P9-AQN-470

LARGE-SCALE STRUCTURES IN THE UNIVERSE

WILEY-PRAXIS SERIES IN ASTRONOMY AND ASTROPHYSICS

Series Editor: **John Mason, B.Sc., Ph.D.**

Few series have been at the centre of such important developments or seen such a wealth of new and exciting, if sometimes controversial, data as modern astronomy, astrophysics and cosmology. This series reflects the very rapid and significant progress being made in current research, as a consequence of new instrumentation and observing techniques, applied right across the electromagnetic spectrum, computer modelling and modern theoretical methods

The crucial links between observation and theory are emphasised, putting into perspective the latest results from the new generations of astronomical detectors, telescopes and space-borne instruments. Complex topics are logically developed and fully explained and, where mathematics is used, the physical concepts behind the equations are clearly summarised.

These books are written principally for professional astronomers, astrophysicists, cosmologists, physicists and space scientists, together with post-graduate and undergraduate students in these fields. Certain books in the series will appeal to amateur astronomers, pre-university college students and non-scientists with a keen interest in astronomy and astrophysics.

For further details of the books listed below and ordering information, why not visit the Praxis Web Site at http://www.praxis-publishing.co.uk

ROBOTIC OBSERVATORIES
Michael F. Bode, Professor of Astrophysics and Assistant Provost for Research, Liverpool John Moores University, UK

THE AURORA: Sun–Earth Interactions, Second edition
Neil Bone, British Astronomical Association and University of Sussex, Brighton, UK

PLANETARY VOLCANISM: A Study of Volcanic Activity in the Solar System, Second edition
Peter Cattermole, formerly lecturer in Geology, Department of Geology, Sheffield University, UK, now Principal Investigator with NASA's Planetary Geology and Geophysics Programme

DIVIDING THE CIRCLE: The Development of Critical Angular Measurement in Astronomy 1500–1850, Second edition
Allan Chapman, Wadham College, University of Oxford, UK

TOWARDS THE EDGE OF THE UNIVERSE: A Review of Modern Cosmology
Stuart Clark, Lecturer in Astronomy, University of Hertfordshire, UK

ASTRONOMY FROM SPACE: The Design and Operation of Orbiting Observatories
John K. Davies, Joint Astronomy Centre, Hawaii, USA

THE DUSTY UNIVERSE
Aneurin Evans, Department of Physics, University of Keele, UK

LARGE-SCALE STRUCTURES IN THE UNIVERSE
Anthony Fairall, Professor of Astronomy, University of Cape Town, South Africa

MARS AND THE DEVELOPMENT OF LIFE, Second edition
Anders Hansson, Ph.D., Senior Science Consultant, International Nanobiological Testbed

ASTEROIDS: Their Nature and Utilization, Second edition
Charles T. Kowal, Allied Signal Corp., Applied Physics Laboratory, Laurel, Maryland, USA

ELECTRONIC IMAGING IN ASTRONOMY: Detectors and Instrumentation
Ian S. McLean, Department of Physics & Astronomy, UCLA, and Director, Infrared Imaging Detector Lab, USA

URANUS: The Planet, Rings and Satellites, Second edition
Ellis D. Miner, Cassini Project Science Manager, NASA Jet Propulsion Laboratory, Pasadena, California, USA

THE PLANET NEPTUNE: An Historical Survey Before Voyager, Second edition
Patrick Moore, CBE, D.Sc.(Hon.)

ACTIVE GALACTIC NUCLEI
Ian Robson, Director, James Clerk Maxwell Telescope, and Director, Joint Astronomy Centre, Hawaii, USA

EXPLORATION OF TERRESTRIAL PLANETS FROM SPACECRAFT, Second edition
Yuri Surkov, Chief of the Planetary Exploration Laboratory, V. I. Vernadsky Institute of Geochemistry and Analytical chemistry, Russian Academy of Sciences, Moscow, Russia

THE HIDDEN UNIVERSE
Roger J. Tayler (*deceased*), formerly Astronomy Centre, University of Sussex, Brighton, UK

Forthcoming titles in the series are listed at the back of the book.

LARGE-SCALE STRUCTURES IN THE UNIVERSE

Anthony Fairall
Department of Astronomy, University of Cape Town, South Africa

JOHN WILEY & SONS
Chichester • New York • Weinheim • Brisbane • Singapore • Toronto

Published in association with
PRAXIS PUBLISHING
Chichester

The White House,
Eastergate, Chichester,
West Sussex, PO20 6UR, England

Published in 1998 by
John Wiley & Sons Ltd
in association with Praxis Publishing Ltd

Wiley Editorial Offices

John Wiley & Sons Ltd, Baffins Lane,
Chichester, West Sussex, PO19 1UD, England

John Wiley & Sons, Inc., 605 Third Avenue,
New York, NY 10158-0012, USA

Wiley-VCH Verlag GmbH, Pappelallee 3,
D-69469 Weinheim, Germany

Jacaranda Wiley Ltd, G.P.O. 33 Park Road, Milton,
Queensland 4001, Australia

John Wiley & Sons (Asia) Pte Ltd, 2 Clementi Loop #02-01,
Jin Xing Distripark, Singapore 12981

John Wiley & Sons (Canada) Ltd, 22 Worcester Road,
Rexdale, Ontario, M9W 1L1, Canada

Library of Congress Cataloguing-in-Publication Data

Fairall, Anthony P.
Large-scale structures in the universe / Anthony P. Fairall.
p. cm.— (Wiley-Praxis series in astronomy and astrophysics)
"Published in association with Praxis Publishing, Chichester."
Includes bibliographical references and index.
ISBN 0-471-96252-X (cloth : alk. paper). — ISBN 0-471-96253-8 (pbk : alk. paper)
1. Cosmology—Measurement. 2. Galaxies—Measurement. I. Title. II. Series.
QB980.F35 1997
523.1—dc20
96-28788
CIP

A catalogue record for this book is available from the British Library

ISBN 0-471-96252-X Cloth
ISBN 0-471-96253-8 Paperback

Printed and Bound in Great Britain by MPG Books Ltd, Bodmin

To my parents

Table of contents

The Atlas of Nearby Large-Scale Structures appears between pages 80 and 81.

Prologue

I have led a privileged life. In my younger days as a graduate student, I worked under the supervision of Fritz Zwicky – one of the great pioneers of extragalactic astronomy. Although often controversial and at odds with his colleagues, many of his ideas and discoveries have proved to be correct. It was Zwicky who introduced me to the large-scale distribution of galaxies, as he viewed the Universe through the wide-angle photographs of his beloved Schmidt telescopes.

Yet I remember being disappointed that I had not been around earlier. It seemed to me then that all the great discoveries had already taken place, and that we had only the details to fill in. How wrong could I have been? Even Zwicky would have been intrigued and amazed at the fabric of the cosmos we have since seen in the large-scale distribution of galaxies.

I have been fortunate enough to have been around at the time when this new view of the Universe was first revealed. I think I would have been appreciative and content to have been a fly on the wall when it all came about; but I, as a minor role player, have made my own small contribution – even seeing some features, voids, walls and so on, ahead of anybody else. It is the same satisfied feeling that must be experienced by an explorer who treads virgin territory. Like most scientific revolutions it came about as the cumulative actions of many players – though there were often certain individuals who had early insight, or first glimpses. I think of John Herschel and Jaan Einasto – two modest gentlemen of science. But the story of the recognition follows in the opening chapter.

It has also been a time of crucial events in human history. Within my lifetime, I have witnessed the first excursions from our home planet – the Apollo Moon landings and the detailed exploration of our Solar System by robotic spacecraft. I have also seen the discovery of the cosmic microwave background, which forms the limit to all we may ever see of the Universe. Through it, our understanding of cosmology has reached a milestone, and we are able to contemplate the history of the Universe as never before. In short, what a wonderful time to be around.

Whether or not I wanted to, I have been forced to face up to aspects of the anthropic principle. In this century, so much has been achieved by the human race in only a few billionths of the age of the Universe. Yet I was here to see it. And it goes further. I once calculated that probably only one part in 10^{40} of this Universe was fit for human habitation; yet, in spite of impossible odds, I managed to get to the right place! Of course the arguments go further – the right sort of physics, or the Sun, Earth and its inhabitants could not have developed as they have. Even if considering that the Earth has, and is harbouring, millions of different species of creature, then might I not have existed as a lower life form? Not so. You and I find ourselves in the ringside seats. We view the Universe from an extremely privileged position.

Yet the human quest to understand everything is far from done, and as if to emphasise this come revelations that confound us – perhaps no more so than the structure of the Universe on the largest possible scale that we can consider. This book is an attempt to convey something of the excitement of its discovery and the character of the cosmos on its largest scale.

Author's preface

This book is written by an observer, with very strong emphasis on observational material. It is concerned with the three-dimensional distribution of galaxies in space – or more correctly in 'redshift space', where the radial component is the measured redshift or 'velocity of recession'. So much so, that the dimensions in redshift space are expressed in 'kilometres per second'. Such dimensions can of course be converted into physical distances by means of the Hubble constant (as explained in Section 1.4), but until a precise value of that parameter is decided it is easier to work in and think in 'kilometres per second'.

A feature within this book is the 26-page Atlas of Nearby Large-Scale Structures (which accompanies Chapter 4, between pages 80 and 81). If you have only five minutes to make a fleeting examination of this book, then turn to the Atlas and look at the fascinating cosmic labyrinth in which we dwell.

There are many things which this book is not. It is not a book about cosmology. Though large-scale structures are intimately tied up with cosmological scenarios, I have not attempted to explain the standard tenets of cosmology. That would be a small book in itself, and in any case there are many such excellent texts available.

This is not a theoretical book. So much of the current understanding of the cosmos on its largest scale is mathematical in nature, but there are other books that approach the topic from the mathematical side – none more so than *Cosmology: The Origin and Evolution of Cosmic Structure* by P. Coles and F. Lucchin (John Wiley & Sons, 1995). The intention is that my observational approach should prove complementary. Mention should also be made of *Structure Formation in the Universe* by T. Padmanabhan (Cambridge University Press, 1993), which has a similar theoretical basis.

Neither is this a book about clusters of galaxies. Though a number of clusters receive mention or even description, they do so as part of large-scale structures – but other important aspects of clusters, such as X-ray emission, morphology of members, sub-groups and internal dynamics, are not covered here. That too is a book in itself.

The text within this book is written as an introduction for the student. However, in the way that an entrance hall leads to a large building, I have provided extensive, and usually more specialised, references in the reading lists at the end of each chapter. Aside from a few historical references, these are taken from the literature over the last 15 or so years, with selection favouring more recent works (up to early 1997). Some chapters, especially those dealing with the more theoretical aspects, seemed to require a large number of references which could not easily be reduced in numbers without bias. The selection is inevitably subjective, but I hope it provides a useful set.

My greatest fear is that I may have overlooked what others would have considered important, or that I have failed to mention the work of a particular investigator who feels it ought to have been given prominence. If so, I offer my apologies. My intention has been to carry out as honest and fair a job as I can.

Anthony Fairall
Cape Town, December 1997

Acknowledgements

My interest in astronomy started in my teenage years, thanks to a book by Patrick Moore, the Bulawayo Astronomical Society and, a little later, Charles Maxwell, who ran the Astronomy Club at Prince Edward School. During my undergraduate years in Cape Town, I received both kindness and guidance from David Evans and Richard Stoy, then both at the Royal Observatory. From my graduate years (Texas and California), I owe particular debts to Fritz Zwicky, Harlan Smith and Dan Weedman.

During the course of preparing this book I have had to call for help from Patrick Woudt (who was completing his Ph.D under my supervision, and therefore could not refuse), Harold Corwin (who provided me with a copy of northern ZCAT data and an updated list of Abell clusters), Margie Walters, Penny Dobbie (for graphical and typographical matters respectively) and Brian Warner. I have also enjoyed positive interaction with Clive Horwood and John Mason of Praxis, and Bob Marriott who was copy editor.

I treasure the friendships I have made within the 'large-scale structure' community. I have had useful discussions with many of its members and, in some cases, fruitful collaborations as well. In particular, I must mention Renée Kraan-Korteweg, Dave Latham, Luiz da Costa, Chantal Balkowski, John Huchra, Jaan Einasto, Brent Tully, Ken-ichi Wakamatsu and Margaret Geller. I have gained much from my past students, especially those involved with large-scale structures – Guinevere Kauffmann, Archie Maurellis, Hartmut Winkler, Yuri Andersson and, of course, Patrick Woudt. I have often benefitted from the close proximity of the 'Cosmology Group', headed by George Ellis, here at the University of Cape Town. Many researchers in the field have also been kind enough to allow me permission to reproduce their diagrams.

Finally, a thank-you to my wife Alex for support and understanding – and for introducing me to 'early mornings', without which I would never have finished this book.

List of illustrations, plates and tables

Chapter 1

Chapter 2

Chapter 7

Chapter 8

Chapter 9

Tables

The Atlas of Nearby Large-Scale Structures appears between pages 80 and 81.

1

The recognition of large-scale structures

1.1 THE DISCOVERY OF NEBULAE

On a clear night we look out from the Earth to the Universe that surrounds us. We see a realm of darkness. The blackness of space is speckled by thousands of stars, and the faintest stars seem to conglomerate towards the luminous band which, by its mythological association, is called the Milky Way. It is as though we had woken up in a maze; it is not obvious where we are, or even what the nature or the distance of what we see is – as if we had been given a gigantic riddle to solve.

It has taken almost all of mankind's existence to solve that riddle. Though some of the interpretations of the Greeks embodied scientific sense, most of the theories were fantasy compounded with religious beliefs. Only since the advent of the telescope in the seventeenth century, have we seen the wandering planets to be worlds like our own. Though the idea had been considered by learned men, only in the last three hundred years or so have the pinpoint stars been properly understood to be distant suns.

Amongst the stars are diffuse patches of light – luminous clouds known in astronomy as 'nebulae' (the Latin word for 'clouds'). A few are bright enough to be faintly visible to the naked eye, and have been known since antiquity. The early users of telescopes, such as Cassini, were able to see more of them.

By 1768 some 68 nebulae were known, 42 of which were southern hemisphere objects located by Nicolas Louis de Lacaille during his sojourn at the Cape of Good Hope. During the same period Charles Messier, in Paris, located more objects while using a small telescope to search for comets, and he therefore recorded them to prevent any subsequent confusion. His original catalogue of 103 objects was published in full in 1771, and even today (now extended to 110 objects) it remains in popular use. Many of Messier's objects lie in the Milky Way, and others lie away from the plane of the Galaxy. In particular, there is a concentration in the constellation of Virgo.

Soon after Messier, the very powerful telescopes and dedication of William Herschel and his son John, in the late eighteenth and early nineteenth centuries, were to revolutionise observational astronomy. Amongst their monumental achievements was the cataloguing of not just hundreds, but thousands of nebulae.

The large Herschelian telescopes were supported and aimed within triangular wooden frameworks (see Figure 1.1). This design made it impracticable to follow objects across

Fig. 1.1. Sir John Herschel was the first astronomer to make a systematic telescopic survey of the entire sky, thereby confirming or discovering the nebulae to be incorporated in his *General Catalogue*. Most of the southern nebulae were found with his 18¼-inch, 20-foot focal length reflector during his time at the Cape of Good Hope (1834–38). The telescope is seen here close to the eastern face of Table Mountain (left of centre). Today, a monument marks the site where the great telescope was erected, and the surrounding estate has given way to urban development. (Much of this book has been written within two kilometres of this site.) (Delineation by Sir John Herschel; lithograph by G.H. Ford.)

the sky, but it was perfect for surveying. The telescope was aimed at a point (on the meridian) above either the northern or southern horizon. It was then gradually raised and lowered (by an assistant) to nod up and down by a small amount as the sky was slowly swept past by the rotation of the Earth. This allowed the observer to scan a strip of the sky – described by William Herschel as a 'sweep'. On a subsequent night, the telescope was set to a slightly different position, and so in time a survey of the entire visible sky could be accomplished.

In his lifetime, William Herschel published three lists, totalling 2,500 nebulae. After excursions in mathematics and chemistry, John Herschel decided to follow up his father's work. Using the most improved of the Herschelian telescopes, he surveyed the northern skies, re-observing every one of his father's nebulae and also adding another 500. In order to complete the work over the entire sky, he relocated the telescope to the Cape of Good Hope for four years, and from there he added another 1,700 nebulae. In 1864 he published a compilation of all known nebulae – the *General Catalogue*, containing 4,630 nebulae discovered by himself and his father, and only 450 discovered by other observers. Even today, the major reference work is the *New General Catalogue* – a recompilation of Herschel's work, with further additions, carried out by J.L.E. Dreyer, Director of Armagh Observatory, and first published in 1888.

1.2 THE EARLIEST RECOGNITION OF LARGE-SCALE STRUCTURES

The Herschels recognised that there were different types of nebula. Some were clearly star clusters, while the 'planetary' nebulae were a new breed of object altogether. But it was the vast majority – removed from the plane of the Milky Way – that caused much debate. Their occurrence seemed to satisfy a philosophical school. In 1750, Thomas Wright had speculated that since we clearly inhabit a flattened stellar system – seen as the Milky Way – there might be other such systems. Immanuel Kant took the idea further as a 'theory of island universes', pointing out that they had already 'been seen by astronomers' – with reference to the nebulae.

William Herschel entertained the idea. Having constructed the 'millstone' model of our stellar system (or Galaxy, a much later term), he was prepared to believe that many of his nebulae could be stellar systems in their own right. Even with his telescopes, they were simply too far away for their member stars to be seen individually; instead their collective light was merged to form the nebulosity. That a few nebulae could be partially resolved into stars supported the theory. However, later in life, Herschel was to reverse his opinion. Seeing that many nebulae had central condensations – and our stellar system apparently did not – he promoted a 'nebular hypothesis', inspired by the suggested origin of our Solar System. This held that a nebula consisted rather of vaporous material, seen in the process of condensing into what would become a star. Herschel himself had demonstrated the universality of gravitational systems by means of his double-star observations. He was quite correct in supposing stars formed in this manner, but to consider that the nebulae were simply stars in formation, and not 'island universes', was in retrospect his mistake. Curiously, his son was to make the same mistaken reversal of opinion, yet was almost unknowingly the first to discover and describe 'large-scale' structure – the topic of this book.

Whilst compiling his *General Catalogue*, John Herschel was able to comment on the sky distribution of nebulae. The concentration in Virgo had grown much more conspicuous; he reported that a third of the nebulae were contained in only an eighth of the sky. More importantly, he saw the nebular system (excluding the obvious star clusters associated with the Milky Way) as being separated from the 'sidereal' system of the stars and Milky Way. He was prepared to adopt the 'island universe' and to consider our stellar system as but one of the many.

Consequently he was the first to describe what we now know as the Local 'Supercluster'. Herschel supposed the nebulae to form a roughly spherical system, centred on the condensation in Virgo. He saw that our own Galaxy lay far from this dense region but was "involved within its outlying members". The distribution of galaxies was far from uniform; branches or protuberances ran outward from the denser core. Our Galaxy "forms an element of some one of its protuberances". This, as we shall see, is an astonishingly accurate description of our supercluster.

The study of large-scale structures in the Universe would have been advanced by more than a hundred years had this correct interpretation been accepted by the scientific community, and even by John Herschel himself! The younger Herschel still revered his famous father, and was reluctant to overthrow the 'nebular hypothesis' with a theory lacking conclusive evidence. His authoritative textbook *Outlines of Astronomy*, reprinted in numerous editions, still promoted his father's interpretation. In addition, the debate on the nature of the nebulae swung against the 'island universe' theory.

In 1864, William Huggins successfully showed that a number of nebulae had discrete line emission spectra, and not the continuous spectra of stars, thereby establishing their fluid gaseous nature. The supposition that the light might rather be the combined emission from faint unresolved stars was excluded for at least these cases.

There were also reports of apparent changes in the appearance of some nebulae – feasible if they really were the flow of gaseous vapours, but quite impossible in the time scale of enormous stellar systems. The apparent changes can now be seen to be spurious; they merely reflect the difficulties of making reproducible sketches after looking through telescope eyepieces.

One author who discussed John Herschel's theory of the Virgo 'Supercluster', but who could not accept it, was Richard Proctor, a Victorian populariser of astronomy. Proctor's counter-argument was that had Herschel's view been correct, the sidereal system – the stars and Milky Way – would show no favoured alignment with the nebular system any more than it does with the Solar System. Yet the general congregation towards Virgo matched the Galactic pole, and the zone most avoided by nebulae matched the plane of the Milky Way. This came at a time when the true magnitude of obscuring material within the Milky Way was far from being understood, and it still seemed as if the Sun were situated close to the centre of the Galaxy.

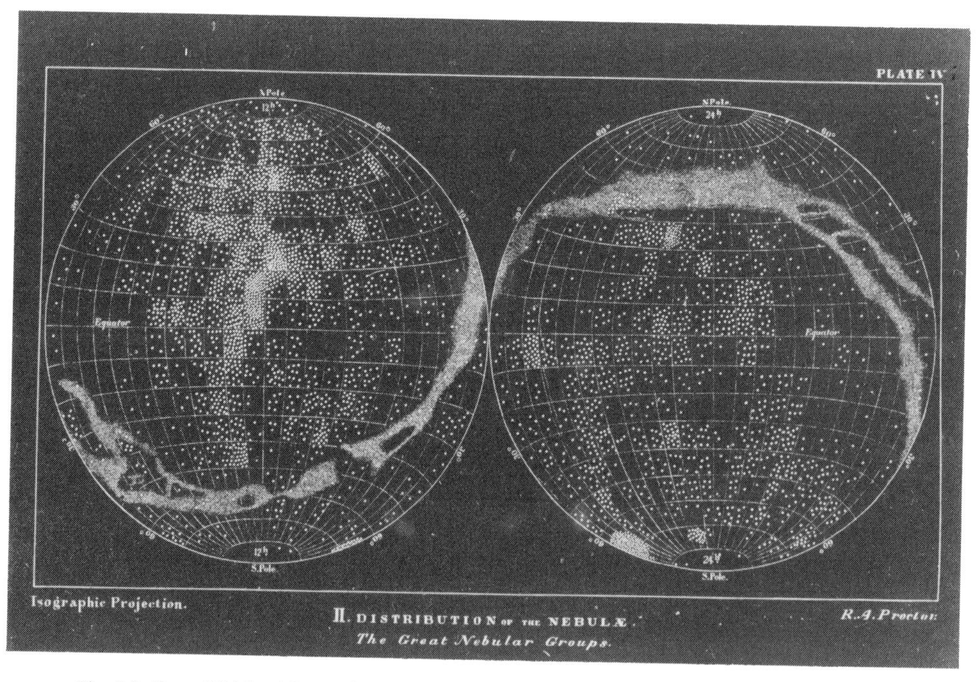

Fig. 1.2. One of Richard Proctor's plots of Herschel's *General Catalogue*, from which various groupings of nebulae were described. His 'great nebular groups' closely match various nearby large-scale structures recognised today. (From R.A. Proctor, *The Universe of Stars*, published by Longmans, Green and Co., London, 1878.)

However, in his book *The Universe of Stars*, Proctor presents some remarkable plots of the distribution of the nebulae in Herschel's *General Catalogue*, one of which is reproduced here as Figure 1.2. Proctor was probably the first to make plots showing the distribution of nebulae (whereas John Herschel had simply counted nebulae in regions of the sky). In the accompanying text, he proceeds to detail the major groupings of nebulae according to the constellations in which they occurred; for example, one of the "streams or clusters" of nebulae "passing over the left hand of Virgo, the tail of Hydra, and nearly the whole extent of Centaurus, to Crux and Musca". Although Proctor's interpretation was incorrect, his descriptions of the distribution of nebulae include the major large-scale features recognised today.

Proctor's book also includes a two-colour plot of nebulae (Figure 1.3) drawn by Sidney Waters, that shows the distribution in finer detail. The plot seems, to this author, quite astonishing in that it accurately reproduces the character of the distribution of galaxies, as only recognised late in the twentieth century. Many filamentary structures are apparent, and since the nebulae are sampled to a limited depth, in some places voids encircled by nebulae are apparent. The plot is purely two-dimensional, but it is a remarkably accurate representation of the structures that form the topic of this book.

Though spiral structure in such nebulae was first seen by the Earl of Rosse in the 1840s, the coming of astronomical photography in the late nineteenth century led to the discovery that many of the nebulae showed elegant and symmetrical spiral forms. David Gill, a respected astronomer (who directed the Royal Observatory at the Cape of Good Hope) was one of the first to capitalise on photography. Early in the twentieth century he recognised

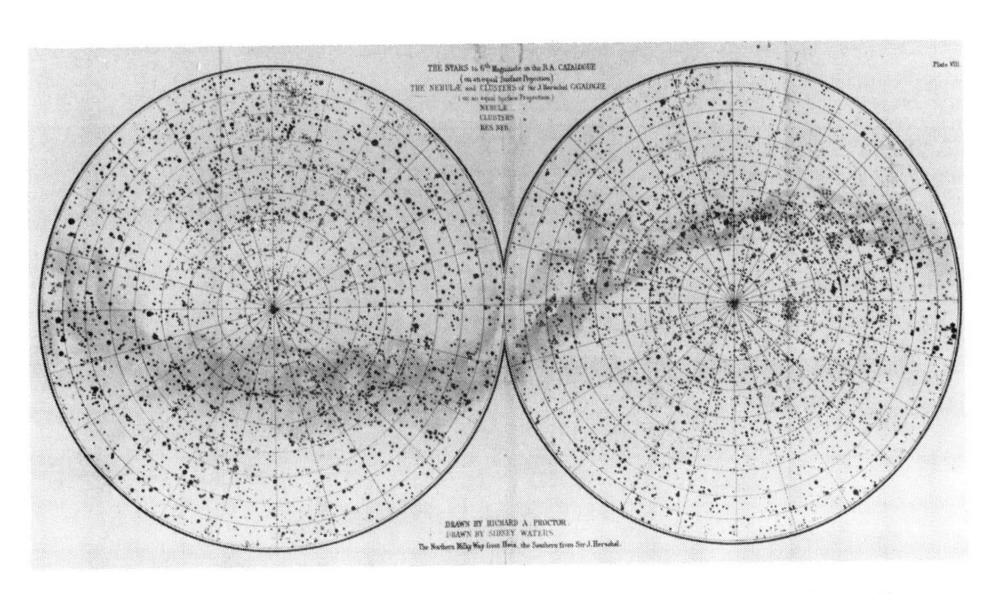

Fig. 1.3. A plot of Herschel's *General Catalogue* by Richard Proctor and Sidney Waters. Although not remarked on at the time, the superior resolution reveals cellular structures – galaxies surrounding voids – particularly in the right-hand panel, though the scale and absence of colour in the reproduction above do not make it possible here. (From R.A. Proctor, *The Universe of Stars*, published by Longmans, Green and Co., London, 1878.

the similarity between the texture of the Milky Way and the spiral structure in the nebulae. He endorsed and thereby revived the 'island universe' theory, even influencing the eminent theoretician, Arthur S. Eddington.

1.3 SUPERCLUSTERS IN THE SKY

The debate on the nature of the 'whirlpool' nebulae again became a great astronomical controversy, though now mainly centred in the United States. Vesto Slipher and others upheld the 'island universe' theory, and opposing it was Harlow Shapley, whose overestimate of the size of our Galaxy suggested that nebulae were much too small for comparison. He also believed the observations of Adriaan van Maanen, who claimed to have measured rotations in spiral nebulae — all in favour of their being stars under formation.

It was not until the 1920s that the 'island universe' theory was at last proved correct. Edwin Hubble, using the world's most powerful telescope – the 100-inch Hooker reflector at Mount Wilson, California – was able to identify Cepheid variable stars within the great spiral nebula in Andromeda. Recognised as 'standard candles' for estimating distances, they provided convincing proof that this, and later other nebulae, were indeed extragalactic.

Our Galaxy was no longer unique. It was much as though Los Angeles had been found to be no longer the only inhabited city on Earth, yet it passed as a relatively quiet scientific revolution. Perhaps the increase in distance scale was simply too much for normal minds, outside of those in the specialised field, to comprehend; few of the general public, even today, know anything about it.

Hubble took it upon himself to communicate the news. In 1935, he delivered a set of lectures at Yale. These lectures were written up as *The Realm of the Nebulae*. This classic book opened up a new perspective on the large-scale character of the Universe. In one chapter of the book, Hubble addressed 'The Distribution of the Nebulae'. In it he reports mainly on his own investigation, whereby 44,000 galaxies were counted in 1,283 small photographed fields, scattered over the sky. Hubble's survey immediately revealed the 'Zone of Avoidance' caused by the foreground obscuration of the Milky Way, and the decrease in galaxy numbers as the Galactic plane was approached. He made quantitative corrections for this obscuration, and then examined the distribution.

Presumably Hubble never read Herschel's description of the Virgo 'Supercluster', as he makes no mention of it in his text. Very likely, he did not bother to research the conclusions of outdated visual observations, particularly from the other side of the Atlantic. In any case, his telescope and equipment was vastly more powerful than Herschel's. His survey went much deeper and saw beyond what we now know are the nearby large-scale structures. Impressive though his endeavour seemed, his deductions were incorrect.

Hubble concluded that there was "no evidence of conspicuous systematic variation in the distribution of nebulae over the sky." He was aware of groups and 'great' clusters of galaxies: "Condensations in the general field may have produced the clusters, or evaporation of clusters may have populated the general field". However, "the tendency to cluster appears to operate on a limited scale. No organisations on a scale larger than the great clusters, and no clusters with as many as a thousand members, are definitely known." This last remark was not thrown out to dispute John Herschel, as no historical reference is given. More likely it was intended to dampen the claims of Hubble's great scientific rival,

Harlow Shapley. Shapley's work was mainly based on shallower, but much wider angle photographs, many obtained with the 24-inch Bruce telescope at the Boyden Observatory in South Africa. Such photography penetrated much deeper than Herschel's visual observations. In the southern skies, he identified a "remote cloud of galaxies" in Centaurus and a "distant metagalactic cloud" in Horologium – systems far larger than clusters. (We shall be discussing these systems in Chapter 5.)

Most of the scientific world chose to believe Hubble. He had after all been right about the 'island universes' (where Shapley had been wrong), discovered the recession of galaxies, and worked with the world's largest telescope. Hubble's legacy remained for decades, and it is only in more recent times that Shapley has been seen to be right; he was one of the first to recognise the existence of large-scale structures in the Universe.

While only some 20 'great clusters' of galaxies were known in Hubble's day, the numbers grew into thousands with the National Geographic–Palomar Sky Survey of the 1950s – wide-angle Schmidt photographs that covered the entire sky visible from Palomar, with a fast (low) f-ratio that readily showed the galaxies. Fritz Zwicky, of the California Institute of Technology, had long championed the Schmidt telescope. Following use of a portable version in the 1930s, the 18-inch Schmidt was built on Mount Palomar, to be succeeded by the 48-inch, the telescope behind the Palomar Sky Survey. Soon after its completion, George Abell scrutinised all its photographs to complete a now famous survey of over 2,700 clusters.

Hubble's assertion that clusters were the largest conglomerations was in line with simple gravitational theory. In a cluster, the typical 'crossing time' for a galaxy to go from one side of the cluster to the other is a billion years or more – already a good fraction of the age of the Universe. If larger entities were to exist, the 'crossing time' would be many times greater still – more than the age of the Universe. Not only would there not be enough time for galaxies to cross, there would also be insufficient time for the entities to form in the first place (unless considerably more mass were present to speed up the formation process, as will be discussed in Chapter 9).

Yet, slowly, observational data revealed the larger structures. By 1953, Gerard de Vaucouleurs was advocating the existence of a Local 'Supergalaxy', a flattened conglomeration of nearby structures centred on the Virgo cluster (see Figure 1.4). John Herschel's original system had been revived, but, as before, it did not gain general acceptance. De Vaucouleurs also claimed the existence of a southern supergalaxy. In 1960 he put the diameter of the local supergalaxy at 30 megaparsec (100 million light years) — far larger than a cluster of galaxies.

In 1967, C.D. Shane and C.A. Wirtanen published the Lick Observatory survey on the distribution of galaxies, in which they counted more than a million galaxies identified on their survey photographs, and prepared contour maps of their sky distribution. They claimed the existence of three apparent 'clouds' of galaxies in Serpens/Virgo, Corona and Hercules, with dimensions comparable to the Local Supercluster. Meanwhile, Fritz Zwicky had used the Palomar Sky Survey to produce an atlas of galaxies and clusters. He drew contour lines around the clusters, but the suggested sizes were vastly extended from the normal cluster diameters; Zwicky referred to them as 'cluster cells'. Still, relatively few researchers were convinced that entities larger than clusters existed; arguments flowed both ways and the situation was unresolved, and so it remained until three-dimensional mapping became possible.

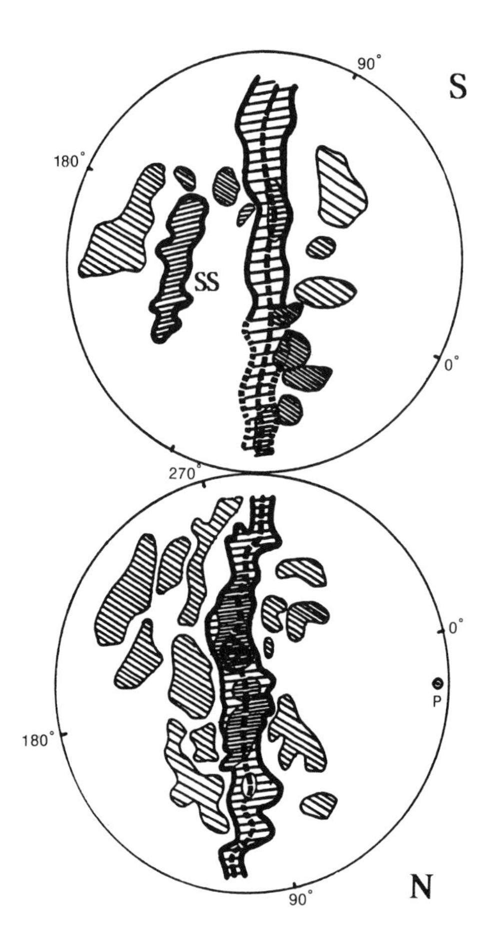

Fig. 1.4. Diagrams presented by Gerard de Vaucouleurs in 1953 to indicate clustering of relatively bright galaxies in the southern and northern Galactic hemispheres, in support of the existence of a Local 'Supergalaxy'. By permission of the *Astronomical Journal*.

1.4 REDSHIFT AND DISTANCE

The distribution of galaxies as seen in the sky is, of course, a two-dimensional projection of their three-dimensional positions in space. Distances to galaxies can be roughly gauged by their angular sizes; one can pick out which are near or far. We now know, however, that galaxies exhibit a range of true physical size, and yet retain the basic spiral or elliptical forms. Consequently, one may sometimes have a small nearby galaxy look as if it were distant, and a distant galaxy look as if it were near. However, this is less likely to be a problem with clusters of galaxies, where the 'average' angular size is a reflection of distance. Abell and Zwicky were able to assign five or six distance classes ranging from the very near to the extremely distant. Consequently, they provided for the first time a 'three-dimensional' view of the cosmos, showing where their clusters were situated in space.

However, Edwin Hubble had opened up an easier way to obtain relative distances to galaxies. The distance scale that he established using Cepheid variable stars not only demonstrated that the galaxies were truly external to our Milky Way; it also showed that distance to a galaxy was in proportion to its velocity of recession.

Since the early twentieth century, measurements of the spectra of galaxies had revealed the spectral features to be slightly displaced towards longer wavelengths; that is, they were 'redshifted'. Since the shift was in proportion to the wavelength of light involved, the interpretation was a Doppler shift caused by the galaxy concerned moving away from our Galaxy. Hubble, using only a flimsy set of data, claimed that the greater the distance, the greater the velocity of recession. This linear relation was subsequently confirmed and extended to far greater distances, and is today known as the Hubble Law,

$$V = Hd$$

where V is the velocity of recession usually measured in kilometres per second (km/s), and d is the distance, generally measured in megaparsecs (1 megaparsec, abbreviated 1 Mpc = 3.26 million light years or 3.08×10^{19} kilometres). H is the Hubble constant, which is somewhere around 70 km/s per Mpc. Hubble's original determination put H as high as 540 km/s per Mpc.

Soon after Hubble's announcement, cosmologists interpreted the finding as the expansion of the Universe. There is an ever increasing separation of the galaxies as the space containing them expands (see Figure 1.5). Hubble himself preferred to see the galaxies in dynamical motion, rather than being separated by what apparently seemed to him a bizarre 'Alice in Wonderland' growth of the Universe. The concept of space itself expanding was only made possible by the cosmological implications of Einstein's famous concepts of General Relativity.

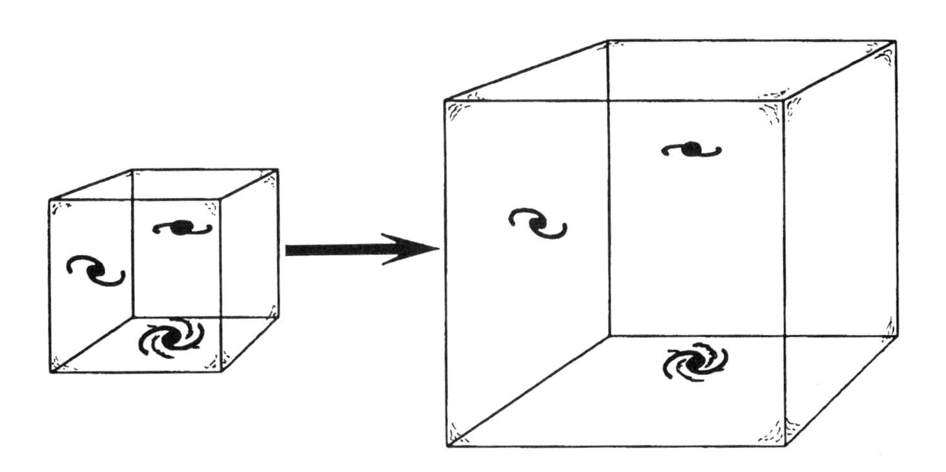

Fig. 1.5. The expansion of the Universe causes the distances between galaxies to increase. A useful analogy is that of a fruit-cake baking. The galaxies are like the raisins in the cake – they do not in themselves grow larger. The space that separates them is like the cake mixture that grows in volume, causing them to be spaced further and further apart.

Within clusters, and even within groups of galaxies such as our Local Group, gravity has arrested and overcome the cosmological expansion. In doing so, it has set the galaxies in motion, such that a more accurate version of the Hubble formula would read as follows:

$$V = Hd + V_{pec}$$

V_{pec} is the galaxies' own peculiar radial motion, which might add to, or subtract from, the cosmological velocity of expansion. V_{pec} may amount to several hundred kilometres per second for motions within clusters, or even large-scale streaming (Chapter 7), but even these velocities are low enough that, for now, the basic Hubble Law can be taken as a good approximation.

In this book, we shall therefore talk of redshift as an expression of distance. We shall deal with redshift space as though it were truly three-dimensional space. Only in rich clusters of galaxies will the difference be noticeable, as there the internal motions will distort the shape of the cluster into elongated radial forms, as will be shown presently in the illustrations. Plots in redshift space are generally in the form of 'slices' in one or the other angular celestial coordinates – usually Declination or Right Ascension†.

We shall also take the licence of expressing redshifts as velocities – a 'redshift of 5,000 km/s', for example. While the value of the Hubble constant would enable us to convert to megaparsecs or even light years, the exact value of that constant is still a matter of debate. For many years, there were two opposing schools – those favouring H = 50 and those preferring H = 100 km/s per Mpc; one of the classic controversies of astronomy. The Hubble Space Telescope has now narrowed the accepted value into the range 60 to 75 km/s per Mpc. In this book, we shall often use velocity of recession as representative of distance. Alternatively, following common practice, we shall sometimes express sizes and distances as 'h^{-1} Mpc', where $H = h \times 100$ km/s per Mpc. The value of h is therefore about 0.7, but can easily be adjusted as a more precise value becomes known. A redshift of 5,000 km/s would therefore correspond to a distance of $50h^{-1}$ Mpc. In the literature, occasional use is made of h_{75} where $H = h_{75} \times 75$ km/s per Mpc, or even h_{50}. In this book h, without any subscript, is always from $H = h \times 100$ km/s per Mpc.

The procedures for measuring redshifts of galaxies will be discussed in detail in the next chapter. Meanwhile, the brief introduction above will serve to show how redshift plots came to be instrumental in the recognition of large-scale structures in the Universe.

1.5 THE FIRST INDICATIONS OF CELLULAR STRUCTURE

For many years, the spectrograms used to measure redshifts were recorded directly onto photographic emulsions. In order to accumulate enough of a galaxy's light to produce a reasonable photographic record, long exposures were necessary. Early this century, some exposures stretched over many nights of observing time. By 1914, Slipher had measured only 13 galaxies. It took until the mid-1930s for the number to exceed 100. Improvements in the speed of photographic emulsions followed; by 1956, when Humason, Mayall and

†The sky is projected onto the interior of a celestial sphere of infinite radius. The extension of the Earth's equatorial plane to this sphere provides the basis for a coordinate system equivalent to that used on the terrestrial sphere. Declination (+ or –) is the direct equivalent of latitude (north or south); Right Ascension corresponds to eastern longitude, but is usually expressed in hours, minutes and seconds, where 24 hours equal 360 degrees.

Sandage published a major catalogue, there were over 800. When de Vaucouleurs' *Reference Catalogue* was published in the mid-1960s, fewer than 1,500 redshifts were available, but since then, image intensifiers, reticon and CCD detectors have revolutionised productivity, such that numbers have doubled every seven years.

The data in the *Second Reference Catalogue*, complete up to 1975, was to provide the avenue towards a major breakthrough. This was to come about at the Tallinn Observatory in Estonia, then part of the Soviet Union. There, Jaan Einasto and colleagues had been exploring dark haloes around galaxies, and the identification of companions to galaxies was of importance. One of the group, Mihkel Joeveer, made redshift plots using data from the *Second Reference Catalogue*. More than just seeing superclusters, the plots revealed filaments of galaxies surrounding apparently empty voids. The breakthrough came in 1977, just before the observatory was about to host a Symposium of the International Astronomical Union. The title chosen for the symposium was 'The Large-Scale Structure of the Universe', apparently the first time such terminology (and the title of this book) was used. At the conference, Joeveer and Einasto presented a paper entitled 'Has the Universe the Cell Structure?' They described the idea that galaxies were typically arranged in filaments surrounding empty voids – a cellular structure such as seen in Figure 1.6. It was a sensational claim.

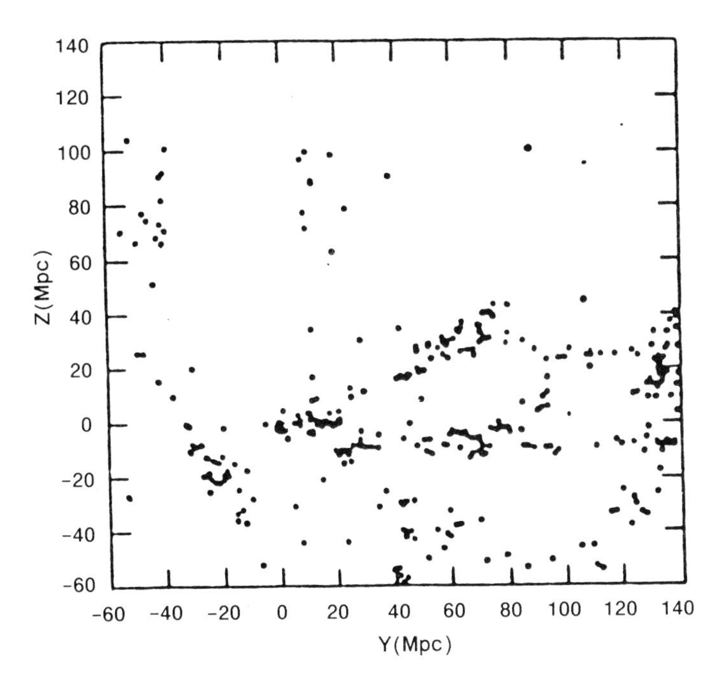

Fig. 1.6. Evidence of the cellular distribution of galaxies; one of many plots presented by Einasto and collaborators. The more luminous galaxies have been plotted in three-dimensional space, and this diagram is a 'slice' through that volume. Note the empty spaces enclosed by walls of galaxies. The distances to the galaxies are based on their redshifts, but all apparent members of the Coma Cluster have been put at a common distance. (Reproduced with permission from J. Einasto and *Nature* (**300**, 407). (Copyright(1982) Macmillan Magazines Ltd.)

News of the claim travelled widely, but converts outside the Soviet Union were few; the idea seemed to overthrow all that was understood in the West about clustering. It was seen to support the Soviet theoretical cosmological view – particularly that of Ya. B. Zel'dovich and colleagues in Moscow – involving 'pancaking', rather than the American views of galaxy formation, at a time when 'cold war' rivalry flourished.

Perhaps the greatest criticism of Einasto's work, and the reason for the Americans' disbelief, was that it was based on uncontrolled data. The redshifts reported in the *Second Reference Catalogue* were only for those galaxies which various investigators had chosen to observe. They were therefore highly selective, and far removed from a uniform homogeneous sample. This raises a vital issue in mapping the Universe – for catalogues are ahead of controlled surveys – and much further discussion will follow later (in Chapter 3). For now, suffice to say that one cannot pre-select in redshift; and, in any case, Einasto was right! Furthermore, the cosmological theories of Zel'dovich are now highly respected.

Soon after, American efforts, whether seeking to spurn Einasto or otherwise, began to show the same picture of galaxies surrounding voids. Figure 1.7 shows a classic plot by Gregory and Thompson that reveals an extended structure running from the Coma Cluster, behind the foreground voids. Plotted in redshift space, the Coma Cluster appears here radially distorted into a 'finger of God', since the internal motions within the cluster have added to or subtracted from the common cosmological velocity of expansion.

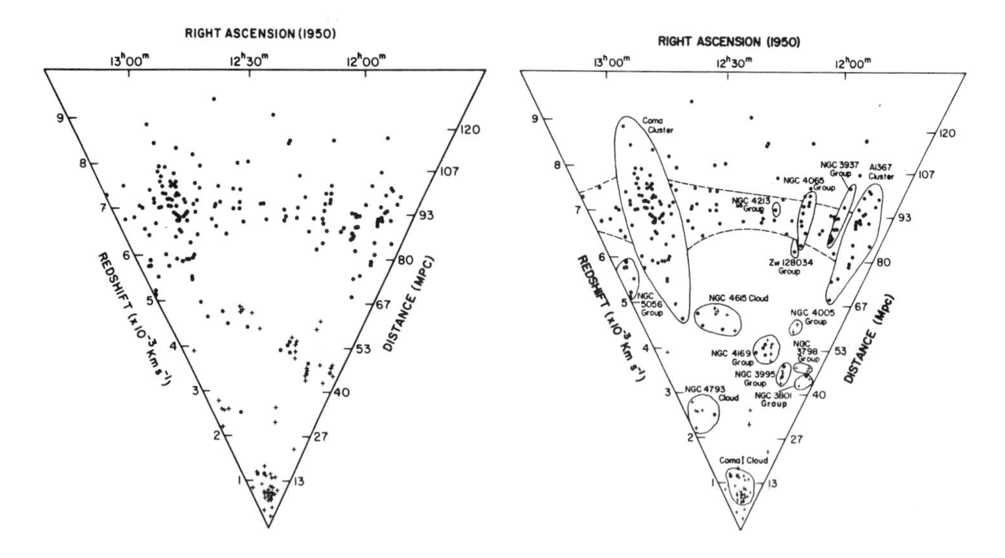

Fig. 1.7. A wedge or pie plot by Stephan Gregory and Laird Thompson showing the interconnections between the Coma and Abell 1367 clusters. The plot is the first of many in this book shown in 'redshift space' – not quite conventional space. As a consequence, the two clusters appear stretched radially (see explanation in 1.4). This survey also revealed empty voids in front of the Coma 'Supercluster'. In this plot, the voids appear stretched horizontally into elliptical shapes, because the angular scale has been exaggerated by a factor of approximately 2. Were this not the case, they would appear roughly circular. The voids are in common with those shown in Figure 1.6, but the cut is roughly perpendicular. (Reproduced with permission from S. Gregory and the *Astrophysical Journal* (**222**, 784, 1978).)

In parallel with this came plots of redshifts obtained by Guido Chincarini, Herbert Rood, Riccardo Giovanelli and Martha Haynes. Their plots confirmed the tendency for galaxies to conglomerate in 'superclusters' – much more extended than clusters – but also confirmed the existence of empty spaces between such structures. In particular they looked at Coma and also the Hercules Supercluster. Chincarini, one of the first to discuss the new-found 'tapestry' of the Universe, pushed the idea that redshifts were 'segregated'.

1.6 DISCOVERIES IN THE 1980s

In 1981, Robert Kirschner, Augustus Oemler, Paul Schechter and Stephen Shectman claimed the existence of a possible 'million cubic megaparsec' void in the direction of the constellation of Boötes. They had found that an underdense region consistently showed up between 12,000 and 18,000 km/s in three 'pencil-beam'-style probes, separated from each other by roughly 35 degrees. The novelty of the claim and its implications meant that it received considerable publicity (enough that some writers, unaware of earlier work, have

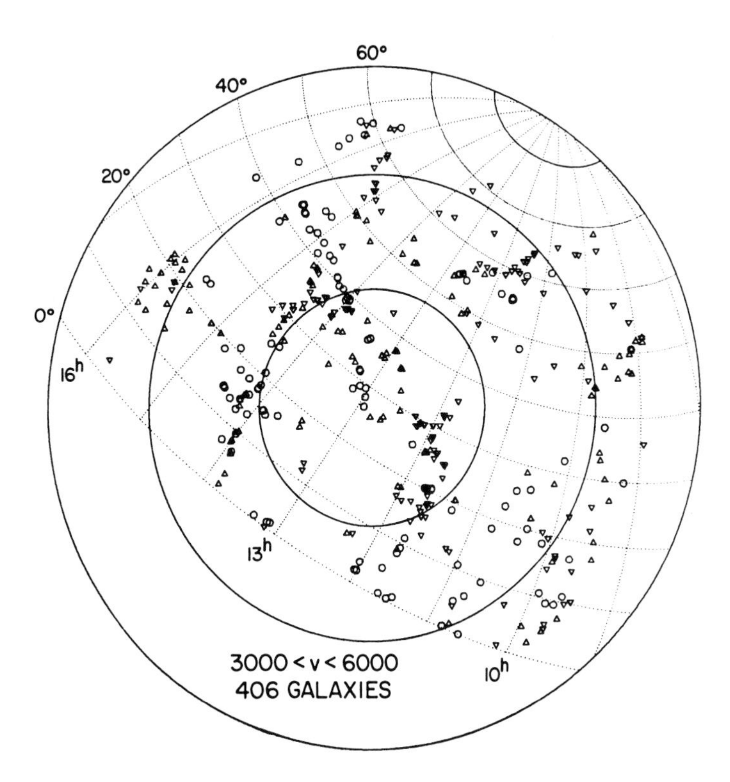

Fig. 1.8. A sample plot from the Center for Astrophysics (CfA1) survey, showing the frothy cellular structure. Data for the redshift interval between 3,000 and 6,000 km/s are projected on much of the sky of the northern Galactic hemisphere. Within the redshift range, galaxies denoted by inverse triangles are closer than 4,000 km/s, and those by circles more distant than 5,000 km/s; the intermediate range is shown by upright triangles. (Reproduced with permission from J. Huchra and the *Astrophysical Journal* (**253**, 423, 1982).)

labelled it as the first void discovered). The initial report implied that the void would have a diameter approaching 10,000 km/s in redshift space, significantly larger than any of the voids in the plots seen so far. However, the Boötes 'Void' is more of a general underdense region; in the years following, various galaxies were discovered within its boundaries, while its overall size was reduced.

The first proper redshift survey with statistically controlled data (to limiting magnitude 14.5; see Chapter 3) was carried out by Marc Davis, John Huchra, David Latham, John Tonry and colleagues at the Harvard–Smithsonian Center for Astrophysics (in Cambridge, Massachusetts). Now known as the CfA1 survey, its data revealed a frothy structure, including large filamentary structures and large voids (a sample is shown as Figure 1.8). The authors saw the frothy texture as presenting a severe challenge to theories of galaxy and cluster formation.

From a personal standpoint, the Gregory and Thomson plot and the CfA1 survey inspired the author to plot data from his list of compact galaxies and his newly established redshift catalogue. Working with my student, Hartmut Winkler, we examined the distribution of southern galaxies and in 1983 tentatively identified several overdense regions and voids, all of which have stood the test of time (see Figure 1.9). We also exchanged mate-

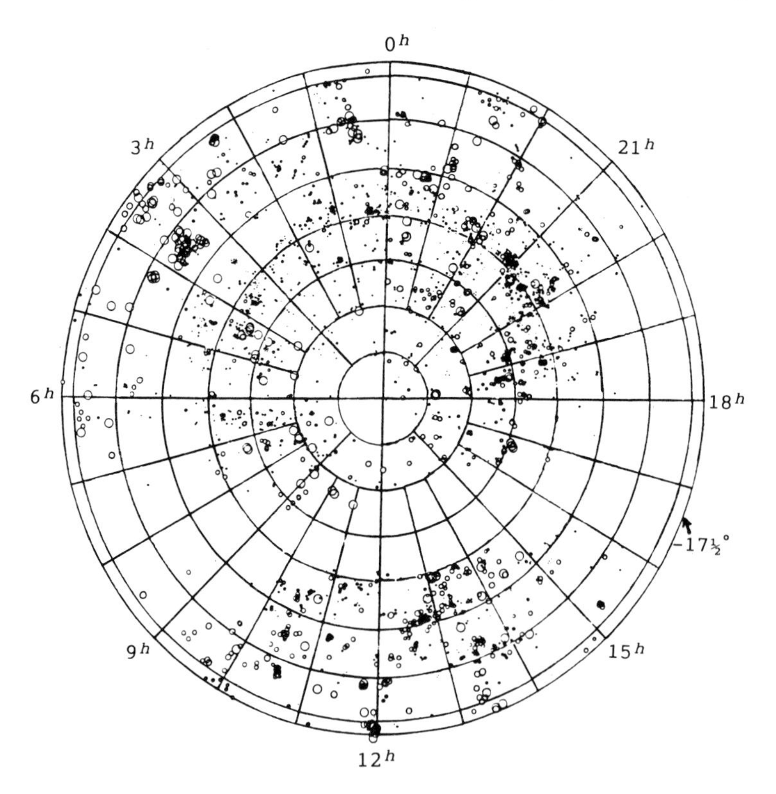

Fig. 1.9. A plot made by H. Winkler in 1983 from the author's *Southern Redshift Catalogue*, showing the redshift distribution in the southern sky (south of Dec. −17.5°). The size of the the galaxy symbols diminishes with increasing redshift. A continuous structure – the 'Centaurus–Pavo Supercluster' – runs over much of the diagram.

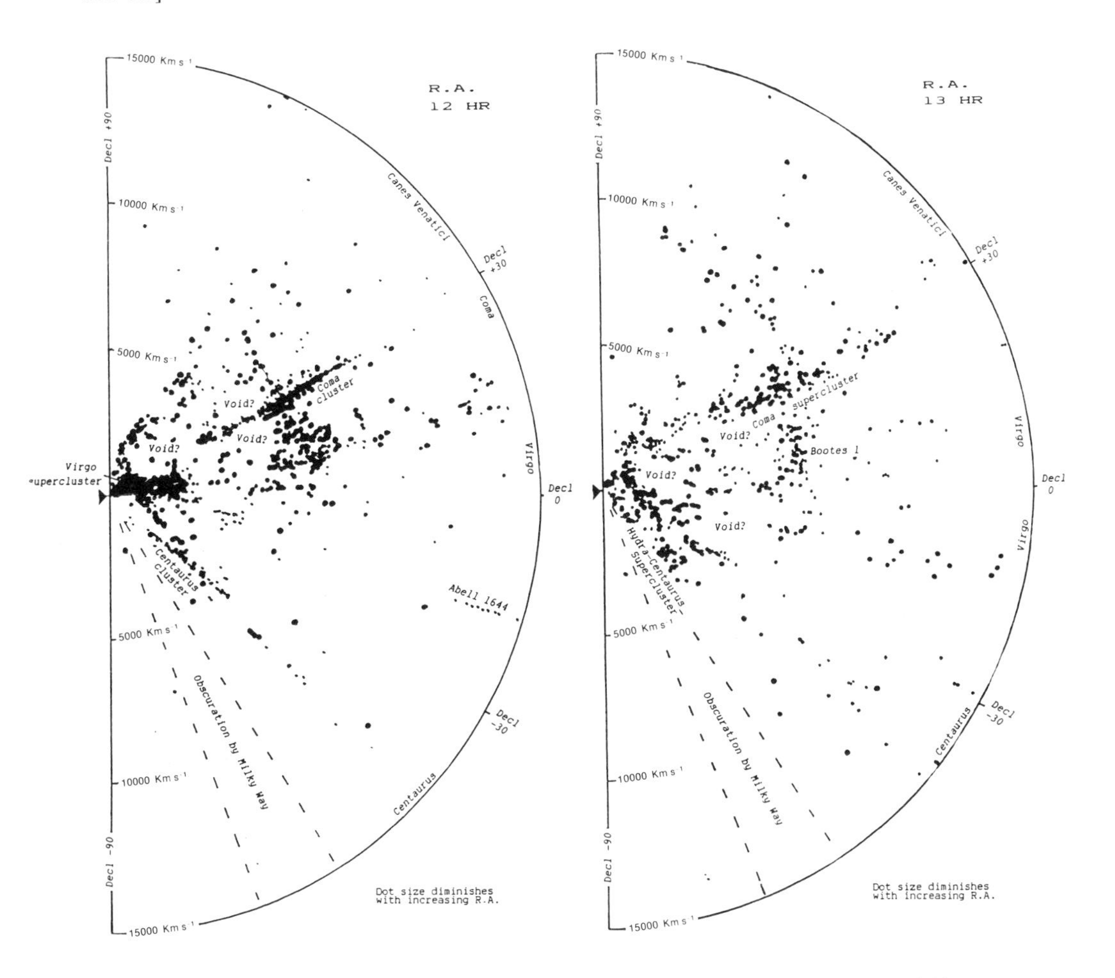

Fig. 1.10. Sample sections from 24 Right Ascension slices, plotted in 1985 by the author and 60 undergraduate students, from Huchra's 1983 ZCAT. All but five of the plots (those most affected by the Milky Way) revealed cellular structure such as seen here. Some of these voids are in common with those appearing earlier in Figures 1.7 and 1.8.

rial with John Huchra's ZCAT redshift catalogue, and global plots of his 1983 catalogue (see Figure 1.10) confirmed the tendency towards cellular structure, wherever the data was dense enough to show it. The plots also identified some 40 apparent voids.

A paper by P. Fontanelli in 1984 added new data to the Coma/Abell 1367 structure; an accompanying plot (shown here as Figure 1.11) extended and updated the original Gregory/Thomson plot, revealing more detailed structure and more distinct voids.

The first portion of the deeper CfA2 survey (to magnitude 15.5) was released as a 'Slice of the Universe' in early 1986 (see Figure 1.12). For a small strip of the sky, it had more data than the equivalent catalogue surveys, yet was statistically controlled and, for many within the scientific community, was therefore believable. It again confirmed the cellular structure, and suggested what the authors – Valerie de Lapparent, Margaret Geller and

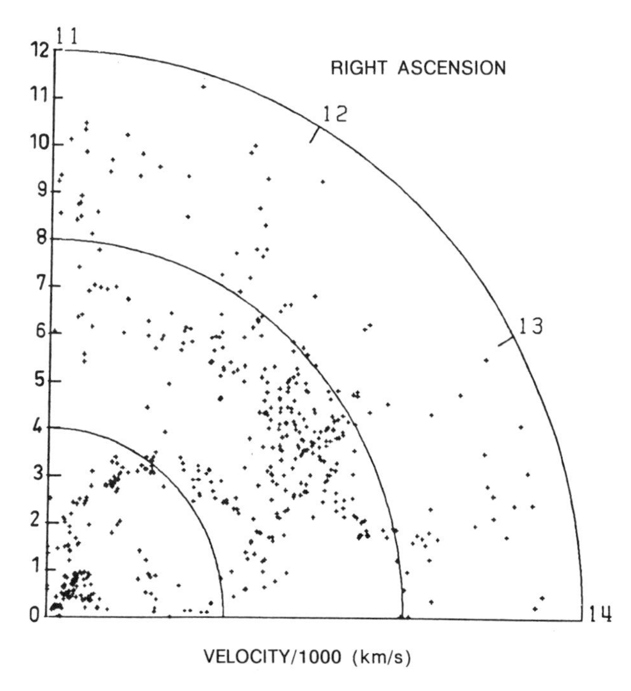

Fig. 1.11. A further pie plot – a slice between Dec. 26° and 39° – of the Coma Cluster region, by P. Fontanelli. Note that the angles of Right Ascension have been exaggerated. (Reproduced by permission from *Astronomy and Astrophysics* (**138**, 85, 1984).

John Huchra – described as a 'soapsud' texture. At the centre of the plot, the 'finger of God' of the Coma Cluster is contained within the 'head' and 'body' of a formation that resembles a stick man, with arms and legs spread (the same stick figure can be seen, for instance, in the Fontanelli plot). Resting on sound credentials, the 'Slice of the Universe' plot was very widely publicised, appearing in *Time* magazine and numerous scientific journals and textbooks. More than all preceeding plots, this diagram had an impact far beyond the field of speciality and brought home to many the new cosmic texture. It has proved the most influential of the pioneering studies of large-scale structure.

In complementary fashion, Martha Haynes and Riccardo Giovanelli obtained a large number of redshifts, using radio telescopes, in the Perseus–Pisces region. With existing published redshifts, they were able to plot the distribution of over 2,700 galaxies in this portion of the sky. Their plots (see sample in Figure 1.13) also showed the characteristic filamentary structures and cellular texture.

Another pioneer of large-scale structures was Brent Tully of the University of Hawaii. During the 1980s, he and colleague Richard Fisher assembled a catalogue and atlas of nearby galaxies (to which we shall make frequent reference in Chapter 4) that has proved a definitive work in mapping the Local Virgo Supercluster. Tully also explored further afield by means of the Abell clusters, using software packages that displayed three-dimensional density contours. On the basis of these, he claimed the existence of conglom-

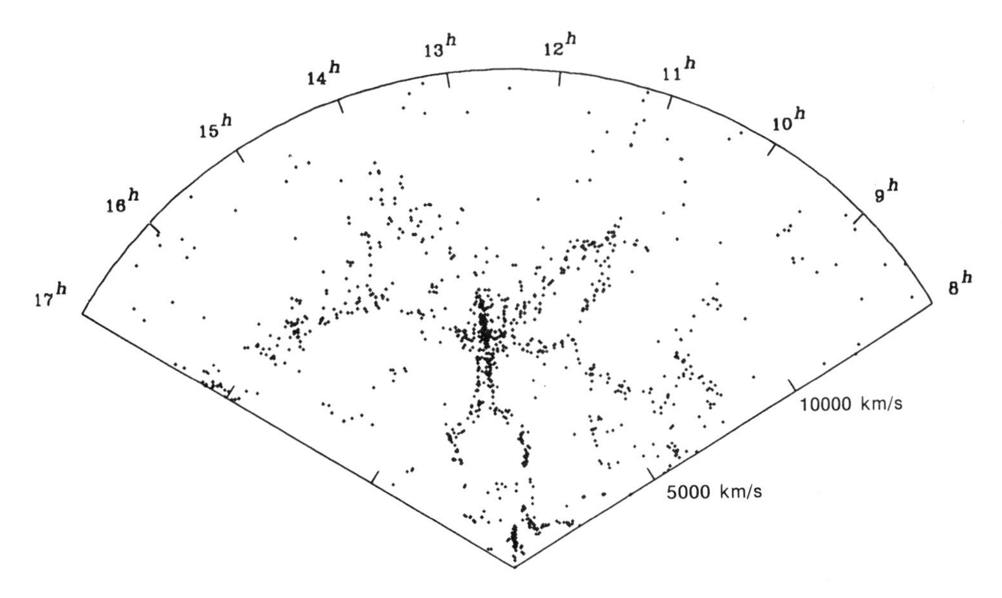

Fig. 1.12. A 'Slice of the Universe'. The first slice of the CfA2 survey, published in 1986 – probably the most widely publicised plot showing the character of large-scale structure. The galaxies form a curious 'stick man' figure in the centre, most of the vertical portion of which is the elongated 'finger of God' of the Coma Cluster. The central portion of the plot is in common with Figures 1.6 and 1.11, but has more data; some of the same voids also appeared previously in Figures 1.6 and 1.10. (The 'arms' of the stick man are later seen to be a cross-section through the Great Wall.) (Reproduced with permission from M. Geller and the *Astrophysical Journal* (**302**, L1, 1986).)

erations of clusters on scales reaching 30,000 km/s. Fig 1.14 shows such large assemblages, though this black and white diagram does not match the splendour of the stunning three-dimensional colour versions that Tully generated for conference presentations. In 1989, Marc Postman and colleagues were to dispute the reality of Tully's 'supercluster complexes'; nevertheless Tully's work stresses the interconnections, such that it is often difficult to say where one structure ends and another begins. His work also showed that the scale of homogeneity in the Universe was clearly far larger than had been anticipated by cosmologists. (Much more will be said about this in Chapter 5.)

The southern equivalent of the CfA1 (and CfA2) survey has been the Southern Sky Redshift Survey (and its extension). It has been a collaborative effort between observatories in South America and South Africa, led by Harvard graduate Luiz da Costa of the National Observatory of Brazil. (The author and colleagues at the South African Astronomical Observatory have contributed). It is a statistically controlled survey that has enabled quantitative comparisons between the structures discerned in northern and southern skies, thereby revealing similarity in texture. A plot is shown in Figure 1.15.

Other interest in the southern skies centred on the Centaurus region. In 1986, a group of researchers investigating elliptical galaxies was able to detect a streaming motion, over and above the cosmological expansion, that might be indicative of the presence of a large

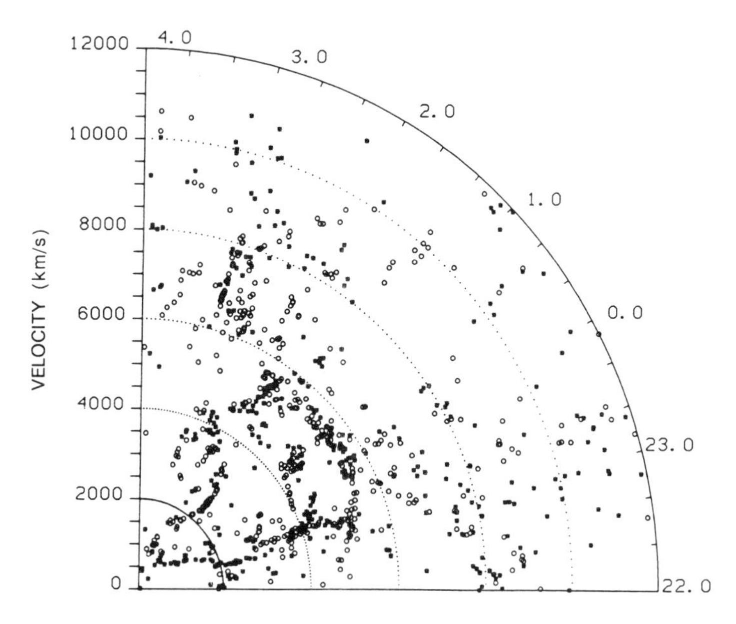

Fig. 1.13. One of the redshift plots published by Haynes and Giovanelli revealing the character of the distribution of galaxies towards Perseus–Pisces, and suggesting interconnections with our own supercluster. The angular coordinate is Right Ascension; galaxies between Declinations 0° and +20° are shown. (Reproduced with permission from R. Giovanelli and the *Astrophysical Journal* (**306**, L58, 1986).)

'overdensity' in the direction of Centaurus. One of their number – Alan Dressler – labelled it the 'Great Attractor', a name that has stuck ever since. It was not immediately clear, and is still not so, what exactly the Great Attractor is, but large-scale motions are a consequence of the mass overdensities brought about by large-scale structures.

Ongoing progress at the Center for Astrophysics continued with the work of John Huchra and Margaret Geller. Further slices in the CfA2 survey led to the recognition that an elongated feature that appeared in the original 'Slice of the Universe' extended into neighbouring slices. It was more of a slab-like concentration of galaxies, seen flat on in the sky, and was soon dubbed the 'Great Wall' (probably by Avishai Dekel). It appears in Figure 1.16. A similar 'Great Southern Wall' was soon recognised in the southern skies – part of the almost rectilinear structure in that region.

Thus was laid the initial knowledge of large-scale structures in the Universe, and the foundations for more detailed mapping. In the 1990s, many further major redshift surveys were to be completed and major undertakings put underway, so that by the time of writing this book (1996–97), there are hundreds of papers in the literature that concern themselves with plots of large-scale structures in one way or another. There are also surveys that now probe deep into the cosmos. In 1990, Tom Broadhurst, Richard Ellis, David Koo and Alex Scalay made the sensational claim of finding periodicity in the structures when pushing deep in the direction of the Galactic poles. Current efforts to gather a million redshifts may unravel such mysteries of the cosmos. These matters will, however, be dealt with in detail in the remaining chapters of this book.

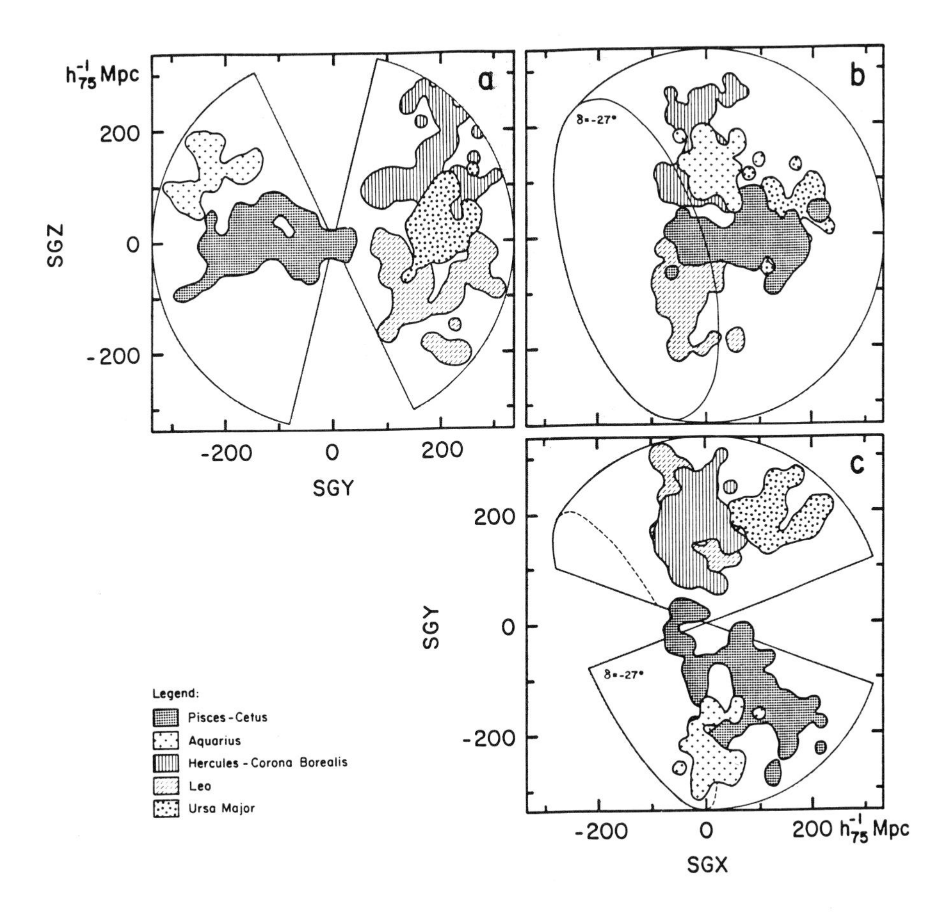

Fig. 1.14. A set of orthogonal views of five major supercluster complexes out to 30,000 km/s, as identified by B. Tully. The rectangular coordinates are 'supergalactic' (see Section 3.3 for explanation). (Reproduced with permission from B. Tully and the *Astrophysical Journal* (**323**, 1, 1987).)

1.7 FURTHER READING

General

Historical perspectives

Chincarini, G. and Rood, H.J., The Cosmic Tapestry, *Sky and Telescope* (1980).

Gregory, S.A. and Thompson, L.A., Superclusters and Voids in the Distribution of Galaxies, *Scientific American*, **246**, 3, 88 (March 1982).

Herschel, J.F.W., *Outlines of Astronomy,* 5th Edition, Longmans, Green and Co., 1858.

Hubble, E., *The Realm of the Nebulae*, Yale University Press, 1936.

Oort, J.H., Superclusters, *Ann. Rev. Astron. Astrophys.*, **21**, 373 (1983).

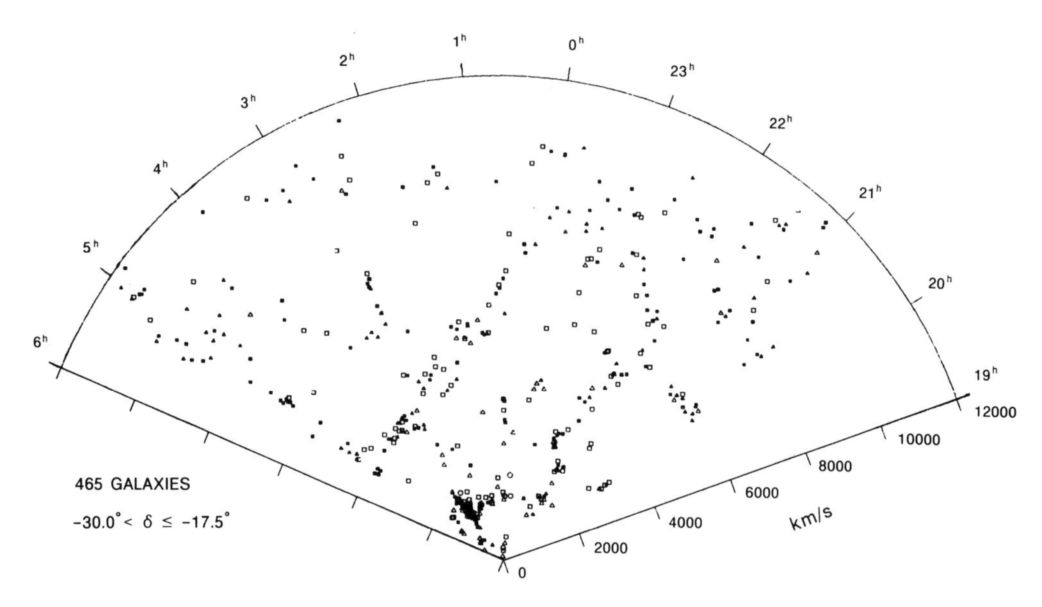

Fig. 1.15. A sample plot from the Southern Sky Redshift Survey confirming that cellular structures also are to be found in the southern skies. The angular coordinate is again Right Ascension; the Declination range is -17.5° to −30°. (Reproduced with permission from L. da Costa and the *Astrophysical Journal* (**327**, 544, 1988).)

Overbye, D., *Lonely Hearts of the Cosmos*, Harper Collins, 1991.
Proctor, R.A., *The Universe of Stars*, Longmans, Green and Co., 1878.

Specialised

Pioneering publications mentioned in the text

Chincarini, G., Clumpy Structure of the Universe and General Field, *Nature*, **272**, 515 (1978).
da Costa, L.N. *et al.*, The Southern Sky Redshift Survey, *Astrophys. J.*, **327**, 544 (1988).
Davis, M. *et al.*, A Survey of Galaxy Redshifts. II. The Large Scale Space Distribution, *Astrophys. J.*, **253**, 423 (1982).
de Vaucouleurs, G., Evidence of Local Supergalaxy, *Astron. J.*, **58**, 30 (1953).
Einasto, J., Hypergalaxies, [in] *The Large-Scale Structure of the Universe*, IAU Symposium 79 (*Ed.* M. Longair and J. Einasto), 51 (1978).
Fairall, A.P. and 60 undergraduate students, Redshift Plots of Huchra's Z Catalogue, *Publ. Dept. Astron. Univ. Cape Town,* Number 8 (1985).
Fontanelli, P., The Coma/A1367 Filament of Galaxies, *Astron. Astrophys.*, **138**, 85 (1984).
Gregory, S.A. and Thompson, L.A., The Coma/A1367 Supercluster and its Environs, *Astrophys. J.*, **222**, 784 (1978).
Haynes, M.P. and Giovanelli, R., The Connection between Pisces–Perseus and the Local Supercluster, *Astrophys. J.*, **306**, L55 (1986).

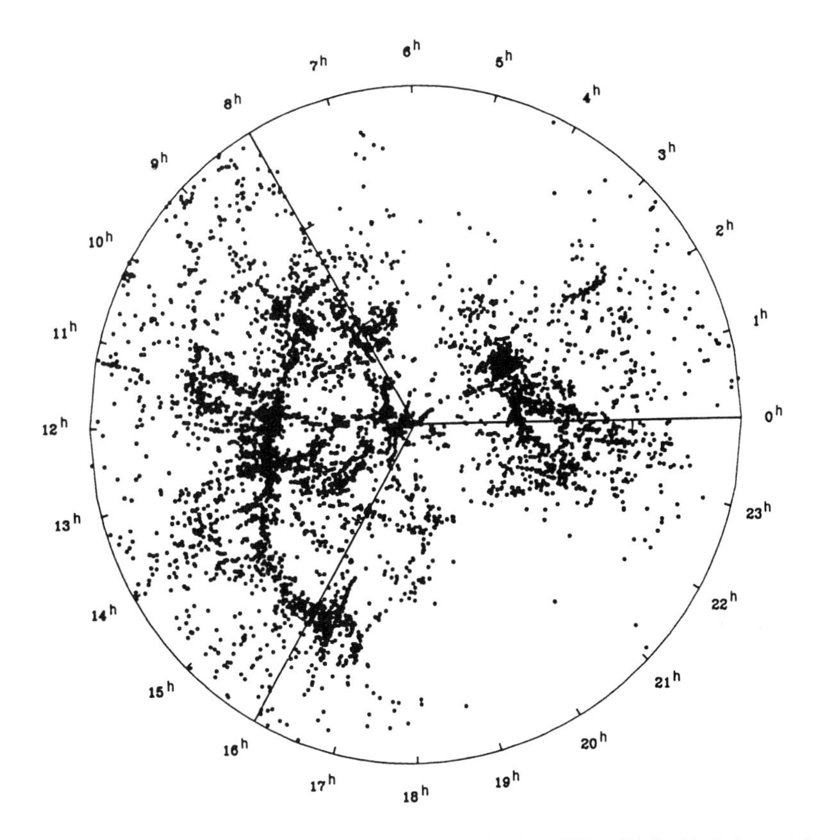

Fig. 1.16. The Great Wall, a plot of ZCAT data, for Declinations 20° to 40°, by M. Geller and J. Huchra, showing the elongated structure that dominates the left-hand side of the diagram. (Reproduced with permission from M. Geller.)

Joeveer, M. and Einasto, J., Has the Universe the Cell Structure?, [in] *The Large-Scale Structure of the Universe*, IAU Symposium 79 (*Ed.* M. Longair and J. Einasto), 241 (1978).

Kirshner, R.P. *et al.*, A Million Cubic Megaparsec void in Boötes?, *Astrophys. J.,* **248**, L57 (1981).

Shane, C.D. and Wirtanen, C.A., The Distribution of Galaxies, *Publ. Lick Obs.,* **XXII,** Part 1 (1967).

Tully, R. Brent, Alignment of Clusters and Galaxies on Scales up to 0.1*c*, *Astrophys. J.*, **303**, 25 (1986).

Winkler, H., The Spatial Distribution of Galaxies in the Southern Sky, *Mon. Not. Astr. Soc. Sthn. Africa,* **42**, 74 (1983)

Zel'dovich, Ya. B. *et al.*, Giant Voids in the Universe, *Nature*, **300**, 407 (1982).

2

Measuring galaxy redshifts

2.1 GALAXY MORPHOLOGY AND LUMINOUS CONSTITUENTS

The recognition and mapping of large-scale structures in the distribution of galaxies (as outlined in the opening chapter) has come about from spectroscopic observations at the telescope, from which redshifts could afterwards be extracted. As its title suggests, this chapter is devoted to the instrumentation and reduction procedures involved in obtaining redshifts. It is this author's 'bread and butter', having probably measured approaching 5,000 individual galaxy redshifts over a 30-year career. Over that period, great advances in technology have improved the speed, reliability and accuracy, and it is a humbling experience to realise that all the observations that went into my Ph.D. thesis could now be done in a single night!

Up to now, all the diagrams have depicted the galaxies as 'points'. That is correct in the sense that, on the same scale, even the largest galaxies would have diameters no more than a tenth of that of the full stop printed at the end of this sentence. But it is as well to remember that galaxies are really gigantic stellar systems in themselves, and this opening section summarises their contents and character.

The description 'island universes' is very apt. Each galaxy harbours a vast population of stars – perhaps a million million stars in a giant galaxy such as our own. There are almost as many stars in a galaxy as there are cells in a human body! Each star is, of course, a sun like our own, and like our Sun may possess a planetary system in orbit about it. Apart from stars and planets, a significant fraction of the apparent mass may be in the form of interstellar matter.

There are also more exotic components. A small percentage of galaxies exhibit active nuclei, usually best interpreted as the manifestations of material being violently sucked towards a black hole. It is likely that all large galaxies possess black holes – with masses measured in millions of that of our Sun – but most lie dormant, with only the mildest displays, like that probably possessed by our Galaxy.

The manner in which many galaxies rotate suggests that they may also contain large amounts of non-luminous material. The mass of such material may be comparable to, or even exceed, that of the stars. It may be that what we see as galaxies are like icebergs, where the visible portion is only a small fraction of the whole. All sorts of suggestions have been made concerning the nature of the non-luminous material, of which the most plausible is 'Jupiter-like' objects. Later discussion in this book will concern the general possibility of there being vast amounts of non-luminous matter in the Universe. Such dark matter could even be separate from the galaxies.

But galaxies are things of beauty, as clearly seen in the photographs shown in Figure 2.1. Redshifts can only be obtained from the 'light' that the galaxy emits, and the pictures here show the portions of the galaxy from where that light originates. The photographs, however, tend to exaggerate the brightness of the outer regions; it is the central nucleus that contributes most of the light. Most, if not almost all, of the light is from stars, but the intrinsic luminosity of individual stars varies enormously. The most luminous stars (such as the three that make up Orion's Belt in our night sky) are approaching a million times greater luminosity than our Sun! The least luminous stars known have more like 1/10,000 of the Sun's luminosity. 'Luminosity functions' suggest that the relative numbers of stars increase dramatically towards the fainter limits. Thus very high-luminosity stars are extremely rare, and very low-luminosity stars are very common. Even so, the small minority of rare highly luminous stars accounts for the bulk of the emitted starlight. And this is especially relevant when considering the appearance of distant galaxies.

Many of the galaxies we see show spiral structure. For example, Figure 2.1 includes a galaxy with two dominant spiral arms. It would be tempting to suppose that the stars are only contained within the spiral arms, and that the spaces between the arms are empty. Yet, judging from our knowledge of our own Galaxy, there are probably just as many stars in the apparently empty spaces as there are in the spiral arms themselves. It is simply that the very high-luminosity stars are confined to the spirals arms.

High-luminosity stars are the most massive stars. While their masses may approach 10^2 times that of our Sun, they attain their extreme luminosity by consuming nuclear fuel at a rate perhaps 10^5 times faster. Consequently their lifetimes may only be of the order of 10^6 to 10^7 years. Against astronomical time standards, this is like saying that their lifetimes are so short that they die before leaving the maternity home! The presence of high-luminosity stars, such as those marking the spiral arms, therefore mark the sites of star formation.

The favoured interpretation of the spiral arms is that of a 'density wave' – a pattern of compression and rarefaction – in the interstellar gas in the disk of the galaxy. The spiral arms represent the regions of compression. The wave pattern rotates about the centre of the galaxy, but not as fast as the constituents – the stars and gas – revolve. In overtaking the wave pattern, the interstellar gas moves through the spiral compression waves. It enters from the inner edges of the spiral arms, is compressed as it passes through, then expands as it leaves the outer edges, and in time catches up with the next spiral arm. It also travels at supersonic speeds. This results in the creation of shock fronts along the inner edges of the compression waves – the inner edges of the spiral arms. The extreme compression within the shock is believed to initiate star formation, hence the presence of the high-luminosity, but short-lived, stars that mark the spiral arms. Dark dust lanes wrapped on the inside of the arms reveal the shock fronts.

The most extreme examples of such high-luminosity short-lived stars will have sufficiently high surface temperatures that they will emit enough thermal ultraviolet radiation to ionise the surrounding interstellar matter. A state of equilibrium where recombination equals ionisation (out to the radius of a 'Stromgren sphere') causes an HII region, which emits a recombination spectrum – an emission line spectrum. Such regions tend to occur butted up along the outside of the dust lanes, where such stars enjoy their brief lives.

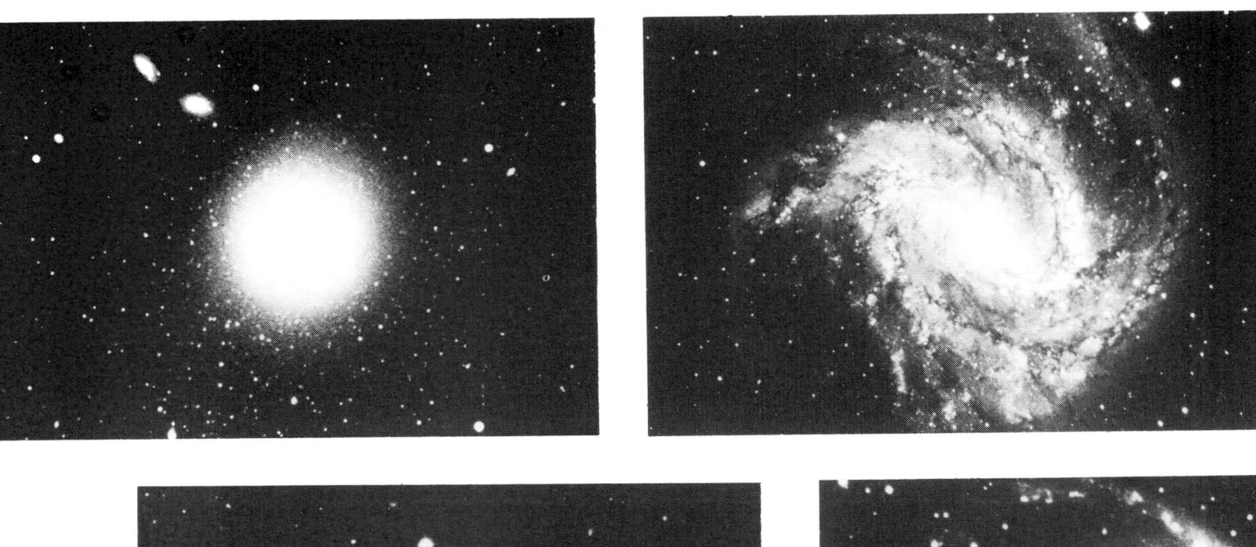

Fig. 2.1. Galaxies exhibit a variety of forms. Spiral galaxies are the most common sort in lower density environments. Elliptical galaxies prefer clusters. (Adapted from slides and reproduced with permission from David Malin and the Anglo–Australian Telescope Board.)

Most of the interstellar matter is, however, neutral hydrogen (designated as HI) which is cold and invisible, at least at optical wavelengths. Fortunately, a transition of the hydrogen atom's spin alignment, between its nucleus and electron, results in substantial emission at low-energy radio wavelengths, known as the 21-cm emission line.

This, however, is only a simple summary of a vast tract of astronomy, and it hardly does justice to an immense subject. For the purposes of this book, galaxies need only be thought of as points of light. The light emitted by them may originate from a high-density population of stars, or from regions of lower density but with ongoing star formation, or partly from HII regions, or even active nuclei. Alternatively, galaxies with a substantial content of neutral hydrogen gas are easily detected by radio telescopes.

2.2 SPECTROGRAPHS

A spectrograph is an instrument that takes the light of a distant galaxy and disperses it into a spectrum, from which the redshift or other information may be extracted. A telescope is used to gather sufficient light to direct into the instrument. The spectrograph is either mounted on the telescope, or placed near the telescope and fed by optical fibres. The majority of professional telescopes today operate as Cassegrain reflectors, where the spectrograph is attached to the lower end (such as shown in Figure 2.2).

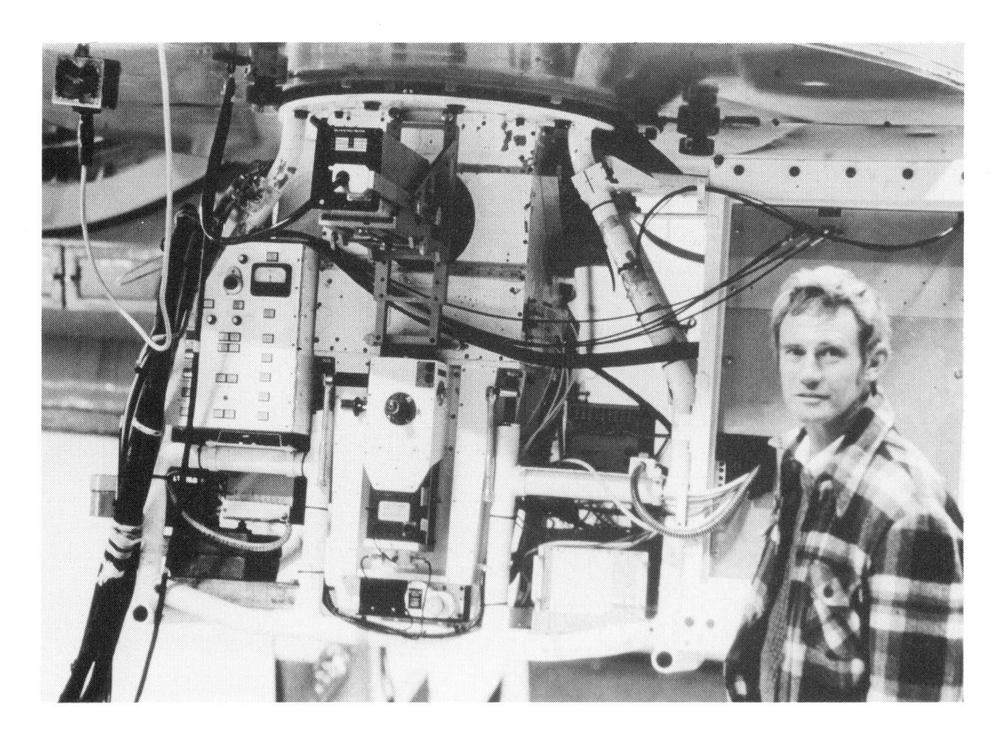

Fig. 2.2. A spectrograph mounted at the Cassegrain focus of the 1.9-m (74-inch) Radcliffe reflector of the South African Astronomical Observatory. (Photograph by Herschel Mair.)

Analogous to the funnel in a rain gauge, a telescope is used simply as a device to funnel light into the spectrograph. Provided it can do this, the actual image quality the telescope produces is not as critical for spectroscopy as it may be for other purposes. This has suggested that specialised spectroscopic telescopes could be built, with less stringent optical specifications, and therefore less expense. The idea has now culminated in a new form of very large telescope, built at a fraction of the normal cost. The Hobby–Eberly telescope at McDonald Observatory uses a fixed spherical primary mirror. The images it produces are tracked by a unit that carries corrective optics and the light so focused is intercepted with a fibre. The long-term success of the system will prove influential in future projects.

The light of the galaxy is isolated from its surroundings by positioning the galaxy's image (in the focal plane of the telescope) on the entrance slit of the spectrograph. Most spectrographs allow for the slit width (and length) to be adjustable. While as narrow a slit as possible is desirable for maximum spectral resolution, too narrow a slit will mean that much of the galaxy's image spills over onto the sides of the slit and light is lost. Obviously, a compromise is necessary.

Atmospheric 'seeing' usually smears star-like images – or the central portion of a distant galaxy – into disks typically 1–2 arcsec across. The better the seeing, the smaller the disk. Seeing conditions where star images are as small as 0.5 arcsec occur under optimum conditions at good observing sites. By contrast, humid air, affected by the temperature variations of condensation and evaporation, may produce very poor observing conditions and 'blow up' the seeing to many arcsec. A typical width set for the spectrograph slit would correspond to around 1.5 arcsec in the sky. Spectroscopy is highly dependent on seeing. Under good seeing conditions, over 90 per cent of the galaxy's light may pass through the slit. However, if seeing deteriorates, the figure could drop far below 50 per cent. Exposure times need to be adjusted accordingly.

Figure 2.3 shows the basic layout of a spectrograph. After the slit, the light is led into a collimator which converts all the light rays emerging from the slit into a parallel beam. Since there is often the need to keep up good response in the violet end of the spectrum, glass lenses are avoided, and instead, reflecting elements are employed. A miniature Cassegrain telescope, used in reverse fashion, is a favoured design for a collimator; normally a telescope takes a parallel beam of starlight and focuses it to a point, but here it is used in reverse mode. The focal ratio of the collimator (its focal length divided by its

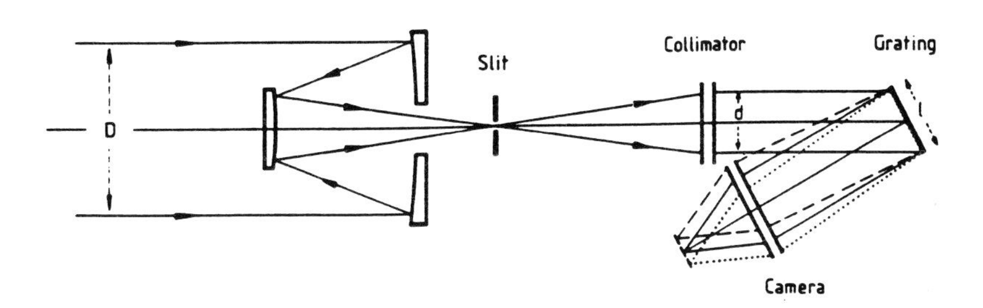

Fig. 2.3. The basic optical layout of a spectrograph. (Reproduced with permission from the *Monthly Notices of the Royal Astronomical Society* (**250**, 796, 1991).)

aperture) must match that of the telescope on which the instrument is mounted, otherwise light will be lost by spilling over the sides of the elements, or obscured by oversize components.

The emerging parallel beam from the collimator passes to a dispersing element. In years gone by, glass prisms were used, but the problems of poor transmission of violet light through glass have long since led to a switch to reflection gratings. Figure 2.4 shows

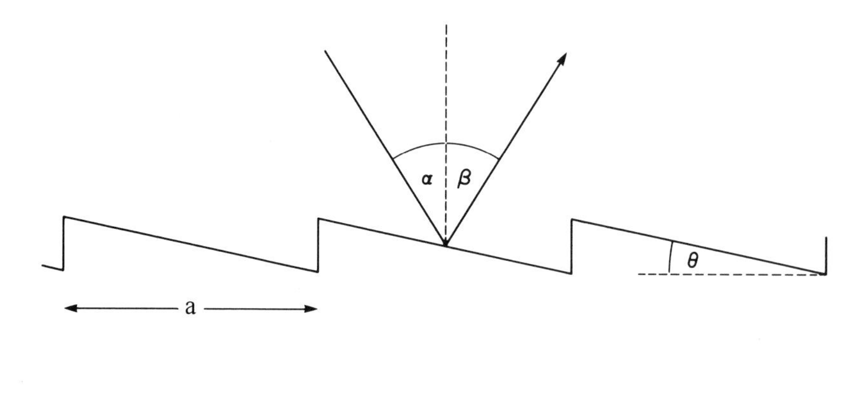

Fig. 2.4. A grossly enlarged segment of a reflection grating, as seen in cross section. A light ray is shown reflecting off one of the many microscopic facets.

light being reflected off the highly enlarged surface of a reflection grating. To make such a grating, a sharp cutting tool, such as diamond, is drawn across a perfectly flat glass plate in a straight line; the tool is then offset slightly and the procedure repeated. The offsets are uniform and typically less than 2 μm, so that there are usually several hundred 'lines' per millimetre. Gratings work by the constructive interference of light waves, given by the expression

$$n\lambda = a\,(\sin\alpha \pm \sin\beta)$$

where n is the order number (an integer, either positive, negative or zero), λ is the wavelength of light, and the angles and spacing are as indicated in the figure. Except for the zero order, the grating thus disperses different wavelengths to different angles (in a plane perpendicular to that of the slit). Only one order can be recorded, and it is wasteful to loose valuable light to other orders. However, the angled facets cut on the glass plate also serve like mirrors to concentrate the light into a particular direction and therefore into a particular order – usually first order. The angle θ is referred to as the blaze angle. (The manufacturing process and the need for strict tolerances make reflection gratings expensive, and although it is possible to make inexpensive plastic mouldings from a master, plastic is physically insufficiently stable for this job.)

The beams reflected off the grating are intercepted by a camera, which reverses the role of the collimator by bringing the parallel beams back to a focus. Were the light monochromatic, the image so formed would be the image of the entrance slit. For a continuum of colours, the image of the entrance slit is smeared sideways to form a spectrum; consequently, resolution in the spectrum is dependent on the narrowness of the slit. The resolution, R, can be described in terms of $\Delta\lambda$, representing the slit width expressed as wavelength, but since this quantity varies with wavelength, the more usual definition is

$$R = \frac{\lambda}{\Delta\lambda}$$

Most modern spectrograph cameras seek an image of maximum brightness, so a fast camera with a low f-ratio is used – typically a variation on the Schmidt design. The image is also usually located physically clear of the camera body to permit an electronic detector to be mounted.

2.3 DETECTORS

Since the 1960s, spectrographs have used electronic detectors of one sort or another, rather than photographic emulsions, to permit them to observe much fainter objects in much shorter times. For example, when I was a student working with Fritz Zwicky, I was loaned photographic spectra taken with the Hale 5-metre (200-inch) telescope, each a 4-hour exposure! Today, using modern detectors, those same spectra could be obtained in little more than 4 minutes with only a 2-metre telescope.

The choice of detector, mounted in the focal plane of the spectrograph's camera, is nowadays almost always a CCD (charge-coupled device) detector. These detectors are often more than 80 per cent efficient in recording photons of light. Within a silicon layer, the energy of the incoming photons elevates electrons from the 'valence' levels (where they are attached to individual atoms) to the 'conduction' layer (where they are free to move). These 'photoelectrons' are, however, trapped and held in a matrix of potential wells, created by a miniature rectilinear array of electrical 'wires' immediately beneath the silicon. Two sets of wires run perpendicular to each other, creating 'columns' and 'rows' of wells. In each set, every third 'wire' is positive with respect to its neighbours, thereby forming a potential to attract and hold electrons. The matrix of potential wells forms the 'pixels' of the image being recorded. The photoelectrons are accumulated in these wells, much as the water from falling raindrops might be collected in an array of buckets.

Once the spectrographic exposure is complete, the CCD is 'read out' (as though the amount of rainwater in each bucket was measured in sequence). By shifting the positive potentials to a neighbouring 'wire', the potential well is displaced sideways within the silicon, carrying the photoelectrons with it. By continuously repeating this action, the potential wells, with their load of photoelectrons, are moved sideways in conveyor belt action. This mechanism is used to bring each well to a corner electrode where its contents can be electronically measured. The wells in the same column are read out sequentially, then the entire remaining array is shunted over one column so that the next column can be read out. Eventually, all the wells are processed. The electron measurements are encoded and passed on to a computer to assemble as an image. Dummy readouts can also be used to 'prepare' the CCD immediately before an exposure begins, by ensuring all the wells have been emptied.

CCDs have one major disadvantage in terms of spectroscopy of galaxies. Usually the most conspicuous spectral features are the stellar H and K lines (discussed later) which occur at wavelengths of 397 and 393 nm – but at the violet end of the spectrum, where the CCD has much poorer response. Attempts to improve the violet response are sometimes made. Fluorescent coatings can re-emit the incoming photons at longer wavelengths. Since the silicon itself is fairly opaque to ultraviolet, CCDs can also be 'thinned' (a very delicate conversion) and used upside-down.

In comparison with CCDs used in camcorders and similar equipment, the CCDs used in spectrographs are highly selected for optimum cosmetic properties (and are therefore expensive to purchase). In addition, much more effort is made to reduce electronic noise. Usually the entire CCD will be cooled and operated at the temperature of liquid nitrogen to reduce thermal noise. Also, 'readout' noise is reduced by stretching the readout to some 10 seconds or longer. Camcorders, which operate with abundant photons, do not need such precautions, but the light of faint galaxies is so feeble that relatively few photons are available, and astronomical CCDs must be able to record a very weak signal.

The image extracted from a single CCD is of course monochromatic; colour is hardly necessary since it is conveyed by the varying intensity along the spectrum. (Nevertheless, those who visit my telescope control room seem unimpressed that a spectrum of colours is represented by a 'black and white' image; or they are totally confused when false colours are used to represent intensity levels on the computer display.)

Over the years, the number of pixels in a CCD has increased with improved manufacturing techniques. Today, arrays of more than 1,000 × 1,000 pixels are readily available. Physically, the CCDs may measure about 30 × 30 mm. Oblong CCDs, with perhaps 2,000 × 300 pixels, are best suited for spectroscopy, with the wavelength dispersion spread over the longer dimension. The shorter dimension then conveys information along the slit. Usually the galaxy spectrum will occupy several central rows of pixels, with the sky-alone spectrum on either side.

The pixel sizes of CCDs are typically about 15—20 μm across. In mounting a CCD as a spectrograph detector, the pixel size must be matched to the resolution of the spectrograph. The pixel size ought to be about a half of the element of resolution – the width of an entrance slit in the focal plane of the camera. The choice of too coarse a pixel size would degrade resolution; the choice of too fine a size would give more pixels than necessary for the resolution, and such wastage would curtail the wavelength coverage.

Prior to CCDs, particularly in the 1960s to 1980s, many spectrographs used image intensifiers (vacuum tube devices) as detectors. In an image intensifier, the faint image is incident upon a photocathode surface, and the photoelectrons so released are accelerated by high voltage. Using either electrostatic or magnetic fields, the electrons are focused onto a fluorescent screen to produce a bright replica of the original image. It is common to use multiple stages. (As with all such electronic devices, background noise is inevitable, and final amplification is a trade-off against signal-to-noise in the system. The spectral response of the entire system is largely influenced by the spectral response of the first photocathode.) A transfer lens then conveys the bright replica created on the image intensifier's fluorescent screen to a recording device. Early image-tube spectrographs still resorted to photographic emulsion for this.

Developments then brought video techniques with digital output. Perhaps the most successful were the Reticon diode arrays – one-dimensional devices aligned with the spectral dispersion. CCDs and Reticons produce digital outputs – literally photon counting – that can be easily digested by an acquisition computer, which serves to show the observer what is being recorded, but does not normally carry out a full reduction of the data. Such detailed reductions normally involve time-consuming interactions that are best carried out once the observer has carried his recorded raw data back to his home institution.

2.4 MULTI-FIBRE SPECTROSCOPY

The discussion so far has implied that the spectrograph is used to observe a single galaxy at a time. This is certainly still the case with many instruments, but it has long been realised that efficiency could be greatly improved if a number of adjacent objects could be recorded simultaneously. The advent of fibre optics has made this possible, so that instead of the image of a galaxy being positioned on the entrance slit, its light is captured by a fibre. The fibres aligned on many such galaxies are then led to a long spectrograph slit, so that all the galaxies' spectra are recorded simultaneously, but are separated along the length of the slit. (In many cases, the spectrograph is no longer mounted on the telescope, but stands nearby.) This arrangement calls for very accurate positioning of the fibres, so that they correspond to the positions of galaxy images in the focal plane of the telescope. Precise positioning is crucial. The best method is to use the same telescope to photograph the field first. Precise positions can then be measured, and – most desirably – photometric measurements of the galaxy's brightness can be obtained at the same time.

To position and support the fibres, most of the 'first-generation' systems use holes drilled in metal plates. A different metal plate is prepared (in a precision machine shop) for each field. At the telescope, the plate is positioned in the focal plane and the fibres are then manually plugged into the holes. The procedure involves 'book-keeping'; that is, keeping track of which fibres plug into which holes. Bright stars are then used to set the aim of the telescope in register with the plate, and to monitor the guiding of the telescope during the exposure. Though the technique is somewhat pedestrian, it is the easiest way to position fibres and is being retained for the forthcoming Sloan Digital Sky Survey (where fibres will be plugged in, in any order, and scanning two perpendicular shafts of light across the plate will enable a computer to sort out the book-keeping).

Plugging fibres into holes means that the fibre intercepts the light end on, each fibre having a polished end face to facilitate entry. An alternative arrangement – in particular practised at the Anglo–Australian Observatory – is to equip each fibre head with a miniature right-angled prism. The fibres can then be led away 'sideways'. For example, in the FLAIR system used with the Schmidt telescope, the fibre heads are manually glued on to an existing photographic plate of the field to be observed. The fibres are then positioned precisely according to the images of the galaxies on the photograph. The photographic plate is then remounted in the telescope in the same way as when it was originally exposed, but this time carrying the fibres for the spectroscopic observations.

'Second-generation' systems have attempted to reduce the need for machining metal plates and the manual labour of plugging in fibres. For example, the MEFOS system, developed by Paul Felenbok and workers at the Observatory of Paris–Meudon, employs robotic arms to carry the fibres. Like fishermen around a circular pond taking care never to strike their rods together, the arms move radially in or out, with limited sideways movement taking only a few seconds to adopt new configurations. The aiming of the telescope and the robot positions are arranged in advance by a software package provided with the coordinates of the galaxies in the field, and the scale (and slight distortion) of the telescope's focal plane. In order that the positioning of the fibres can be precise, each robot arm carries an imaging bundle of fibres (as used extensively in the medical world) that first provides an image of the galaxy and a tiny piece of surrounding sky. An array of such images is displayed (a CCD exposure) and interactive software then allows the observer to tweak the positioning of each robotic arm. This facility proves very useful; for example, when observing in crowded fields close to the Milky Way. The arms are then repositioned slightly so that the spectrograph fibres are set exactly on the galaxy. In fact, each arm carries two such fibres so that the second fibre generally falls on a blank piece of sky and can be used for 'sky subtraction' (see below).

In the MEFOS system, every arm has its own robotic control system, and the physical size of the arms limits the number that can be employed. By contrast, the Anglo–Australian Observatory uses a single robot to position fibres sequentially. As described above, the fibre ends have miniature right-angled prisms so are led in sideways.

Fig. 2.5. The '2dF' system on the Anglo–Australian Telescope. Magnetic buttons have been accurately positioned where images of galaxies will fall in the focal plane. Each button carries the tip of an optical fibre, to carry the light away to the spectrograph.

Each fibre head further carries a small magnetic base, so that it can adhere to a metal base plate. A single robot picks up each fibre head and places it in the correct position. The new 2dF system is the most ambitious yet in employing 400 fibres over a 2-degree wide field (see Figure 2.5). As the time taken by the robot to set the fibres in position is considerable (and one would hardly wish to sacrifice observing time) two metal base plates and two sets of fibres are involved. Whilst one set is being exposed, the robot works on setting up the next field.

The use of fibres is well established, or even essential for the acquisition of a viable number of redshifts. The future is obviously going to bring even more sophisticated developments.

2.5 OBSERVING PROCEDURES AND REDUCTION

2.5.1 Introduction

Observations with such equipment as described above consist of sequences of exposures or 'integrations', each of which is read out into the acquisition computer. Whilst some of these are exposures (or multiple exposures) using the light of the galaxies to be measured, it is also necessary (as will be described in this section) to make separate exposures of calibration arcs, preceding and following each galaxy integration. There are exposures for 'flat fields', focusing, standard radial-velocity stars, and so on.

2.5.2 Photon noise

Detection or measurement in astronomy is invariably a case of signal versus noise. For each exposure, photons are counted for each pixel element, or channel, of the detector. Since these photons do not arrive at uniform intervals of time, but rather in random fashion, they introduce noise. Even if nature intended the same number of photons to enter each pixel, the random arrival times mean that at any moment, some pixels are 'ahead' and some are 'behind' in their counts. Such noise is dealt with by Poissonian statistics, whereby the standard deviation is given by the square root of the number of photons counted. Consequently the signal-to-noise ratio also varies as the square root of the counts. The larger the number of counts, the better the signal-to-noise. For this reason, it would be desirable if all exposures or 'integrations' were as long as possible, so as to collect as many photons as possible and to have as good a signal-to-noise ratio as possible. Obviously, when observing faint galaxies, where photons are scarce, a compromise is necessary. For calibration arcs or flat-fields, which have much higher counts, it is hardly a problem.

2.5.3 Focusing

Before a spectrograph can be used, the instrument must be accurately focused, usually by physically shifting one of the camera elements. A quantitative indication of focus is needed; that most commonly used is a pair of Hartmann shutters. Each of these two shutters can obscure half of the wide beam of light emerging from the collimator. A pair of exposures of the calibration arc are made, first with one and then with the other shutter in place. If the camera is not in focus, then spectral features (the arc lines) are displaced from one exposure to the other. This displacement is nowadays normally measured by a suitable software routine (cross correlation, as described below). The focus can then be adjusted until there is no significant displacement. Over and above focusing, some spectrographs may call for other adjustments (e.g. rotation and tilt of the CCD) prior to observing, but these will be peculiar to the instrument concerned.

2.5.4 Cosmic ray events

CCDs not only detect photons; they also detect cosmic rays. CCD images are usually dappled with a light sprinkling of cosmic ray spots. However, while their intensity can be quite high, they are contained within single pixels. Since the pixel size is smaller than the resolution in the image, they are readily identifiable as cosmic rays, rather than information in the image. As such, they can be located by software routines, and the intensity level within the affected pixel can be replaced by an average value from neighbouring pixels. This 'tidying up' is the first step in reducing the image.

2.5.5 Flat fields

The pixel elements of a CCD array (or the older Reticon arrays) do not have identical efficiencies in recording incoming photons. The first step in an observing session is to establish the relative sensitivities. Separate 'flat-field' exposures are made (usually at the beginning and end of the night) with even illumination over the detector. This illumination

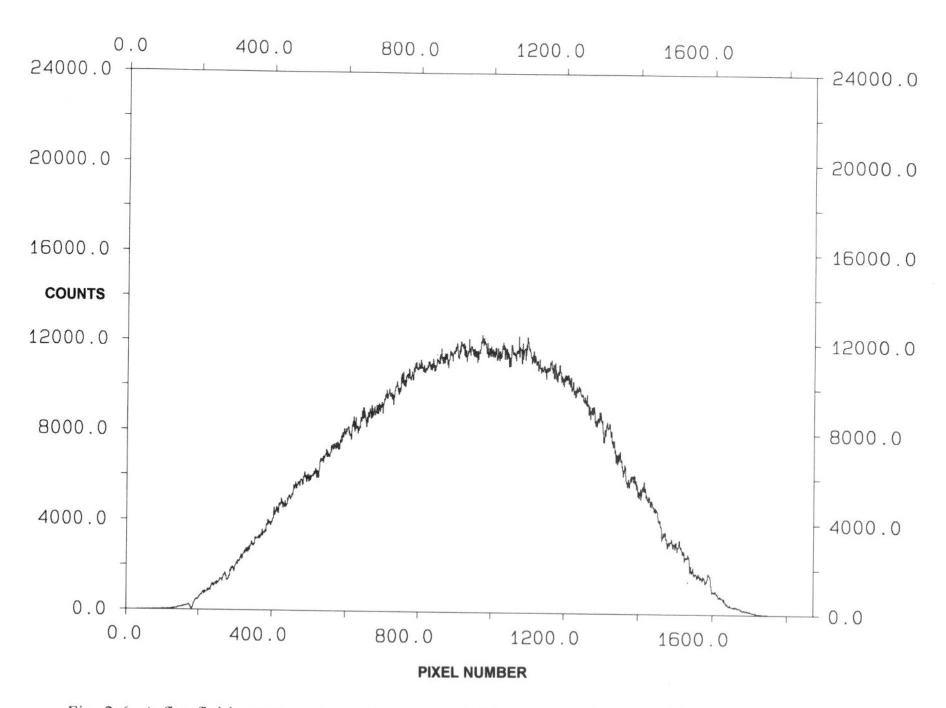

Fig. 2.6. A flat-field exposure (one-dimensional). The general shape of this spectrum is that of a smooth 'thermal' continuum from an incandescent lamp, curtailed at either end by the instrumental response of the spectrograph. The 'noise' superposed on the continuum is due to the uneven response of individual pixels. As described in the text, an exposure such as this is used to evaluate individual pixel sensitivities.

is typically provided by a simple incandescent lamp shining on the slit of the spectrograph; or the telescope could be aimed at an illuminated white screen within the dome. Additional exposures could also be made of the twilight sky; the sky should give even illumination along the slit (but not along the spectrum since it would have the spectral variations and features of diluted sunlight). The recorded output ought to be an even smooth featureless spectrum, but instead the variations in sensitivity from pixel to pixel create a lot of noise (see Figure 2.6). The flat-field exposure also contains the photon noise, already described, but this is minimised by a high count rate, or by averaging multiple exposures.

The correct spectrum would be a heavily smoothed version of this recorded spectrum. Such a spectrum can be generated by the software (of both the acquisition and reduction packages). Then, by dividing the original noisy spectrum by this heavily smoothed version, the pixel-to-pixel variations, or 'flat-field corrections', are extracted; i.e. a correction factor for each individual pixel. For less efficient pixels, the flat-field correction is slightly more than unity; for over-efficient pixels, it is slightly less than unity. If the original noisy spectrum were to be multiplied through by the flat-field corrections, it would then be restored to the even smooth featureless spectrum that it should have been. It is a reflection on the quality of the detector as to whether the flat-field corrections are stable, but it is obviously safest if frequent flat-field exposures are carried out as part of the observing procedures.

Ideally, the flat-field exposure should be made with the same reflection grating, set at the same angle, as used for the galaxy observations. If so, then the spectrum would be precisely positioned on the detector where the galaxy spectrum would fall. Changing the grating dispersion or angle may shift the spectrum very slightly sideways on the detector, due to the minute tolerances in the mechanical engineering of the spectrograph and the mounting of gratings. Unfortunately, it is often necessary to disturb the grating, because the temperature of an incandescent lamp filament is nowhere as hot as a stellar surface, so the peak colour of its light is quite different. If adequate flat-field photons are to be recorded where the blue end of the galaxy's spectrum is to fall, then a change in grating angle is necessary. If the grating position is to be disturbed, then one might as well change the grating to a higher dispersion. That way the number of photons registered will not tail off at the red or blue ends of the spectrum; rather, there will be good photon counts over the entire array.

2.5.6 Calibration arcs

Once observing is in progress, exposures of galaxies (and stars) must be bracketed by short 'arc' exposures. These serve for wavelength calibration; they are used for establishing the spectrographic central wavelength for each individual pixel. The slight mechanical flexure of the spectrograph (as the telescope slowly tracks the galaxy), as well as thermal and electronic variations, necessitate arc exposures before and after each galaxy exposure. This is particularly necessary if accurate radial velocities are to be extracted. When lengthy exposures are undertaken, they may have to be interrupted at intervals for arc exposures. The interval usually depends on the experience with the spectrograph, but in general the interval must be shortened if the telescope is observing relatively low down towards the horizon, where the mechanical stress and flexure of the spectrograph are greatest.

The arc itself is a low-pressure gas discharge lamp, located inside the spectrograph, that produces a set of sharp emission lines, of precisely known wavelengths. Argon, neon and helium (or combinations of these) are common choices (see, for example, Figure 2.7). Note that each arc line has a small width to it, even though the light responsible was monochromatic. Rather, the width of the arc line reflects the width of the entrance slit of

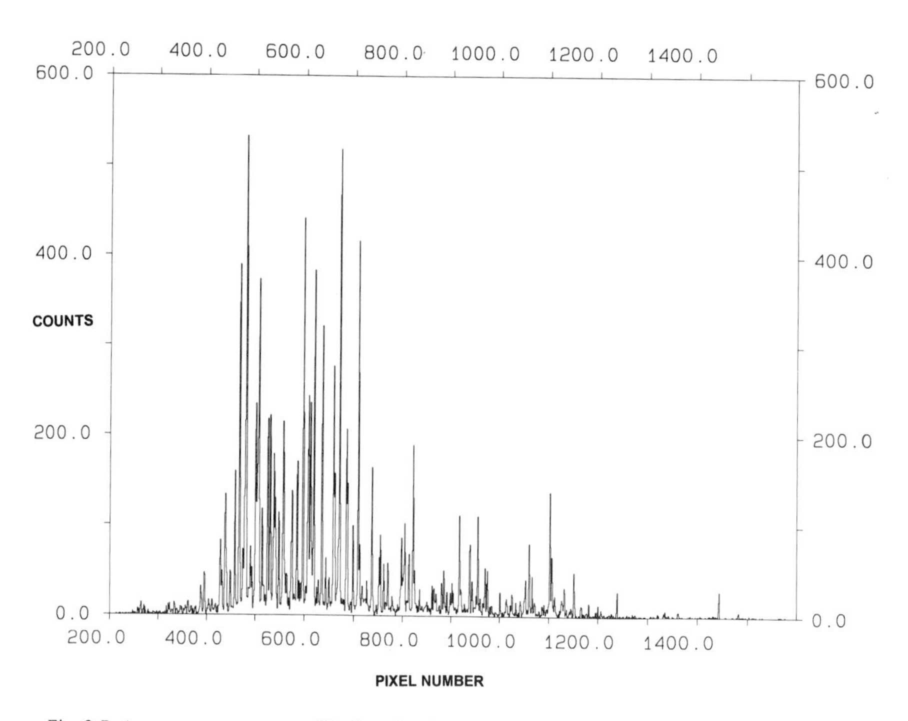

Fig. 2.7. An argon arc exposure. The low-density gas in a discharge lamp gives rise to an array of emission lines rather than a continuous spectrum. The known wavelengths of each of these lines can be used for calibration of the spectrograph.

the spectrograph, and to a lesser degree the quality of the focus of the camera. Opening the slit wider would make the lines fatter; narrowing the slit would make the lines more precise. As discussed earlier, the slit width is set as a compromise, allowing in as much of the galaxy's light as possible, without too serious a degradation of spectral resolution. The spectral resolution of the detector is, of course, a single pixel or channel. As mentioned earlier, there is no purpose in narrowing a slit such that an arc line is narrower than a single pixel. Instead, the system is matched such that for the slit width chosen, arc lines are marginally resolved; that is, recorded in two or three pixels. Typically, if an arc line fell in the exact centre of a pixel, then the two adjacent pixels would register about 25 per cent of the photon counts in the central pixel. It almost goes without saying that the slit width must not be changed between observing the galaxy and recording the arc, particularly since only one jaw of the slit is movable, and opening the slit will shift the centroids of the arc lines slightly.

During reduction, each pair of arc exposures is summed, and a one-dimensional arc spectrum, corresponding to the position of the galaxy spectrum, extracted. Usually the galaxy spectrum will have occupied the middle rows of the CCD image. The arc spectrum is assembled from the same middle rows. Gaussian curves are then fitted to each individual arc line profile, and its centroid (relative to pixel numbers) determined. A mathematical polynomial function is then derived that relates pixel numbers to wavelengths. A fifth-order polynomial would be a typical choice. The procedure for doing this is usually an iterative one, that tends towards the best fit. With each iteration, the software will normally show how much each arc line's wavelength deviates from the polynomial fit. Each iteration then tweaks the polynomial to reduce the least squares of the deviations, until no significant improvement can be made.

Arc fitting is the most demanding aspect of data reduction. In the process of fitting Gaussian curves, many arc lines – especially those with low counts or close neighbours – may be rejected, because noise has made it too difficult to fit a satisfactory profile. To increase the chance of a good fit being obtained, a set of minimum levels is selected, and the computer then discards all data below these levels. An ideal arc spectrum would have only strong lines evenly spaced along the spectrum. However, in reality, there are portions of the spectrum only serviced by weaker lines. Compromise levels then have to be selected, so that enough background is subtracted without leaving too few counts in the lines for Gaussian profiles to be fitted. The loss of strategic arc lines could seriously affect the accuracy of radial velocity measurements. Consequently there has to be a system for monitoring which lines are being recognised; and where critical lines are lost, the arc fit has to be reworked with different levels.

In order for the software to recognise the lines, an initial set of identifications versus approximate pixel number has normally to be provided for the software package. The package usually has a library list of arc line wavelengths. This list may be varied according to the grating dispersion used. For high dispersions, a very detailed listing may be required, whereas low dispersions need a library of lines without close neighbours so that the software is not ambiguous in making identifications. A minimum of around forty identified lines spread along the spectral range covered is a generally accepted rule. Again interaction is necessary; tolerances for identifying lines need to be chosen and correct identifications need to be monitored.

This author's experience is that it is usually best to work with extreme caution in establishing an initial good arc fit. Thereafter, during the rest of the night's observing, (and as long as the grating is undisturbed) the software need only tweak the fit. One still monitors to ensure that the repeated fits do not deviate by more than expected. A standard problem to watch is whether the software loses identifications at the extreme ends of the spectrum; should it do so the repeated iterations will have an adverse effect, causing the polynomial fit to 'whip about' at either end, with ripples throughout.

When the detector has registered more than one target object along the slit, independent arc fittings (that again differ only slightly) may have to be made for the different portions of the detector concerned. The polynomial coefficients derived from the arc fitting can then be used to convert a galaxy or star spectrum, originally recorded against pixel number, to one according to wavelength. The process effectively 'rebins' the photon counts into channels or pixels of uniform width in wavelength. Thus a wavelength-calibrated spectrum is obtained.

It is also best to check the initial arc fit using the coefficients to produce a wavelength-calibrated version of the arc spectrum itself. The correct identification of arc lines can then be easily confirmed, giving one full confidence that no errors have crept in.

2.5.7 Radial velocity standard stars

As a further check that nothing can go wrong, it is good policy to observe radial velocity standard stars, of similar spectral type (G and K) to that of the galaxies observed. (Such stars are listed in the *Astronomical Almanac*, for instance.) They can be observed during evening and morning twilight, without any penalty in galaxy observing time. As with the galaxy spectra, the stellar spectrum is normally contained in the middle rows of the CCD exposure, so that suitable summing produces a one-dimensional spectrum. Thereafter, polynomial coefficients extracted from the adjacent arc exposures can be used to calibrate the spectrum against wavelength.

A sample spectrum is shown in Figure 2.8. The prominent spectral features labelled there are the same as those expected in galaxy spectra (since the galaxies are assemblages of similar stars). The spectra of these standard stars should be processed in exactly the same manner as that intended for the galaxies, and radial velocities extracted (as detailed

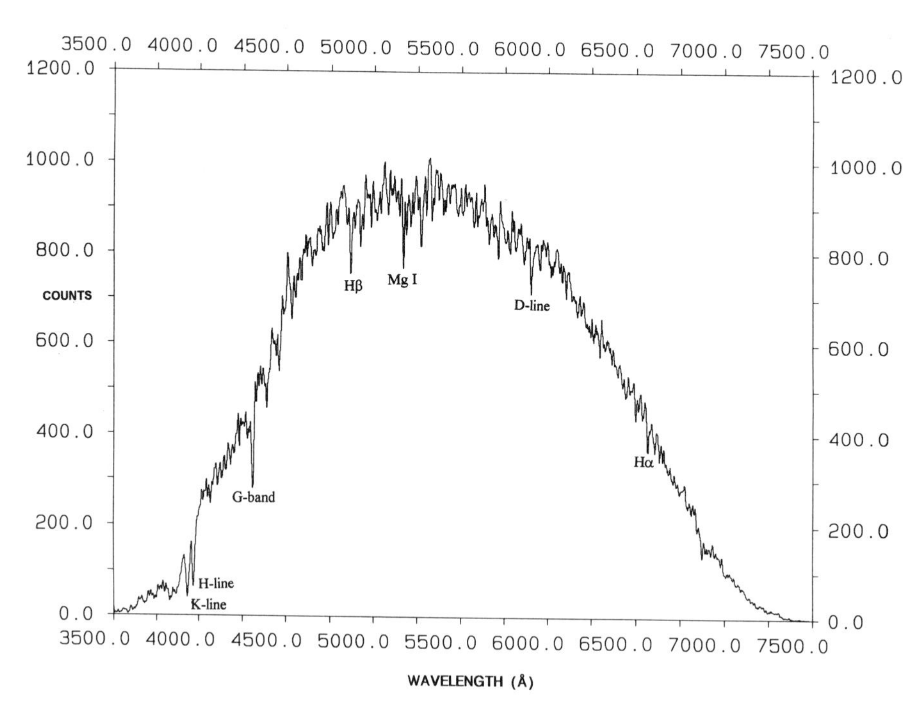

Fig. 2.8. A spectrum of a star of early G spectral type. The principal absorption features are labelled. This star is a standard star with a precisely known radial velocity. Measuring that velocity from this spectrum is a way of checking that there are no systematic errors in observation and reduction. (Wavelengths are calibrated in Ångstroms, where 10 Å = 1 nm.)

below). The values so measured should, within the established uncertainties, match the known radial velocities. It is confirmation that the system – both hardware and software – is working correctly. Some researchers may feel this an unnecessary precaution, but given the complexity of the spectrograph and the reduction software, it is very easy to overlook or be unaware of certain aspects, limitations or even systematic errors. Observing radial velocity stars is the most sensible precaution.

2.5.8 Sky subtraction

By comparison with the standard radial velocity stars, the overriding problem in obtaining spectra of galaxies is their general faintness. In a spectrum recorded in a realistic exposure time, the total number of galaxy photons registered may be as little as a few thousand. At this level of faintness, there is also a significant contribution from the (foreground) night sky – recorded together with the galaxy. There needs to be a means whereby the sky alone is also exposed, preferably simultaneously with the galaxy. The sky contribution can then be subtracted from the combined spectrum to leave only the recorded light from the galaxy.

In the case of a CCD exposure, the sky alone will be recorded on either side of the galaxy spectrum. The spectrum of the night sky has a number of emission lines towards the yellow and red – particularly [OI] at 557.7, 630.0 and 636.4 nm from the Earth's upper atmosphere – which will be seen running across the whole spectrum. Towards the blue end, the sky continuum rises. Before subtraction, the sky segments will also have to be independently calibrated for wavelength (i.e. have their own set of arc coefficients derived). Since the sky emission lines are relatively strong, only precise wavelength calibration will give perfect subtraction, except for photon noise.

2.6 GALAXY SPECTRA AND REDSHIFTS

As described above, an exposure of a galaxy spectrum must first have cosmic ray events removed, then be flat-field corrected, recalibrated for wavelength by means of the arc coefficients, and sky-subtracted. The spectrum will still not be corrected for the spectral response of the instrument (which involves observing standard spectrophotometric stars), but if only radial velocities are desired, there is no need to do so.

A sample spectrum of a galaxy is shown in Figure 2.9. This is an example of an absorption line spectrum – the integrated spectra of the stars in the galaxy. In order to make the absorption features more visible to the eye, some smoothing has also been carried out. This is necessary because the average number of photons recorded per channel is low. The photon noise (as discussed earlier) can therefore be reduced by increasing exposure time, but a compromise must be reached as the noise only decreases as the square root of the exposure time. In our example, the maximum count per channel is around 50 photons, which is adequate for the purpose. All too often, a galaxy will prove fainter than expected, and one may have to accept lower maximum photon counts per channel and not spend an unreasonably long time observing the galaxy. This, however, may affect the reliability of the redshift extracted, as will be discussed at the end of this chapter.

Table 2.1 lists the principal absorption line features found in galaxy spectra and their 'rest' wavelengths; many of them can be seen in our sample spectrum. It is possible to

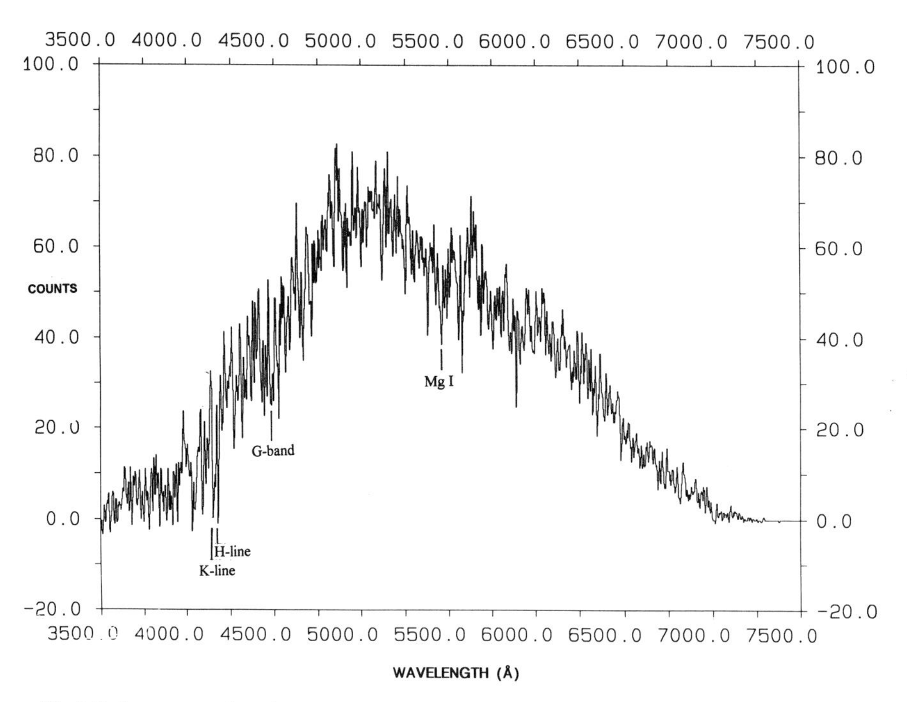

Fig. 2.9. A spectrum of a galaxy where there is a substantial amount of noise due to the relatively low number of photon counts. Some of the features indicated in the previous figure can be identified.

Table 2.1. Principal absorption features in galaxy spectra

Wavelength (Å)	Identification
3933.7	Ca II (K-line)
3968.5	Ca II (H-line)
4101.7	Hδ
4226.7	Blend (Ca I)
4304.4	G-band
4340.5	Hγ
4385.0	Blend
4861.3	Hβ
5175.4	Mg I
5889.9	Na I (D-line)
6562.8	Hα

measure their wavelengths, either by interaction off a computer screen, or on a paper hard copy. The measured wavelengths will almost invariably be higher than the rest wavelengths. The features have been Doppler shifted to the red due to the galaxy's velocity of recession.

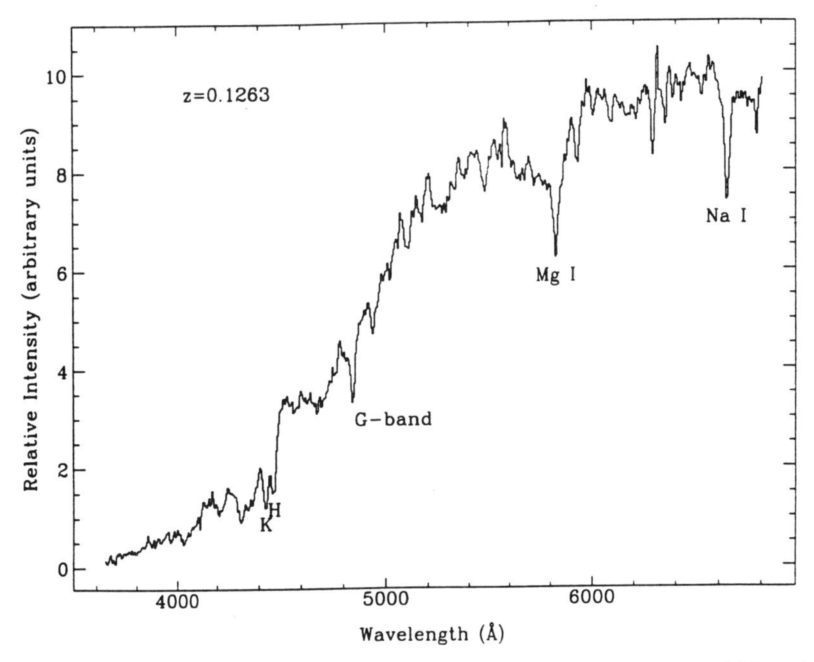

Fig. 2.10. A spectrum of a galaxy with high signal-to-noise, as observed with EFOSC on the 3.6-m telescope of the European Southern Observatory. Note that the features identified in Figure 2.8 have been shifted by some hundreds of Ångstroms towards the right due to the substantial redshift of the galaxy. (Reproduced with permission of C. Collins and the *Monthly Notices of the Royal Astronomical Society* (**274**, 1071, 1995).)

The object of the observation is therefore to measure the redshift, defined as

$$z = \frac{\lambda}{\lambda_0} - 1$$

where λ is the measured wavelength and λ_0 is the rest wavelength. It should, of course, be the same for all the features in the same galaxy, but uncertainties in measurement will obviously produce some scatter that can be expressed as a standard deviation. Such uncertainties reflect 'internal' errors (see discussion in 2.9). If the velocity of the galaxy is fairly low, the redshift can be converted into the velocity of recession by the simple relationship

$$V = cz$$

where c is the velocity of light (299,790 km/s). However, if the velocity of the galaxy is relativistic, then the relationship is

$$V = c\frac{(z+1)^2 - 1}{(z+1)^2 + 1}$$

The difference between the two relationships is expressed in Table 2.2. It is not often appreciated that at velocities as low as 1,000 km/s, the difference is already 3 km/s; at a few thousand km/s it has climbed to 50 km/s, while by 20,000 km/s or so it is several hundred km/s.

Table 2.2. Differences in velocity when a relativistic correction is applied

z	cz km/s	With correction km/s	Difference km/s
0.005	1,499	1,495	4
0.01	2,998	2,983	15
0.02	5,996	5,936	60
0.03	8,994	8,859	135
0.06	17,987	17,449	538
0.10	29,979	28,487	1,492

The convention in astronomical publications is to use cz, irrespective of the velocity. However, it is apparent that errors have crept into some of the velocities published in the literature, due to the authors of some software packages thinking the relativistic formula should be used, when the users of the software did not realise it.

The other adopted convention is to publish velocities with a heliocentric correction. The Earth moves at some 30 km/s in a near-circular orbit about the Sun which affects the measured velocity. One therefore corrects the measured velocity to that relative to the Sun.

In the past, further corrections were often made; in particular that advocated by de Vaucouleurs in his Reference Catalogues, which corrects to a local frame within the Local Cluster of galaxies:

$$V_0 = V + 300 \sin l \cos b$$

where l and b are galactic longitude and latitude (where the centre of the Galaxy lies towards l = 0° and b = 0°). However A. Yahil, G. Tammann and A. Sandage have suggested that the 300 km/s velocity should be towards l = 107°, b = –8°, rather than l = 90°, b = 0°, as in the formula above.

Radio astronomers (see 2.7 below) are often fond of correcting velocities to the 'local standard of rest', after removing the Sun's local motion relative to neighbouring stars, which is believed to amount to 19 km/s in a direction RA = 18 hrs, Dec. = +30°.

The author, as a cataloguer of redshifts, is all too aware of the different corrections that can be found with published redshifts, which complicate comparisons. Given the confusion that arises, the preferred recipe is simply to provide 'heliocentric cz' – the convention followed in this book.

The modern-day alternative to the 'manual' measurement of spectral features is to conduct a mathematical 'cross-correlation'. The spectrum is first rebinned into channels, of equal width in 'log wavelength'. The reason for doing this is that the Doppler shift is then the same for all channels. The spectrum is filtered to remove large-scale variations and very small-scale variations (smaller than the spectral features); a form of Fourier filtering is commonly employed in the software packages. The spectrum is then 'slid' against a template (prepared from a known bright galaxy, or bright stars) and a correlation function, such as that shown in Figure 2.11, derived. The peaks in this function show where the observed spectrum best fits the template, and one of them (hopefully the highest) will be the correct offset due to the Doppler shift; the result can be extracted with velocity and

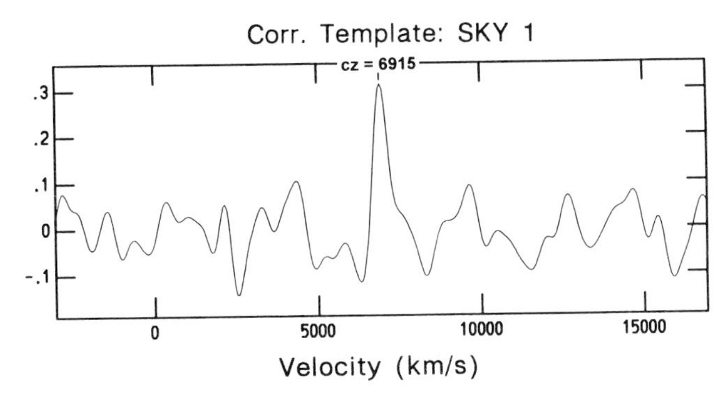

Fig. 2.11. A correlation function between the spectrum of an observed galaxy and a template spectrum. The peak identifies the velocity offset – the Doppler shift – between the two.

error. It has become common to express the uncertainty of such results by the 'contrast factor r' – the vertical scale in Fig 2.11. Not many investigators realise that the contrast factor can be increased or decreased by changing the size of the channels when rebinning; consequently one cannot intercompare r factors between different investigators.

Most software packages incorporate routines to determine the uncertainties of the measurements, based mainly on the r factor, but sometimes taking into account the width of

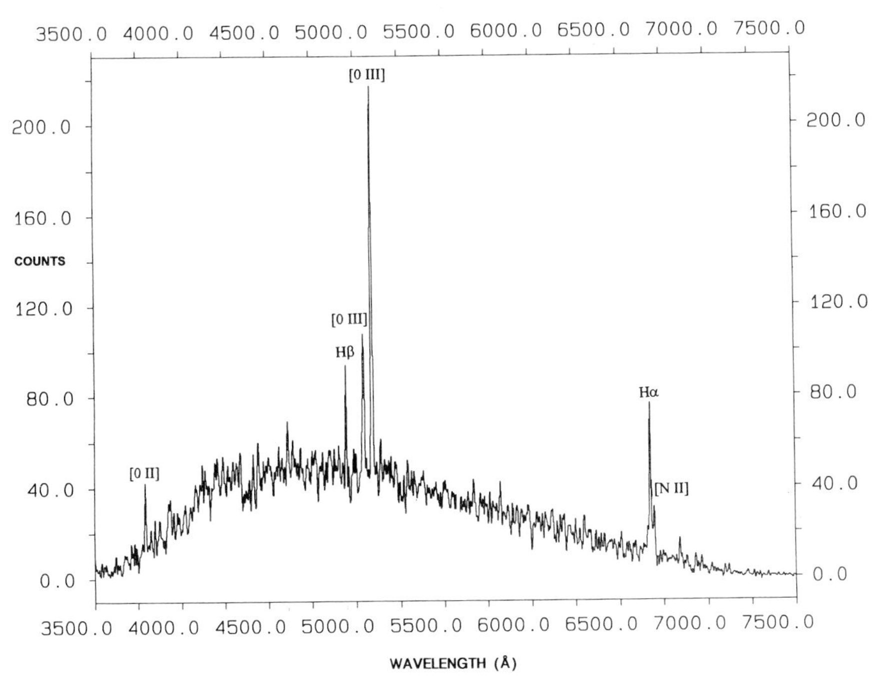

Fig. 2.12. An example of an emission-line galaxy. The emission lines, superposed on the continuum radiation, are identified.

the correlation peak. There is no standard recipe in common use for doing this. Some software packages are overly optimistic as to what the errors should be. In general the internal errors assessed by these routines are much smaller than the true 'external' uncertainties – apparent when results from different observers are intercompared.

A spectrum, such as that shown earlier in Figure 2.10, (where each channel corresponds to 0.28 nm wavelength) can be used to extract a value of cz to better than 100 km/s precision. It also shows the entire optical spectrum of the galaxy. Using a higher dispersion grating in the spectrograph, that may only allow a portion of the spectrum to be recorded, may give velocities of precision 40 km/s or better, as has been the case with the 'z-machine' employed in the Center for Astrophysics surveys. However, since the same number of photons may now have to be shared over twice the number of channels, exposure times need to be longer. It is a trade-off between precision and productivity.

About 10 per cent of galaxies have significant emission lines in their spectrum, caused by interstellar gas at various levels of excitation. The example shown in Figure 2.12 is that of an HII galaxy, the emission lines reflecting an excitation comparable to that in a galactic HII region. At lower excitations, the [OIII] may not be present; at very low excitations only [OII] appears. Seyfert galaxies and quasars – forms of 'active' galaxies – have similar emission lines to those shown here, but the lines themselves are broadened, revealing turbulent motions in the active nuclei. In some cases the broadening of the hydrogen lines indicates internal motions of thousands of kilometres per second. A full list of emission lines appears in Table 2.3.

Table 2.3. Emission lines in galaxy spectra

Wavelength (Å)	
3727.3 [O II]	4861.3 Hβ
3869.7 [Ne III]	4958.9 [O III]
3970.1 [Ne III] + Hε	5006.8 [O III]
4068.6 [S II]	6548.1 [N II]
4076.2 [S II]	6562.8 Hα
4101.7 Hδ	6583.6 [N II]
4340.5 Hγ	6717.0 [S II]
4363.2 [O III]	6731.3 [S II]
4685.7 He II	

In terms of extracting a redshift, emission lines are a blessing. Their relative strength and narrow width means that a reliable redshift, to the precision the system operates, can be readily extracted. As before, the scatter in redshift measurements from individual lines can be handled in mathematical fashion. Some galaxies may give both absorption and emission redshifts. Even with precision measurements, these may differ by a few tens of kilometres per second; in occasional cases even higher.

2.7 REDSHIFTS WITHOUT SPECTROGRAPHS

In contrast to the detailed spectra shown above, it is possible to obtain very crude spectra without using a spectrograph. Telescopes can, of course, be used for conventional photography, with a photographic plate or CCD detector in the focal plane. Placing a large but very thin prism in front of the telescope changes the images of stars and galaxies into small elongated spectra. As there is no spectrographic slit, the resolution of these spectra is limited by the seeing conditions, as well as the very low dispersion of the thin objective prism.

Depending on the field of the telescope, hundreds if not thousands of spectra may be recorded simultaneously. Historically, this technique has enabled the mass classifications of stellar spectra. It has also proved very worthwhile in the discovery of emission-line galaxies and quasars. But could radial velocities be extracted? If so, this would revolutionise the production of redshifts and the mapping of large-scale structures. Investigators at St Andrews University and at the University of Muenster have attempted to extract velocities from the objective prism plates of the UK Schmidt Telescope. Finding large errors, and uncertain reliabilities, the Scottish efforts ceased. Muenster, however, has persisted, even though the researchers will admit to errors of 4,000 km/s in a sample of tens of thousands of redshifts, going to a depth of 60,000 km/s. Regrettably, the errors are too large for mapping large-scale structures.

2.8 RADIO REDSHIFTS

In a somewhat different manner, redshifts of galaxies can also be measured with radio telescopes, such as that shown in Figure 2.13. The principles are the same as with optical telescopes, but instead of dispersing light into a spectrum, the radio telescope gains spectral coverage by means of multi-channel receivers, where each channel is tuned to slightly different frequencies. As with an optical telescope, the radio telescope will track a galaxy and integrate the signal for a long enough period of time for an acceptable signal to be discerned.

While some active galaxies are sources of radio continuum (synchrotron emission), such continua lack spectral features and cannot therefore have redshifts extracted. Instead, it is a single emission line that provides a redshift. Various emission lines exist in the radio and microwave spectrum, though few are strong enough to detect in galaxies other than our own. The exception, and by far the most widely used, is the 1,400 MHz emission (21-cm wavelength) from widespread neutral hydrogen (HI). It comes from the very low-energy transition between two slightly different spin levels within the ground state of hydrogen. Fortuitously, there is abundant cold hydrogen (with electrons in the ground state) available, particularly in spiral galaxies.

Cold interstellar hydrogen gas is found throughout the disk of our Galaxy and others. The spiral structure of our Galaxy has been mapped by means of Doppler shifts of the HI 21-cm line – slight blueshifts and redshifts caused by the rotation of the gas about the centre of the Galaxy. The same line emission may also readily serve to measure redshifts of other galaxies. With increasing distance, the signal is of course weaker; in general it has been possible to measure redshifts in this way out to a few thousand km/s. An alternative

Fig. 2.13. The Parkes radio telescope in Australia. (Photograph by R.H. Stoy.)

to HI emission is CO – carbon monoxide, and a number of CO redshifts have now appeared in the literature.

This book is, however, largely concerned with optical redshifts, and we will not attempt to describe the procedures in the same detail as above (partly because the author is not a radio astronomer!); we shall rather summarise as follows. The radio receiver used for this purpose is a multichannel device, preset to record a certain range of frequencies simultaneously. The radio telescope is aimed at the galaxy and, as with the optical observations, an integration begun. The HI (or CO) emission builds up as a discernible peak above the background until an acceptable signal-to-noise ratio is achieved (examples are shown in Fig 2.14). The central wavelength or frequency of the peak can then be established and the redshift derived as before. The width of the emission peak comes about from the motions

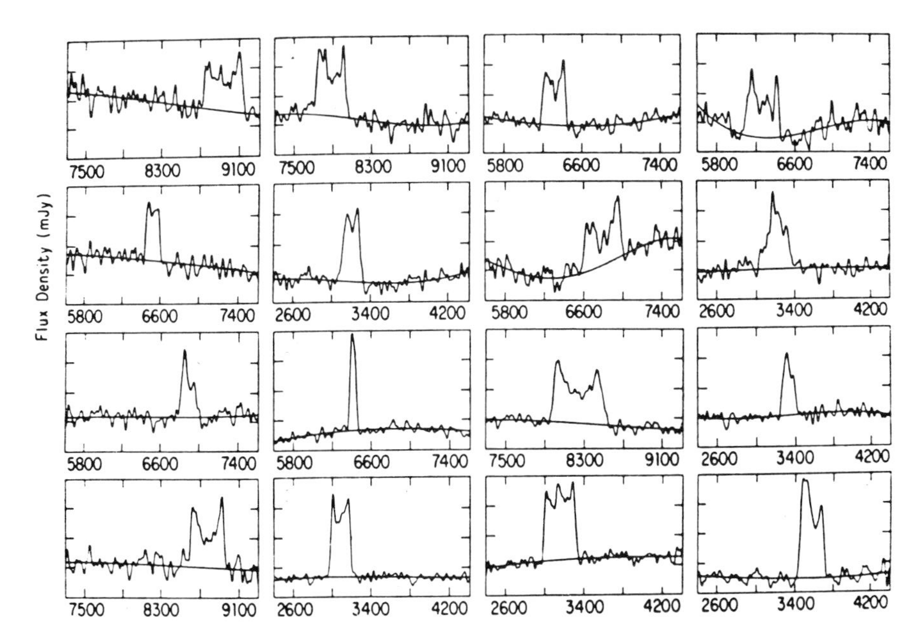

Fig. 2.14. Neutral hydrogen emission as detected by a radio telescope. Each diagram is for a different galaxy. The centroid of the emission peak corresponds to the galaxy's redshift; the width of the emission peak is a measure of the internal motion within the galaxy. The horizontal scales are in km/s. (Reproduced with permission from *Astronomy and Astrophysics* (**138**, 85, 1984).)

of the gas within the galaxy concerned. It is a useful by-product, serving to indicate the total luminosity of the galaxy and its mass via the Tully–Fisher relation (see Chapter 7).

In general, radio velocities are an order of magnitude more accurate than their optical counterparts. Typical uncertainties are around 5 km/s. There is, of course, generally good, if not absolutely precise, agreement between optical and radio measurements for the same galaxy, even though the former is based on the stellar component and the latter on the interstellar material. Very detailed examinations of nearby galaxies, carried out in both radio and optical, do reveal small offsets in velocity (a few km/s) between the optical and radio dynamical centres.

Unfortunately, not every galaxy can be observed in HI. Attempts to cover all galaxies in a fixed sample will invariably result in numerous non-detections. This means the galaxies concerned are deficient in neutral hydrogen, and effectively too faint for radio observation. Crowded fields may also present a problem, for the beam of the radio telescope is usually many minutes of arc in width, and all galaxies that fall in the beam will be recorded. The most common error that can be made is to confuse the contributions of galaxies in the same field.

Radio and optical redshifts have generally proved complementary. Radio works best on spiral galaxies, and the same spiral structure may often prove too faint for optical observation.

Some new receivers employed are multi-beam devices, enabling the radio telescope to observe adjacent parts of the sky simultaneously. That is not to say that each beam could be set on a different galaxy, but in cases of blind searches – for instance looking for galaxies behind the Milky Way – the productivity could be greatly increased.

2.9 QUALITY, DISCREPANCIES AND DISSEMINATION

It would of course be nice to report that, using the procedures outlined above, error-free redshift measurements are made, but unfortunately this is not always the case. Experience shows that of the redshift measurements published in the literature, around 3 per cent on average are erroneous. This is a small but nevertheless significant enough portion to justify discussion here.

But how can mistakes be made? The problem goes back to the necessary compromise concerning signal-to-noise. Statistical fluctuations in noise (with perhaps a touch of Murphy's law thrown in) occasionally contrive to create spurious spectral features. So convincing are these that the person carrying out the reductions – or even the cross correlation software – can latch on to these false features and extract erroneous velocities.

Experience is usually the best preventive. Optically, one likes to see a number of recognisable features to be convinced of correct identifications, as anything less could arouse suspicion. If the blue end of the spectrum is recorded, then the distinctive H and K lines need to show up clearly with the correct spacing and with the distinct rise in the continuum to their redward side. The general character of the spectrum has to ring true. Spikes resembling emission lines are another problem. A single emission line is treated with caution; other lines are searched out to check its authenticity. Astrophysically, the only emission lines that can occur alone are Hα or [OII] at 373 nm – unfortunately at either the red or blue end of the spectrum, where numbers of photons are often lowest. Usually such an emission line must be supported by absorption features to gain acceptance. Even when using cross-correlation software, the sharpness of the correlation peak (see Fig 2.11) is assessed by the observer. Experience increases with years of practice. Almost every observer who measures redshifts finds that the reliability of his or her early work is lower than that for later reductions.

Most observers weed out the uncertain measurements. Either they are not published or they are published as only tentative results. Tentative optical results are still useful as they will often prove correct, and will therefore help verify other observations or greatly save time on follow-up observations on radio telescopes. But tentative results cannot of course be accepted for analysis.

Excluding tentative measurements, the percentage of erroneous redshifts, though small, obviously varies with the researcher or research team. Even with the most careful reduction work possible, and the best reputation in the business, a list of measured redshifts is unlikely to be better than 99 per cent correct. More often, the average publication is, as suggested earlier, around 97 per cent correct, while there are many published papers that barely make better than 90 per cent correct. Clearly, the record could be improved if the investigators would be willing to halve their productivity and observe every galaxy twice. Even so, the author is still aware of some isolated cases where repeated observations have produced repeated erroneous velocities!

The only way for the reliability of a redshift measurement to be assessed is where the same galaxy has been observed by one or more independent researchers, using completely

Table 2.4. A fragment of the *Southern Redshift Catalogue*

```
NGC 3464/E569-G22
 10 54 40 -21 03 53  10 52 14 -20 47 54  269.6  34.1
  3742  17 R    *     R150 Quoted in Bottinelli et al 1990 AASupp 82,391
  3729     O EM *     R260 Mathewson et al 1992 ApJSupp 81,413
  3750     O    *     R285 Zaritsky et al 1993 ApJ 405,464
  3740  13 B EM **    Weighted mean

NGC 3459
 10 54 46 -17 03 59   10 52 18 -16 48 00  267.0  37.5
  2663  10 R    LD    R143 Richter Huchtmeier 1987 AASupp 68,427
  4253  28 O    LD    R261 Strauss et al 1992 ApJSupp 83,29
                      NB Large difference

NGC 3463/E502-G02
 10 55 13 -26 09 29  10 52 49 -25 53 30  272.8  29.8
  3980     R    *     R181 Aaronson et al 1989 ApJ 338,654
  3810 100 O    *     R076 Fairall et al 1989 AASupp 78,269
  3950     O EM *     R260 Mathewson et al 1992 ApJSupp 81,413
  3972  13 B EM **    Weighted mean

MCG-2-28-21
 10 55 49   -9 51 35  10 53 19  -9 35 35  261.8  43.6
  8301  89 O    *     R138 Fairall et al 1992 AJ 103,11
  8288     X          R176 Metcalfe et al 1989 MNRAS 236,207

E437-G72
 10 56 14 -31 56 19  10 53 52 -31 40 18  276.1  24.9
  3379  94 O    *     R138 Fairall et al 1992 AJ 103,11
  3331     O EM *     R260 Mathewson et al 1992 ApJSupp 81,413
  3341  27 O EM **    Weighted mean

MCG-3-28-24
 10 56 22 -15 53 01  10 53 54 -15 37 00  266.5  38.7
  4418  44 O    *     R138 Fairall et al 1992 AJ 103,11

NGC 3469
 10 56 57 -14 18 31  10 54 28 -14 02 30  265.6  40.1
  4613  25 O    *     R138 Fairall et al 1992 AJ 103,11

vdB 36
 10 56 49 -47 39 26  10 54 36 -47 23 24  283.7  10.9
  5670     O    *     R265 Visvanathan vd Bergh 1992 AJ 103,1057

GNX025
 10 57 17   -9 08 01  10 54 46  -8 51 59  261.6  44.4
 11519     O    *     R176 Metcalfe et al 1989 MNRAS 236,207

GNX040
 10 57 22   -9 43 27  10 54 51  -9 27 25  262.1  43.9
 16185     O    *     R176 Metcalfe et al 1989 MNRAS 236,207

E264-G?53
 10 57 16 -46 10 26  10 55 02 -45 54 24  283.1  12.3
 11505     O EM *     R217 Acker et al 1991 AASupp 87,499
```

different equipment. Intercomparison of measurements is also invaluable for assessing the accuracy of the assigned uncertainties. Just as many researchers are over-optimistic about the reliability of their work, so are they similarly over-optimistic about the smallness of their uncertainties. Most uncertainties quoted in the literature are mainly based not on external but on internal derived uncertainties. On average, the uncertainties published are about half their true external values, as is apparent when intercomparisons are made.

The ultimate worth of a redshift measurement lies in its being recorded as such in the established scientific literature, or otherwise preserved for posterity. Efforts to assemble databases of redshift measurements, as extracted from the literature or otherwise made publicly available, are being made. Currently, radial velocities are incorporated into the NASA Extragalactic Database (NED) based at the California Institute of Technology, but accessible worldwide via Internet. The Strasbourg Data Centre and LEDA (Lyons Extragalactic Database) are also repositories for such material. For many years, John Huchra, at the Harvard-Smithsonian Center for Astrophysics, has arranged for the ongoing assembly of ZCAT; for each galaxy listed, it provides what is considered to be the most reliable redshift measurement. Similarly, since 1981 this author has kept a *Southern Redshift Catalogue* (SRC) of radial velocities of galaxies south of the celestial equator. A fragment of this catalogue is displayed in Table 2.4, as an indication of how multiple observations of galaxies identify and sort out erroneous velocities. We shall also be making use of ZCAT and SRC redshifts in Chapter 4.

With the dissemination of large numbers of velocities, even clerical errors are apparent. The efforts of the small cataloguing community are not really independent endeavours so much as they are independent checks – striving towards as error-free data as possible.

In the next chapter, we shall encounter dedicated redshift surveys, each of which has produced considerable data. Regrettably, in many cases the data is still to be released in tabular form, and there are also vast numbers of unpublished redshifts that have not yet made their way into the literature and the major databases.

2.10 FURTHER READING

Specialised

Heavens, A.F., Galaxy redshifts: improved techniques, *Mon. Not. R. astr. Soc.,* **263**, 735 (1993).

Le Fevre, O. and Vettolani, P., Survey Spectrographs for Cosmology at the ESO–VLT, [in] *Clustering in the Universe* (*Ed.* S. Maurogordato, C. Balkowski, C. Tao, J. Tran Thanh Van), p.545, Edition Fontieres, 1995.

Schuecker, P., The Muenster Redshift Project: improved methods for automated galaxy redshift measurements from very low-dispersion objective-prism spectra, *Mon. Not. R. astr. Soc.,* **279**, 1057 (1996).

Yee, H.K.C., Ellingson, E. and Carling, R.G., The CNOC Cluster Redshift Survey Catalogs. I. Observational Strategy and Data Reduction Techniques, *Astrophys. J. Suppl. Series*, **102**, 269 (1996).

Zuiderwijk, E.J., Methods for unsupervised arc-line identification, *Astron. and Astrophys. Suppl. Ser.*, **112**, 537 (1995).

3

Redshift surveys

3.1 THE NEED FOR SELECTION

Large-scale structures in the distribution of galaxies are only apparent once a considerable number of redshift measurements are available. Obviously, the measurement of the redshift of a single galaxy provides but a single datum point in the distribution. Dedicated redshift surveys are clearly necessary. This chapter concerns itself with the fashion in which such surveys are conducted.

Suppose, as we shall do in the next chapter, that we were to take a redshift limit of 10,000 km/s. This corresponds to a radius of approximately 150 Mpc (or 500 million light years) and therefore a spherical volume of space of some 10^7 Mpc^3. Since the density of galaxies (dwarfs excluded) is in the region of 10^{-2} Mpc^{-3}, there would be 10^5 galaxies within that volume. To make a survey, we would like to obtain redshifts of all these galaxies. Unfortunately, we cannot tell beforehand which galaxies do have redshifts less than 10,000 km/s, and which are more distant. Consequently, we would have to observe at least a factor of ten more – spread all over the sky – which is quite impossible given our current technology. We can therefore only hope to measure redshifts for a very small fraction of all those galaxies, and we have no choice but to be selective.

Of course, we could restrict our mapping to a much more modest distance, such that it would be feasible to obtain every redshift; a true 'volume-limited' survey. Such an attempt has been made in Brent Tully's *Atlas of Nearby Galaxies*, which goes out to a velocity of recession of only 3,000 km/s, but, nevertheless, still proves to be incomplete, particularly in terms of dwarf galaxies.

However, 3,000 km/s is smaller than the sizes of the structures we are to encounter. So we do need a much larger sample volume, and for now, at least, we have no hope of observing every galaxy. We have to select, and the manner in which we select is cause for discussion. Much like predicting the outcome of a national election from a limited poll of local voters, we have to know what the relationship is between the selected galaxies and galaxies as a whole. It is a question of statistics.

From a statistical point of view, the best selection would be to take every tenth galaxy – or every hundredth galaxy – because the galaxies so selected would then have the same population characteristics as the entire sample. Unfortunately, from a practical standpoint, this is never possible.

The overriding constraint is luminosity. We shall see below that galaxies have a wide range in intrinsic luminosity. Moreover the situation is exacerbated by the inverse square law – the well-known relation between apparent luminosity and distance. The most luminous galaxies can be seen to great distances; the least luminous only very close by.

Apparent luminosity, in one way or another, is the overriding selection factor. The more light, the more easily a redshift can be obtained. We shall presently see how different luminosity criteria are employed for selection.

3.2 REDSHIFT SPACE VERSUS NORMAL SPACE

There is another obvious constraint in carrying out redshift surveys. We have to work in 'redshift space' rather than normal three-dimensional space. As described earlier (Sections 1.4 and 2.6), redshift is used to represent distance. Although later (Chapter 7) we shall look at independent indicators of distance, there is no alternative means of measuring the distance of any individual galaxy to better than about 30 per cent, except for very nearby galaxies

Like it or not, we have to accept redshift space. If a galaxy has any peculiar motion of its own, whether through being in a cluster or through participating in large-scale streaming, that motion will either add to, or subtract from, its cosmological velocity of recession. It will therefore be represented in redshift space as being either closer or further than its true distance. Consequently, the distribution of galaxies in redshift space may show 'distortions' compared with the true distribution of galaxies in normal three-dimensional space.

Redshift space is, however, mapped in spherical coordinates. The distortion is only in the radial coordinate, and not in direction. Hence it may be possible to recognise such distortions, or even extract them mathematically. The most obvious is the elongated distortion of clusters of galaxies into 'fingers of God', shown here in Figure 3.1, and in the illustrations that accompany Chapter 1 (especially Figures 1.7 and 1.10). It is the only noticeable distortion.

Even in redshift space, voids appear to be generally circular. The outward motion of galaxies around their peripheries cannot be more than a few hundred kilometres per second, or they would appear stretched radially in redshift space, as illustrated in Figure 3.2.

Great walls of galaxies look roughly similar, whether seen almost flat-on or edge-on in redshift space. Again there cannot be very large infall motions, or they would appear noticeably thinner when seen flat-on than edge-on; see Figure 3.3.

Systematic large-scale streaming will nevertheless cause systematic displacements in redshift space. If an entire large-scale structure is in motion, then its centroid in redshift space would be different from that in corresponding three-dimensional space, and its apparent distance would be affected. The motions of galaxies, that concern the difference between redshift space and real space, form the topic of Chapter 7. Such motions may, of course, even be responsible for the formation of large-scale structures.

However, that such distinct large-scale structures are seen in redshift space shows that the peculiar velocities of galaxies are at least an order of magnitude smaller than the typical cosmological velocities of recession. Were this not the case, the pattern of interlacing

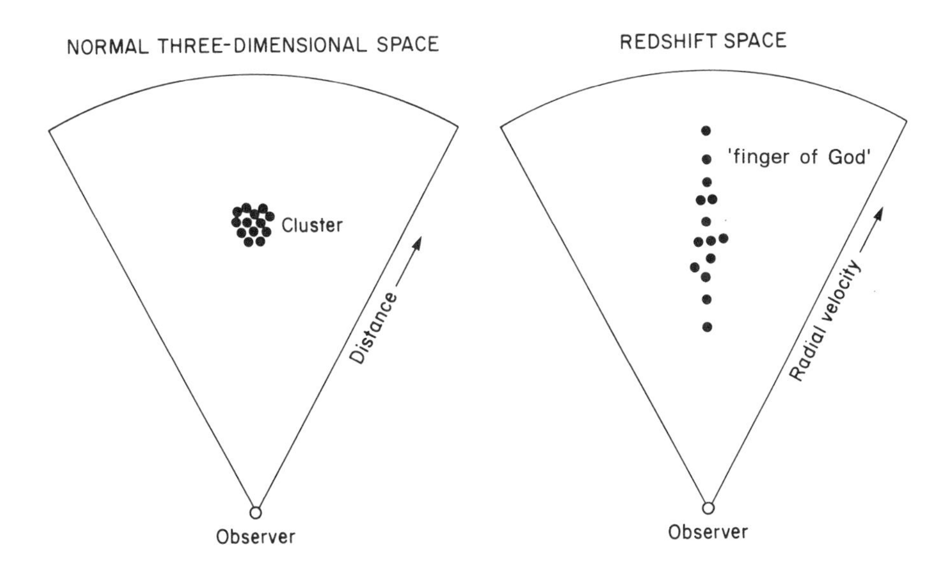

Fig. 3.1. The motions of members within a cluster of galaxies may either add to, or subtract from, the line-of-sight cosmological velocity, so that in redshift space the cluster appears stretched into a radial 'finger of God' (so called because it points towards us).

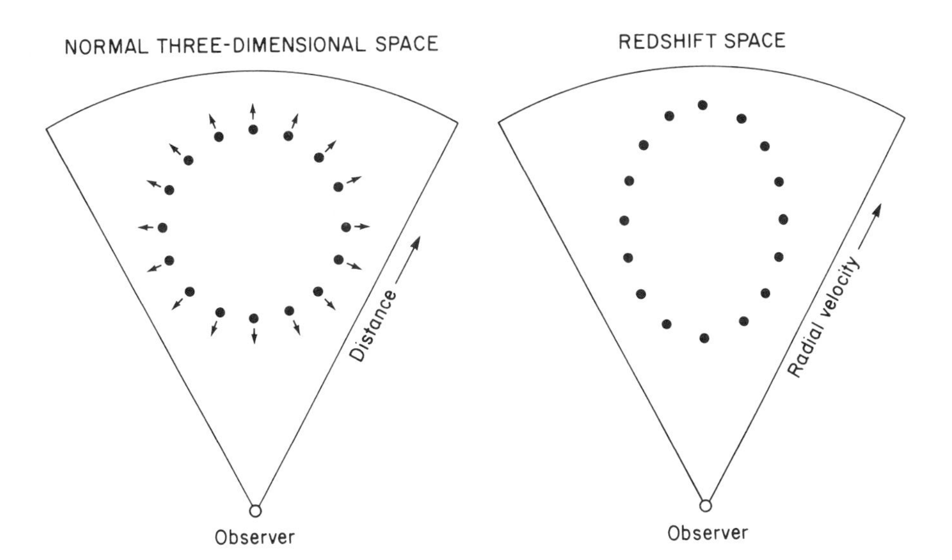

Fig. 3.2. If the galaxies around a circular void were moving radially outward from its centre, such motion might also add to, or subtract from, the line-of-sight cosmological velocity. If so, the voids might appear radially elongated in redshift space; but this is not found to be the case.

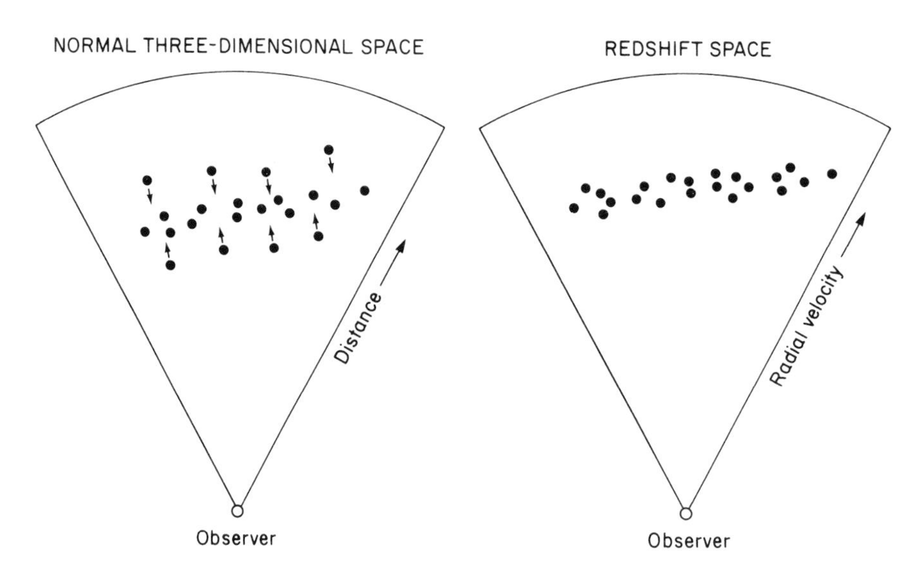

Fig. 3.3. If galaxies were falling inward towards a great wall, their motion would add to, or subtract from, the line-of-sight cosmological velocity, making the wall in redshift space appear thinner than it really is.

structures that apparently exists in normal three-dimensional space would be totally smeared into confusion in redshift space; clearly it is not.

3.3 COORDINATE SYSTEMS

The discussion above makes it clear that the radial coordinate is always to be redshift; we remind the reader of our licence of expressing it directly in km/s; similarly, any linear dimension in redshift space, not necessarily radial, can also be expressed in km/s. Astronomy has various polar coordinate systems for indicating 'directions' into space. In this book we shall encounter three of them, two of which have already been mentioned: The first is the equatorial system, based on the extension of the Earth's equatorial plane into space, where Right Ascension (RA), measured eastwards (0 to 24 hours) is the equivalent of longitude. Declination (Dec.), positive or negative, is the equivalent of northern or southern latitude. Due to a very slow precession of the Earth's axis, the coordinates gradually change, so precise coordinates are always expressed for a particular epoch, usually 1950 or 2000. The second is 'Galactic latitude and longitude' (l and b) based on the plane of the Milky Way. The 'North Galactic Pole' is at RA = 12 hrs 49 min, Dec. = 27° 24′ (1950), and the origin is directed towards the Galactic centre at RA = 17 hrs 42.4 min, Dec. = −28° 55′. The third system is 'Supergalactic latitude and longitude' introduced by Gerard de Vaucouleurs, based on a preferred plane within our Local Supercluster (Chapter 4). The North Supergalactic Pole is at l = 47.37°, b= 6.32°, and the origin is directed at l = 137.37°, b = 0°. (In the next chapter, various maps will be presented where 'great circles' denoting the equatorial plane (the celestial equator) and the Galactic and Supergalactic planes appear.)

Whilst dealing with spatial coordinates, it would also be useful to introduce the concept of 'comoving' coordinates. The Universe is expanding and the galaxies are generally moving apart from one another (as depicted earlier in Figure 1.5). Consequently, the spatial coordinates of galaxies are constantly changing. While this is negligible within our time span of observation, it is not negligible if we look to great redshifts and therefore back in time, to when the galaxies ought to be significantly closer to one another.

The expansion, however, can be represented by a 'universal' scaling factor, which acts on all galaxy separations. In cosmology, it is designated R(t). This being so, we could designate a set of 'comoving coordinates' that expand along with the Universe, according to R(t). Comoving coordinates do not change as the Universe expands. A galaxy that is stationary, relative to the local standard of rest, would therefore not change its comoving coordinates. Only galaxies with peculiar motions of their own would change positions in comoving coordinates.

We could then of course choose any era with which to define the 'scale' of the comoving coordinates. The most obvious choice is the present.

3.4 MAGNITUDE-LIMITED REDSHIFT SURVEYS

3.4.1 The magnitude system

In astronomy, the apparent luminosity of an object is represented by a magnitude, a logarithmic scale whereby a difference in magnitude is related by a ratio of luminosity, as follows:

$$m_1 - m_2 = -2.5\log\left(\frac{L_1}{L_2}\right)$$

The scale was originally designed to grade stars from the brightest (first magnitude) down to the faintest visible to the naked eye (sixth magnitude). The formulation above allows decimal subdivisions and the extension to fainter or brighter magnitudes. Note that apparent magnitude is effectively a measure of faintness, rather than brightness, in the sense that the fainter the object, the higher the numerical magnitude. Also, as with logarithmic scales (decibels for example), adding or subtracting magnitudes is the equivalent of multiplying or dividing luminosities.

The absolute magnitude of an object is the apparent magnitude it would have if viewed from a standardised distance of 10 parsecs (32.6 light years, or 3.09×10^{14} km). It represents the true or intrinsic luminosity, regardless of distance.

Luminosity relates to distance by the well-known inverse square law, so the absolute magnitude M, apparent magnitude m, and distance d in parsecs are related by

$$M - m = 5 - 5\log(d)$$

Although set up for stars, the apparent and absolute brightnesses of galaxies can be similarly represented – assuming that the light over the disk of the galaxy can be integrated. In terms of apparent magnitude, galaxies range from 5th magnitude (the Andromeda Galaxy) to fainter magnitudes. For instance the Shapley–Ames Catalogue shows there are approximately 1,200 galaxies in the sky brighter than 13th magnitude.

In terms of absolute magnitude, a typical giant galaxy is about –20. Highly luminous galaxies might be a few magnitudes brighter (e.g. –23), while faint dwarf galaxies might be a few magnitudes fainter (e.g. –17). To those unfamiliar with the system, the use of large negative numbers, that become numerically more negative with increasing luminosity, is highly confusing, to say the least. Unfortunately the system is entrenched in astronomy and widely used.

Precise magnitudes are established with the use of standardised filters (blue, visual etc.) or standardised photographic emulsions (e.g. Kodak IIIa-J) and must be expressed as such.

As suggested above, galaxies, like stars, come in a wide range of intrinsic luminosities. The brightest stars visible in the night sky are a mixture of nearby stars of modest luminosity and distant stars of high luminosity. The same is true with galaxies. The brightest galaxies in the sky – those most readily observed – show a great range in redshift, so they too consist of nearby galaxies of relatively low luminosity as well as distant high-luminosity galaxies.

3.4.2 The luminosity function

Like stars, it is believed that the variation in the space density of galaxies, according to their luminosity (absolute magnitude), follows a standard form. It can be fitted by a mathematical function; the most widely-accepted form as given by Schecter, i.e.

$$\phi(M) = (0.4\ln 10)\phi^*\left(10^{0.4(M^*-M)}\right)^{1+\alpha} \exp\left(-10^{0.4(M^*-M)}\right)$$

where typical values, for 'blue' absolute magnitudes, would correspond with a space density of $\phi^* = 0.02\ h^3\ \mathrm{Mpc}^{-3}$, $M^* = -19$, and $\alpha = -1.0$. The function correctly reflects the tendency for large numbers of intrinsically faint galaxies to occur. However, if integrated towards ever fainter magnitudes, the number of galaxies would rise towards infinity, whereas the total number of galaxies per unit volume must obviously be finite. Consequently, our knowledge of luminosity functions is uncertain with regard to the numbers of intrinsically faint galaxies. We have limited statistics from our Local Group of galaxies, where a handful of very low-luminosity galaxies surround our Galaxy; these galaxies are only detected because their individual stars are visible. The situation may be further complicated by the relatively recent discovery of a population of galaxies, detected only by their HI radio emission.

The key question, and cause of debate, is whether the luminosity function is universal. Is the manner in which galaxies form, and their stellar contents evolve, the same throughout the Universe, so that a given sample volume will contain the same proportions of galaxies according to their luminosity? Investigations have shown that by and large, this is the case; yet at the same time, various claims of 'luminosity segregation' have also been made. (We shall review the situation later in Chapter 6.) If for now we assume that a universal luminosity function does operate, then a statistically controlled sample of galaxies can be selected on the basis of luminosity or magnitude criteria. 'Statistically controlled' implies that we can extract information about the entire population of galaxies from the selected sample.

3.4.3 Magnitude-limited surveys

From the observational point of view, it is obvious that one would like to start by obtaining redshifts for the brightest galaxies in a region of the sky, and then to observe to fainter and fainter limits. This is known as a magnitude-limited survey. The limit is, of course, in apparent magnitude and not absolute magnitude. But, not only must the redshifts be measured, the apparent magnitudes of the galaxies must clearly also be known.

In the northern skies (Dec. > −2.5°), this is less of a problem. A classic catalogue by Fritz Zwicky and co-workers measured magnitudes photographically, using the 18-inch Schmidt telescope on Mount Palomar in the 1960s. A 'Schraffier' technique was used, whereby small movements of the telescope smeared out the galaxy images to permit microdensitometer measurements. The *Catalogue of Galaxies and of Clusters of Galaxies* includes all galaxies down to photographic magnitude $m_B = 15.5$. Here the subscript B refers to 'blue', the ordinary photographic emulsion used being blue-sensitive. Unfortunately, the magnitude measurements are not without significant uncertainties; the estimated standard deviation of the apparent magnitudes in the Zwicky catalogue is 0.3 magnitudes – enough to cause concern.

The Zwicky catalogue is nevertheless a monumental work; I cannot help recalling how on its completion, one of the observer's concerned, somewhat exhausted, remarked to me 'Is anybody ever going to use all these magnitude measurements?' It may have seemed like that at the time, but ten years later the Zwicky catalogue was to provide the basis for the first controlled wide-angle redshift surveys, as carried out at the Harvard–Smithsonian Center for Astrophysics (CfA).

The Zwicky catalogue shows that the numbers of galaxies, of any given magnitude, decline dramatically towards the plane of the Milky Way. This is because of the obscuration caused by dust clouds within our own Galaxy. The true number N may be recovered according to

$$\log N = \log N_b + B \operatorname{cosec} b$$

where N_b is the observed number at Galactic latitude b. The value of B averages 0.24 magnitudes. However, this is a general rule and, because of small-scale irregularities in the obscuration, cannot be applied with precision to individual galaxies. A better way is to use the 'column densities' of neutral hydrogen, as observed by radio telescopes, to assess the extinction. However, to avoid the inevitable complications, most magnitude-limited samples are restricted to higher Galactic latitudes, where extinction is considered insignificant.

The first magnitude-limited survey was the CfA1 (Center for Astrophysics) survey, mentioned in Chapter 1. It covered all Zwicky galaxies brighter than $m_B = 14.5$, and at Galactic latitudes higher than 40° (in the northern Galactic hemisphere) or 30° (in the southern Galactic hemisphere), where extinction by the Milky Way could be ignored, to a Dec. of 0°; a total of approximately 2,400 galaxies. Since then, the CfA2 survey has sought to extend the limit to $m_B = 15.5$.

Such surveys can in turn be used to derive the luminosity function. Assuming the redshift to represent the distance, absolute magnitudes can be derived. For each interval of absolute magnitude, all galaxies would have been detected out to a distance limit corresponding to the limit in apparent magnitude. Consequently the space density can be derived, and hence – for the range of absolute magnitudes – a luminosity function obtained.

If need be, the galaxies can be divided into distance (redshift) increments. Within each increment, the luminosity function can be examined to see if it agrees with the universal function.

The luminosity function is the relation between the observed sample and the population as a whole. For any particular distance it can be used to correct to the true numbers of galaxies (dwarfs probably excluded). The true numbers then reflect the true density distribution of galaxies, and probably the true distribution of mass.

This is, in essence, the *modus operandi* of controlled redshift surveys. Since only a finite number of galaxies are observed, statistical errors apply throughout and accumulate. The luminosity function can only be verified with reasonable accuracy over an extended region, whereas critics may suggest that it fluctuates on a smaller local scale. The topic has been addressed by numerous analyses published in the scientific literature (and further discussion follows in Chapter 6).

In spite of these drawbacks, apparent magnitude and the luminosity function offer the best avenue for establishing a statistically controlled sample of galaxies, from which the distribution of all galaxies may be inferred. Unlike many other parameters, the apparent magnitude can be determined without ambiguity.

Nowadays, magnitudes can be measured from CCD exposures (Section 2.3) much more accurately than they could in the past by photographic emulsions or multiple aperture photometry. However, CCD fields are usually very small, and often only a single nearby galaxy can be measured at a time; a process more drawn out than even measuring redshifts, since observing conditions have to be 'photometrically' clear. However, 'drift scans' using CCDs are able to calibrate a large number of galaxies within a narrow strip in Declination (see Section 3.12.5 below).

If one is to improve on the Zwicky catalogue magnitudes, or go much fainter, a survey must both measure apparent magnitudes as well as redshifts. Since only galaxies to a certain magnitude limit are to be observed spectroscopically, the magnitude measurements have to be made first. Sequences of CCD exposures, or drift scans, can be used to cover small regions of the sky. From these exposures, galaxies can be identified and their magnitudes and positions measured. The positions are useful, if not essential, if multi-fibre spectroscopy is to be used in the follow-up.

Furthermore, the magnitudes alone can be used as a first assessment of the sample. If the luminosity function and average density are unchanged with distance, then the number of galaxies counted down to a particular magnitude limit should vary with that magnitude limit according to

$$\log N(m) = 0.6m + \text{constant}$$

where N(m) is the number of galaxies brighter than magnitude m (and 'Euclidean space' is assumed). Consequently, deviations from the 0.6 slope can indicate fluctuations in the average density, as may be caused by some large-scale structures. With faint enough magnitudes, the systematic departure from the 0.6 slope shows where the selection in magnitude is incomplete; that is, where faint galaxies have been overlooked.

3.5 DIAMETER LIMITS

There is no equivalent to the Zwicky catalogue available for the southern skies. Photographic magnitudes have been extracted by A. Lauberts and E. Valentijn from the fields of the European Southern Observatory Quick Blue Survey. However, the images are not smeared especially for this purpose, and the photographic emulsion in the central cores of galaxies is often saturated, and the extracted luminosities therefore subject to error. This work came, in any case, only after the selection of galaxies for the Southern Sky Redshift Survey (SSRS - mentioned in Chapter 1). Instead, that survey made use of 'diameter limits'.

An earlier work of Andris Lauberts (the ESO/Uppsala Survey of the ESO(B) Atlas) had identified all galaxies with angular diameters of 1.0 arcmin or greater. Luiz da Costa and colleagues used the diameter limit as the next best thing to a magnitude limit. The main justification is that the 1.0 arcmin limit identified a similar number of galaxies per unit area of sky as did the $m_B = 14.5$ limit of the CfA1 survey. Though similar, diameter limits and magnitude limits are not the same thing. A diameter limit tends to pick up a far greater proportion of low surface brightness galaxies (which are also difficult to observe), and miss a number of compact high-luminosity galaxies, especially ellipticals. Further, at low galactic latitudes, the decline in apparent magnitude, due to extinction by dust clouds, is steeper than that of the diameters. Diameter-limited samples seem to be more immune to galactic extinction until an abrupt cut-off at a certain Galactic latitude; for the Lauberts Catalogue, it is around $b = 10°$.

Unlike magnitudes, diameters are ambiguous. Galaxies do not have sharp boundaries to them, and their apparent diameters are largely a property of the system used to photograph them. There is no tight relationship available that gives a space density of galaxies according to diameter, that could be used to relate the observed sample to the galaxy population as a whole, though one could attempt to extract a version from the SSRS. The SSRS was an attempt to do in the southern Galactic hemisphere in the southern skies, what the CfA1 survey had done in the north. The use of a diameter limit was seen to be the closest choice possible to the magnitude limits used in the north.

The extension of the SSRS – SSRS2 – has in any case been able to employ magnitude limits, as extracted from machine based surveys (see Section 3.8 below). Expanding the original SSRS to some 3,600 galaxies, it goes to $m_B = 15.5$ and is directly comparable to the northern CfA2 survey.

Sometimes, combinations can be used. The more recent Optical Redshift Survey of Basilio Santiago and collaborators, consists of overlapping samples – one limited in apparent magnitude ($m_B<14.5$) and one in angular diameters $\theta_B > 1.9$ arcmin, such that it covers 98 per cent of the sky. The survey is not totally new, but was formed by supplementing existing surveys with additional redshift measurements.

3.6 INFRARED GALAXY SURVEYS

The concerns over possible systematic differences and inconsistencies between the magnitude systems in the northern sky and those in the southern sky, plus the marked extinction effects towards lower Galactic latitudes, led some collaborations to look to the IRAS survey for more uniform selection criteria for controlled surveys.

In the early 1980s, the highly successful IRAS (InfraRed Astronomical Satellite) had carried out a survey over virtually the whole sky at wavelengths of 12, 25 and 60 μm. The IRAS *Point Source Catalogue* provides flux brightnesses for some 15,000 galaxies at these wavelengths. As with radio observations, fluxes are measured not in magnitudes but in Janskys (1 Jy = 10^{-26} watt m^{-2} Hz^{-1}). Because the same spacecraft observed both northern and southern skies, its flux brightnesses – the equivalent of magnitudes – were seen as a more reliable criterion for selection of galaxies for redshift surveys. Furthermore, the infrared fluxes are far less affected by the absorption of the Milky Way, so the survey could probe the much lower Galactic latitudes that the CfA maps and SSRS had excluded.

The main negative aspect of IRAS selection is that the galaxies detected by the survey are predominantly spirals, the infrared emission coming mainly from interstellar dust in the disk of the galaxies. Elliptical galaxies, having little such interstellar material, are less likely to be detected and are therefore usually overlooked, even though they are an important component – and by far the most dominant component of rich clusters – and represent the greatest concentrations of mass.

Historically, two separate collaborations have been investigating IRAS galaxies. One is known as 'QDOT' (for the home institutions of its members: Queen Mary College, Durham, Oxford and Toronto). The other is based in the US, mainly around Michael Strauss and John Huchra. The original QDOT survey sampled one in six of the 60-μm sources with $S > 0.6$ Jy at $|b| > 10°$. By contrast the Strauss survey covered some 5,000 galaxies with $S > 1.936$ Jy at $|b| > 5°$, and where $[S(60\ \mu m)]^2 > S(12) \times S(25)$ – considered to be a good discriminant between galaxies and stars which might be almost coincident in position. More recently, K. Fisher and colleagues have extended it down to 1.2 Jy (using data from the literature and 1,300 new redshifts). The QDOT survey has also since been extended to cover the full sample of some 15,000 galaxies, both by observation and by making use of redshift measurements from other surveys.

Two new near-infrared surveys, from Earth-based observatories, promise to provide galaxy databases for future redshift surveys. The DEep Near-Infrared Survey (DENIS) of the southern sky is a European endeavour to map at 2 μm, while the 2MASS is its complementary American counterpart.

3.7 SELECTION BY GALAXY TYPE

Other global surveys have concentrated on particular galaxy types. Given well known spatial segregation effects according to galaxy type, such surveys cannot easily be used to map the distribution of galaxies as a whole, and are not intended as such. However, they do serve for independent distance indicators (the Tully–Fisher and Faber–Jackson relations) that reveal peculiar motions of galaxies, as will be extensively discussed later (Chapter 7). We shall make mention of those surveys then, and for now restrict our coverage to surveys that directly map the overall distributions of galaxies.

3.8 MACHINE-BASED SURVEYS

As described in Chapter 1, the National Geographic–Palomar Observatory Sky Survey was the first deep survey to cover all the sky visible from that northern observatory. Almost all galaxies selected for northern redshift surveys were originally derived from that survey, picked out by visual scans of the photographs. So successful was the Palomar Survey, that some 25 years later, both the European Southern Observatory and the Royal Observatory, Edinburgh, sought to extend it to the southern hemisphere using similar Schmidt telescopes. The British (UK Schmidt) telescope, sited in Australia, was based directly on the Palomar Schmidt, but with an improved achromatic doublet corrector. The British SRC (Science Research Council) Sky Survey also used IIIa-J photographic emulsion, an improvement over what was available for the Palomar Survey. This emulsion has its greatest sensitivity in the green region of the spectrum, the portion least affected by night sky emission. The result was a survey that went some two magnitudes deeper than the Palomar Survey. So successful was this survey that by the late 1980s, the Palomar telescope was fitted with a British achromatic corrector in preparation for a second northern sky survey.

In parallel to the sky survey, the British also developed automatic scanning machines to survey the photographs: the COSMOS machine in Edinburgh and the APM (Automatic Plate Measuring) machine in Cambridge. The machines scanned the plates, making microdensitometer measurements with a very fine raster. The accompanying software derived positions and magnitudes for the many millions of images detected in each Schmidt photograph. More important, it compared the diameters of the images with their magnitudes to discriminate between stars and galaxies. Consequently it has led to the first reproducible methods for locating galaxies, independent of visual biasing and subjective opinion. It enables a complete sample of galaxies to be selected, together with their magnitudes, over wide portions of the sky. As such, it is an ideal basis for redshift surveys, and groups at Oxford and Durham in particular have used it for this purpose. The only limitations are its success (the Oxford survey around the South Galactic Pole listed two million galaxies) – way beyond the observing capabilities for spectroscopic follow-up, at the time.

Similar scanning machines exist at the Space Telescope Science Institute. They served to extract guide stars for the Hubble Space Telescope from the Palomar and SRC Sky Surveys. In the process, they also extracted magnitudes and identified non-stellar objects, much of which has provided spin-offs for redshift surveys.

There is still the same problem with extracting magnitudes from photographic plates, where the central cores of galaxies have saturated the emulsion (as mentioned earlier). The magnitudes are nowhere near as accurate as CCD observations, but it would be impossible to cover so wide a field of sky. Also there are large numbers of confused images (overlapping of star images may be confused as a galaxy) and these the human eye can sort out more readily than can the software. The reverse can also occur; galaxies can mimic stars, and in extreme cases, nothing short of a redshift can sort it out. Apart from quasars (an acronym of Quasi-Stellar Radio Sources), where the light from the active nucleus so dominates the object that it looks more like a star than a galaxy, there is a population of 'compact galaxies'. (The latter was the topic of the author's Ph.D. thesis, following the promotion by Zwicky in the 1960s.) On the sky surveys, extreme compact galaxies may show images indistinguishable from those of stars. There is still no perfect device, whether human eye or machine, that can identify absolutely every galaxy on a survey photograph.

3.9 CATALOGUE REDSHIFT SURVEYS

As discussed in Section 3.4 above, apparent magnitudes are used generally as the control in redshift surveys since they are the most accessible. However, if luminosity segregation does take place, or if there really is a large population of galaxies consisting mainly of neutral hydrogen rather than stars, much of the statistical control of the sample will be lost.

Moreover, errors in apparent magnitude and in distance (often only represented by the Hubble relation) can be significant. If too, as some sceptics advocate, the visible portion of a galaxy is akin to the visible tip of an iceberg and there is no easy way of determining how much mass is hidden, then much of the efforts in control that have so far concerned us in this chapter may simply be in vain.

We have already seen, in Chapter 1, that the distribution of galaxies reveals distinctive, even conspicuous features; voids, great walls and so on. The situation is very different from the more subtle distribution of stars in the local portion of our Galaxy where, for example, it is difficult to locate spiral structure. It is therefore not surprising that the first revelations of large-scale cellular structure, from Mihkel Joeveer and Jaan Einasto, came from catalogue data and *not* a statistically-controlled magnitude-limited sample. It took ten years before a magnitude-limited sample (the well known 'Slice of the Universe') confirmed their finding.

Catalogue data is simply data contained in databases (e.g. ZCAT, SRC, LEDA, NED; see Section 2.9) that have been published or otherwise released publicly; Joeveer and Einasto had used the redshift data in the *Second Reference Catalogue* (by G. and A. de Vaucouleurs and H. Corwin). Catalogue data is in any case to a large extent controlled in apparent magnitude. The obvious tendency is to observe the brightest galaxies first, then to work down to the fainter ones – except that a fixed magnitude limit is not rigorously enforced. But the obvious advantage of catalogue data is the quantity of data available; simply, all available redshifts from many different researchers. In that way, catalogue surveys will always be years ahead of controlled surveys. A further advantage of catalogued data is that many of its redshifts are based on multiple measurements. Thus the inevitable small percentage of erroneous measurements (see discussion in 2.9) is significantly reduced.

The greatest criticism that can be levelled at catalogue data is that it is subject to 'selection effects'. The mention of 'selection effects' in astronomy (or even science in general) readily conjures up doubts as to the worth of a piece of research. We need, however, to consider how selection effects apply to redshift surveys. By including this or that galaxy, and excluding another, selection is made in direction. But selection *cannot* be made in redshift. A galaxy has to be observed before its redshift is known. One might use the angular diameter of a galaxy to make a very crude estimate of its general redshift, but there is no way that only galaxies of a particular narrow redshift range can be pre-selected. Redshift is therefore immune to selection effects.

Since mappings are made in redshift space, it is possible to consider cases where there may be selection effects in direction, but not in redshift. Some possibilities are suggested in Figure 3.4. In the first case, an elongated feature that runs radially with the line of sight is suspect. It could be real (a cluster or filament running outwards) but it could also have been created by a favoured small survey region in the sky (a selection effect in direction) where neighbouring regions are grossly under-represented. The second case, however, shows an elongated feature that runs across the radial direction. It has to be real, since it is created by a sequence of distinct redshift peaks over different directions of the sky. A similar argument can be applied to the reality of voids, if they are enclosed on their far side by a significant number of galaxies.

The topic, however, is not without controversy. Joeveer and Einasto's findings were not believed by many, and were often ignored – to such an extent that some authors infer it

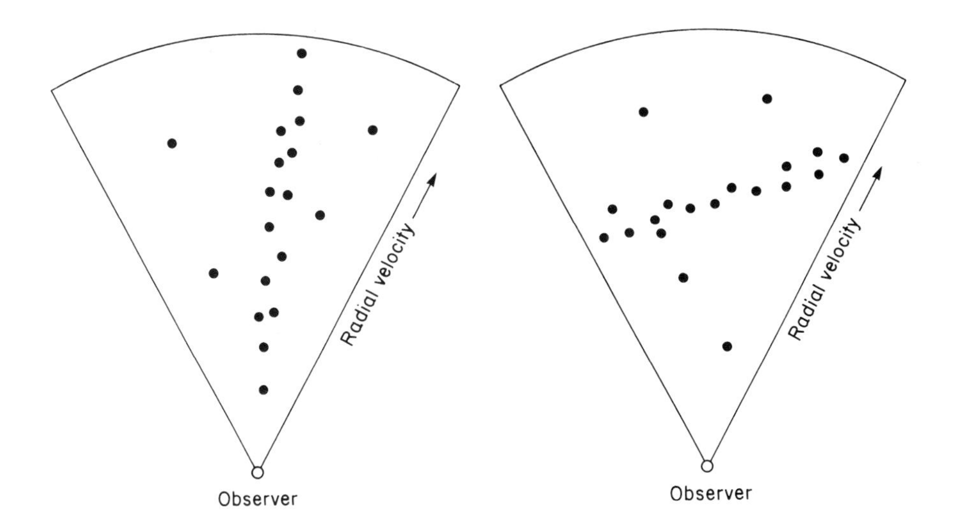

Fig. 3.4. Apparent large-scale structures in redshift space that might be due to selection effects. The discussion in the text argues that while the case on the left may be spurious, that on the right cannot be so.

was a controlled survey, the CfA 'Slice of the Universe' that first 'discovered' that galaxies were concentrated around the peripheries of voids. (The CfA Slice is a landmark in the field, but its authors made it quite clear that they confirmed the cellular structures suggested by Einasto.) In similar fashion the author has found that plots of his catalogued data have on a crucial occasion been rejected by a referee, and more generally ignored at other times. This is because they appeared to go against the scientific wisdom of conducting 'controlled' redshift surveys.

However, it is the author's opinion that catalogue surveys, because of their greater *quantity* of data, and because *redshifts are immune to selection effects*, serve as forerunners in the *mapping* of cosmic structures.

The proof of the pudding is in the eating. All of Joeveer and Einasto's structures were confirmed by the CfA surveys. All the voids tentatively identified by the author and his co-workers in the southern sky (in 1983) have stood the test of time. Figure 3.5 also shows an example of the concordance between controlled and catalogued surveys.

As more and more redshifts are measured, so the data pool increases. The progression of the catalogued plots since this author started in 1982 has been much like watching a photograph appear in a developing tank. First the gross features appear, then as time goes by, finer and finer details are filled in. Eventually, when every galaxy within the sample volume has been observed, we will have the complete picture. (We are not there yet, but in the two chapters following, we shall see what can be discerned so far.)

Catalogue redshift surveys have developed in parallel to the statistically controlled surveys (mentioned in preceding sections). In 1982, Einasto and R. Miller presented data from Huchra's ZCAT, as a 'walk-around' video sequence to examine distributions. Similarly, the author and some sixty students made hand plots of the 1983 ZCAT which revealed major large-scale structures and identified some forty voids. A 1990 update (based

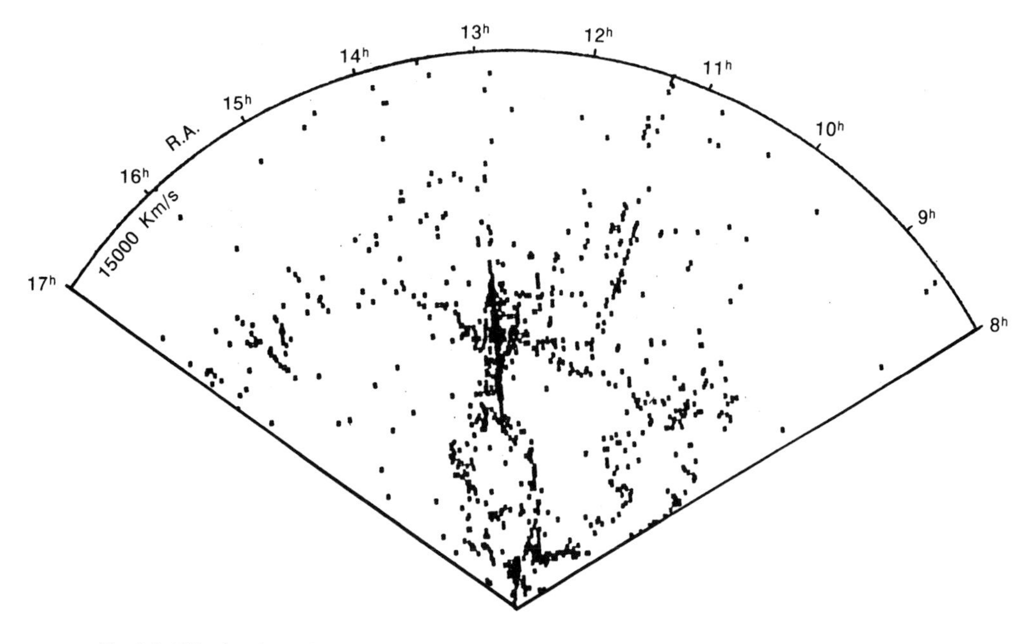

Fig. 3.5. This plot shows the same features as the well-known 'Slice of the Universe' plot (Figure 1.12), but it is drawn from catalogued, rather than from statistically controlled, data.

on the Bologna catalogue and the author's *Southern Redshift Catalogue*) confirmed the features and showed the extent of some of the great wall structures then recognised. These are in addition to the author's ongoing updates of plots of the southern data since 1982. A set of ongoing updates in Fig 3.6 shows how catalogued data has revealed further and finer detail in large-scale structures, without the structures originally recognised being disproved.

Michael Hudson has pioneered an approach that marries redshift catalogue surveys with controlled surveys. A large sample of galaxies can be selected with statistical control – magnitude or diameter limits. The redshift catalogues can then be consulted (Hudson used ZCAT and my SRC) to see what redshifts are available. Of course, the redshift data will prove incomplete for the chosen sample. However, for each small region of the sky, a completeness factor can be established. In this way, corrections can be made for the undersampling of the redshift catalogues. Hudson has been very successful in extracting galaxy densities from the redshift catalogues to map the nearby Universe, and we shall see a sample plot of his work in the next chapter.

3.10 CLUSTER SURVEYS

Clusters of galaxies are considered to mark the density peaks of the distribution of matter in the Universe. Consequently, some redshift surveys – particularly those seeking to probe deeper into extragalactic space – have restricted themselves to clusters rather than to individual galaxies. The distribution of the clusters then reflects gross overdensities, without

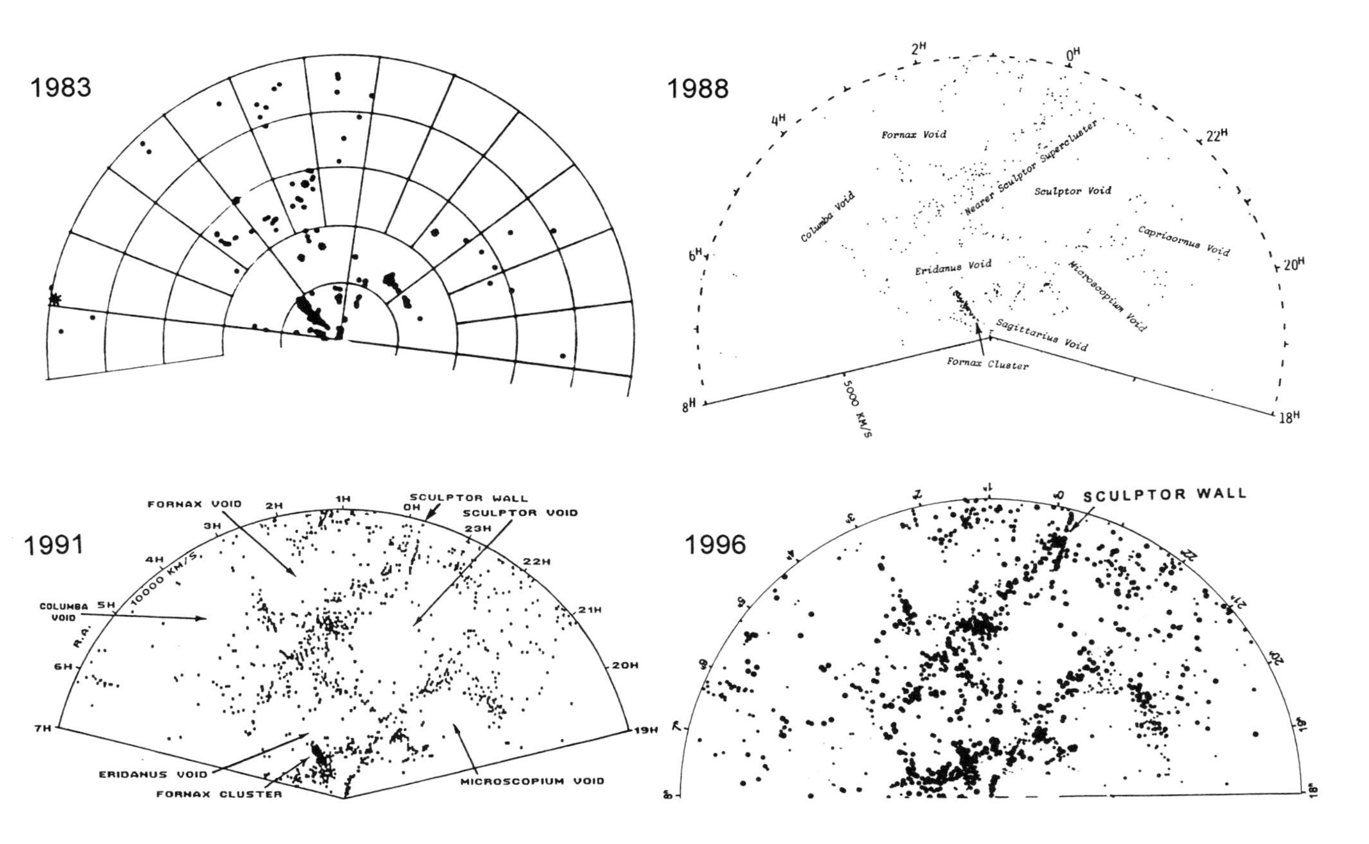

Fig. 3.6. Plots from the author's *Southern Redshift Catalogue* compiled in 1983, 1988, 1991 and 1996. As more data has been added to the catalogue so greater detail is seen; but the same large-scale structures are apparent. The later plots confirm, rather than alter, the structures apparent in the early plots.

the finer detail that surveys of individual galaxies might bring. In this regard, much more will be said in Chapter 5, when we look at the distribution of Abell clusters.

Some early investigators of clusters chanced getting a single redshift, usually that of the brightest galaxy in the cluster. While this was often satisfactory, there were nevertheless all too many cases where the supposed brightest galaxy in the cluster later proved to be a superposed nearby galaxy instead. Many investigators now consider it essential to obtain a minimum of about four or five galaxies in each cluster, to extract a reliable redshift of the cluster as a whole.

The most popular clusters are those from George Abell's catalogue and its extension to the southern skies, started by Abell before his untimely death and completed by Harold Corwin and Ron Olowin. The catalogue is one derived from visual inspection of the sky surveys. In spite of the best intentions, it is impossible for an extended visual survey to be consistent in its selection criteria. The distribution of Abell clusters in the northern skies at least suggests this; such as an excess of clusters at Dec. 70° north. When compared to a computer algorithm employed to identify clusters in the data from the automatic scanning machines, there is probably only about a 75 per cent overlap, even with the same photographic material. Abell would find clusters that the algorithm would not identify, and vice versa.

There is also the problem that some apparent clusters are coincidental overlaps of weaker concentrations, though redshift observations should reveal when this is the case. The large-scale distribution of galaxies reveals many filamentary structures. Such features seen end-on easily mimic clusters.

In spite of the misgivings expressed here, there are many observers who believe that accurate quantitative analyses can be carried out from cluster samples, as though they were statistically controlled data – even more so than the straight catalogue data discussed in the previous section. The Abell clusters alone have given birth to a small industry of research papers. There is no doubt that they serve well for mapping purposes, but this author has doubts on their validity as a controlled survey.

3.11 SKY COVERAGE

Aside from differences in selection effects, redshift surveys also differ in the portion of the sky they choose to cover. For example, in the preceding discussion we have seen how the northern Center for Astrophysics maps were complemented by the Southern Sky Redshift Survey. We also saw how IRAS-selected surveys covered almost the whole sky, though redshift observations would then have to be carried out from separate northern hemisphere and southern hemisphere observatories. Redshift surveys that cover a significant fraction of the sky, if not the entire sky, are obviously best suited for the recognition of nearby large-scale structures. Clearly such surveys involve the gathering of large numbers of redshifts.

By contrast, more modest surveys may concentrate on just a small region of sky, but usually one which is of special interest. Historically, it was the Virgo–Coma region, having the richest collection of nearby galaxies, that drew most attention (as illustrated in Chapter 1). Indeed it was the first portion of the sky to reveal distinct cellular structure.

Similarly the historic Boötes Void was another regional survey. Here, however, the approach was different. Instead of a complete portion of the sky being sampled, only three small areas were examined, and these seemed to show very similar results. Only later did a more extensive survey confirm the findings. The situation is rather like deducing the extent of underground formations from a few drill holes; and we will see this strategy again in action in Chapter 5. So many 'regional' surveys have been carried out that there are already far too many to report here individually, but we shall rather discuss efforts directed at particular portions of the sky.

The Perseus–Pisces region has long since drawn attention, and the early work of Riccardo Giovanelli and Martha Haynes using radio observations is well known. Aside from the fact that the Perseus Cluster is one of the three nearest rich clusters, it lies superposed on what appears to be a massive filament of galaxies – conspicuous since the days of the Herschels – seen on the sky. It is the most obvious indication of the existence of large-scale structures – apparent even without redshift measurements. Perseus–Pisces does however lie at very low Galactic latitudes, so once again magnitude control is a problem.

The Virgo and Coma Clusters, aside from their overall prominence, lie close to the North Galactic Pole, where extinction is minimal. There are no such comparable clusters towards the South Galactic Pole, but obviously this relatively clear window has now been used extensively. One such example has been the Oxford-based survey, using the APM machine at Cambridge (mentioned above).

Clearly, the most difficult region of all the sky to work is close to, or even behind, the Milky Way. In general, our foreground Galaxy heavily obscures about 25 per cent of the extragalactic sky. Moreover, some of the most influential of nearby structures (such as the Great Attractor region, discussed extensively in Chapters 4 and 7) happen to lie in this part of the sky. If we are to get a reasonably complete mapping and understanding of local large-scale structures, then we have to do something about this region.

The obscuration of the Galaxy reduces both the brightness and the diameters of galaxies. Consequently, catalogues of galaxies, selected according to normal criteria, become increasingly incomplete as they approach the Milky Way. We have already noted how magnitude-selected samples are hardest hit and diameter-selected samples somewhat less so. One of the advantages of the IRAS-selected galaxies was that infrared radiation was affected to a smaller degree than normal visible light. Nevertheless, all conventional surveys show the classic Zone of Avoidance (the term coined by Hubble) due to the obscuration of the Milky Way.

The Zone of Avoidance can however be made narrower if the selection criteria are suitably adjusted. In optical terms, this means accepting galaxies with smaller diameters and lower magnitudes. Nevertheless, the overriding complication is the greatly increased star density of the foreground Galaxy. Photographs taken close to the plane of the Milky Way are very heavily crowded with stellar images, making it extremely difficult to discern the faint images of galaxies. The situation would be easier were it possible for the automatic scanning machines to do the job, but the confusion between galaxy images and overlapping and superposed star images makes it apparently impossible for any software algorithm to sort out.

For many years, the author has had a close collaboration with Renée Kraan–Korteweg, one of a small number of cataloguers who have adopted the very pedestrian but effective approach of visually scrutinising the sky survey photographs very slowly and carefully under high magnification. Operating with reduced acceptance criteria, painstaking searches can uncover large numbers of partially obscured, or even heavily obscured, galaxies.

The follow-up redshift observations are of course somewhat more difficult than they might have been, due to the general faintness of the galaxies. There are also some cases of superposed stars lying directly on top of the nucleus of a galaxy, which are very difficult if not impossible to untangle spectroscopically.

There is still a portion of the Milky Way, a few degrees wide, that remains totally opaque to optical surveys. The only possible means of detecting galaxies there lies with radio observations, mainly HI. However, since the galaxies cannot be seen, the radio telescopes have to be aimed blind. Such blind searches may consume vast amounts of observing time while detecting very few galaxies. However, prospects of using multibeam instruments may make this much more viable in the future.

3.12 REDSHIFT SURVEYS – PAST, PRESENT AND FUTURE

A number of important surveys have already been described briefly during the course of this chapter, and others will be mentioned during the next two chapters. Some specific surveys and trends, not otherwise covered, are featured below. It is not the intention of this book to name every redshift survey completed, in progress, or planned (there are far too many) but rather to provide an overview of the major surveys and a representative sample of many others.

3.12.1 Surveys at the Anglo–Australian Observatory

The Anglo–Australian Telescope has a long tradition of redshift surveys based around multiple-fibre instrumentation. In the past, these have included FOCAP, where fibres were hand plugged, and AUTOFIB, where a robot set up fibres in sequence. As described in Section 2.4, the 400-fibre 2-degree field is presently (1997) coming into operation and is set to help revolutionise the acquisition of redshifts. With the availability of the new system, a major survey will get under way that will scan Declination strips, mainly in the vicinity of the south Galactic cap. As before, the system will depend on the databases of galaxies extracted from the automated scans of the UK Sky Survey plates. The FLAIR system operated on the neighbouring UK Schmidt Telescope will continue to survey in complementary fashion, since it can cover much wider fields.

3.12.2 The Stromlo–APM Redshift Survey

Like the surveys at the Anglo-Australian Observatory, this survey (already completed) attempts to survey the redshift distribution of the APM galaxies (Section 3.8) to a modest magnitude limit of $b_J = 17.15$. Even so, the numbers of galaxies to be covered are too high, so a policy of randomly sampling 1 in 20 galaxies was adopted, resulting in a workable survey of nearly 1,800 galaxies. We shall see a sample plot later (in Chapter 5).

3.12.3 The Las Campanas Redshift Survey

The largest survey attempted in recent years has been the Las Campanas Redshift Survey, which has made use of the 2-degree field of the 2.5-m telescope at the Las Campanas

Observatory. It used CCD (drift-scan) R-band imagery to locate and measure magnitudes for galaxies in six Declination strips: three in the north Galactic hemisphere and three in the south. Follow-up spectroscopy, with plug-in fibres, has provided more than 26,000 redshifts, with median redshift of about 30,000 km/s. The results of this major survey will be presented in Chapter 5.

3.12.4 The ESO Slice Project

A collaboration has used the 3.6-m telescope of the European Southern Observatory to obtain some 5,000 redshifts in a Declination strip adjacent to the Las Campanas strips in the south Galactic cap. The objects were selected from the Edinburgh–Durham *Southern Sky Galaxy Catalogue*, from the COSMOS machine scanning the UK Sky Survey Fields. (See Chapter 5)

3.12.5 The Sloan Digital Sky Survey

The success of the Las Campanas survey led some of its participants to propose a far more ambitious survey that is expected to measure a million redshifts. Named the Sloan Survey after its principal benefactor, and involving a number of participating institutions, it uses the same approach as Las Campanas. The accommodation of a much larger number of galaxies is achieved in various ways.

First a dedicated 2.5-m telescope, with a 3-degree field, has been built for the project, so that observations can be carried out full-time (and not, as is usually the case, to be slotted in with numerous other investigations that use the same telescope).

Second, instead of using a single CCD detector for photography, a matrix of 30 large CCDs, that enable simultaneous observations in five different wavelength bands, is used. This camera operates in drift-scan mode: the image of any particular star of galaxy passes across five CCDs, spending about 55 seconds in each. There are a further 24 small CCDs on either side of the main array; these are used for astrometric calibration (accurate positional measurements). It is expected that the system will eventually record accurate magnitudes to as faint as 23rd magnitude in each of the wavelength bands for some 5×10^7 galaxies, in Declination strips in the northern Galactic cap, covering a twelfth of the entire sky.

Third, two double spectrographs – one covering the blue and one the red portion of the spectrum – will operate with 640 fibres, and the spectral resolving power will be about $\lambda/\delta\lambda = 1{,}800$. As mentioned earlier, the fibres will simply be hand-plugged into pre-drilled plates (using dextrous hired help, rather than tired and impatient astronomers). There will be ten fibre harnesses, so ten fields can be set up ready for a full night's observing. In this way, it is hoped to cover about 10^6 galaxies brighter than 18th magnitude.

This is a massive undertaking that has already been many years in planning, and which will still be many years in execution. It is clearly a quantum jump in the acquisition of redshifts, and will set the standards well into the 21st century.

3.12.6 The Canada–France Redshift Survey

Conducted on the CFH (Canada–France–Hawaii) Telescope, this survey has taken advantage of a new spectrograph that enables low-dispersion spectra of relatively faint objects to be obtained. Target galaxies have been selected in the near-infrared I band down to magnitude 22.5. The redshift survey is therefore a deep survey of over 1,000 galaxies, with a median redshift of z = approx. 0.6, that explores evolutionary effects (and about which more will be said in Chapter 5).

3.12.7 Very deep surveys with very large telescopes

The world's largest operating telescope – the 10-metre Keck Telescope – is being used for a redshift survey that will ultimately cover 15,000 galaxies to $m_B = 24.5$, in a Deep Extragalactic Evolutionary Probe (DEEP). Aside from exploring galaxy distributions at much higher redshifts, it too will probe evolutionary effects to even higher redshifts (see Chapter 5).

It may also be that the astronomical world will see a complete new generation of telescopes intended only for spectroscopy. Spectroscopy does not require so sharp and perfect an image as might be desired for direct photography. In many cases, light is simply captured and directed down an optical fibre. Also, a large-aperture telescope often suffers from atmospheric seeing conditions that blur and degrade images.

Much will depend on the long term success or otherwise of the Hobby–Eberley Telescope (again named for its sponsors) at the McDonald Observatory in Texas. This telescope has an array of some 91 hexagonal mirrors, assembled as a primary mirror 11 metres in diameter. Its design is derived from the Arecibo Radio Telescope – the largest-aperture single radio dish in the world – in that the primary mirror is spherical and aimed at a fixed point in the sky. While the Arecibo telescope is aimed directly upwards, the Texas telescope is aimed at 60 degrees, but the whole telescope structure can be raised, rotated to a different position, and set down. While the telescope lacks normal mobility, the advantage of a spherical mirror is that it has no fixed optical axis. Consequently, celestial objects can be tracked, as they move slowly across the sky, by moving the detection apparatus instead of the telescope.

However, as any student of geometrical optics would know, images produced by spherical mirrors suffer enormously from spherical aberration, especially with the relatively short focal length of the Texas telescope. It can nevertheless be overcome by a multiple-mirror corrector that rides with the detector unit. The detection apparatus must, however, be moved and orientated with extreme precision. The attraction of this design is of course the great reduction in cost by using only spherical mirrors that stay in a fixed alignment within a simple framework structure. The mirrors do not require active alignment, as is the case with other multiple-mirror telescopes. Deep probes with this telescope will enable up to 20 objects to be observed simultaneously. This will be best suited for clusters.

Another revolutionary design in spectroscopic telescopes is a vertical telescope whose primary mirror is a spinning dish of mercury. Clearly the future holds much promise in terms of new telescopes that will carry redshift surveys deep into the cosmos.

3.13 FURTHER READING

General

Wide-field Spectroscopy and the Distant Universe (The 35th Herstmonceux Conference), *Ed.* S.J. Maddox and A. Aragon-Salamanca, World Scientific, 1995.

Specialised

Bahcall, N.A. *et al.*, Clustering and Large-Scale Structure with the Sloan Digital Sky Survey, *Publ. astr. Soc. Pacific,* **107**, 790 (1995).

Glazebrook, K. *et al.*, An imaging *K*-band survey – II. The redshift survey and galaxy evolution in the infrared, *Mon. Not. R. astr. Soc.,* **275**, 169 (1995).

Koo, D.C., First Keck Results for Deep (Deep Extragalactic Evolutionary Probe), [in] *Wide-field Spectroscopy and the Distant Universe* (The 35th Herstmonceux Conference), *Ed.* S.J. Maddox and A. Aragon-Salamanca, p.55, World Scientific, 1995.

Lilly, S.J. *et al.*, The Canada–France Redshift Survey. I. Introduction to the Survey, Photometric Catalogs, and Surface Brightness Selection Effects, *Astrophys. J.,* **455**, 50 (1995).

Loveday, J. *et al.*, The Stromlo–APM Redshift Survey. IV. The Redshift Catalog, *Astrophys. J. Suppl.*, **107**, 201 (1996).

Loveday, J., The APM Bright Galaxy Catalogue, *Mon. Not. R. astr. Soc.,* **278**, 1025 (1996).

Mamon, G.A., The Denis 2 Micron Survey and its Cosmological Applications, [in] *Wide-field Spectroscopy and the Distant Universe* (The 35th Herstmonceux Conference), *Ed.* S.J. Maddox and A. Aragon-Salamanca, p.73, World Scientific, 1995.

McGill, C., The redshift projection – I. Caustics and correlation functions, *Mon. Not. R. astr. Soc.,* **242**, 428 (1990).

Raychaudhury, S., A Flair Redshift Survey in the Direction of the Motion of the Local Group, [in] *Wide-field Spectroscopy and the Distant Universe* (The 35th Herstmonceux Conference), *Ed.* S.J. Maddox and A. Aragon-Salamanca, p.110, World Scientific, 1995.

Rowan-Robinson, M. *et al.*, A sparse-sampled redshift survey of IRAS galaxies – I. The convergence of the *IRAS* dipole and the origin of our motion with respect to the microwave background, *Mon. Not. R. astr. Soc.,* **247**, 1 (1990).

Santiago, B.X. *et al.*, The Optical Redshift Survey: Sample Selection and the Galaxy Distribution, *Astrophys. J.,* **446**, 457 (1995).

Santiago, B.X. *et al.*, The Optical Redshift Survey. II. Derivation of the Luminosity and Diameter Functions and of the Density Field, *Astrophys. J.,* **461**, 38 (1996).

Saunders, W. *et al.*, The Point Source Catalog Redshift Survey, [in] *Wide-field Spectroscopy and the Distant Universe* (The 35th Herstmonceux Conference), *Ed.* S.J. Maddox and A. Aragon-Salamanca, p.88, World Scientific, 1995.

Shectman, S.A. and Landy, S.D., The Las Campanas Redshift Survey, *Astrophys. J.,* **470**, 172 (1996).

Strauss, M. *et al.*, A Redshift Survey of IRAS galaxies. IV. The Galaxy Distribution and the Inferred Density Field, *Astrophys. J.,* **385**, 421 (1992).

Wegner, G. *et al.*, The Peculiar Motions of Early-Type Galaxies in Two Distant Regions. I. Cluster and Galaxy Selection, *Astrophys. J. Suppl.*, **106**, 1 (1996).
Willmer, C.N. *et al.*, A Medium-Deep Survey of a Minislice at the North Galactic Pole. II. The Data, *Astrophys. J. Suppl.*, **104**, 199 (1996).
Zucca, E. *et al.*, Statistical analysis of the galaxy distribution in a diameter-limited sample in the northern sky, *Mon. Not. R. astr. Soc.,* **253**, 401 (1991).

4

Mapping the cosmos – nearby structures

4.1 INTRODUCTION

'There are more things in heaven and earth, Horatio, than are dreamt of in your philosophy' - so reads the famous line in *Hamlet*. Clearly, Shakespeare must have been anticipating the large-scale distribution of galaxies! After all, on a cosmic scale, galaxies are effectively 'particles', with plenty of space between them. The intuition of physics therefore suggests that they ought to behave as particles. The same could be said of stars – and stars do behave as expected: they may cluster in places under mutual gravitation, but are otherwise scattered at random. One would hardly expect them to assemble into formation, like aircraft at a show, spelling out a message; and they do not.

But galaxies are different. They are assembled in formations that spell out a message to those who research cosmology. No scientist, even in his wildest dreams, could have imagined it. The formations were completely unanticipated! This cosmological windfall has opened up a wonderful insight into the workings of the Universe. But, like the early readings of Egyptian hieroglyphs, we are still trying to decipher its meaning.

We shall see in this chapter that galaxies are arranged in wall-like or filament-like condensations that interconnect in a three-dimensional labyrinth. The intervening spaces possess relatively few galaxies or may even be completely empty. The largest such voids are around 5,000 km/s (75 Mpc, or 250 million light years) in extent. Voids appear to range in size down to perhaps only 2 Mpc, the small voids seemingly permeating the condensations of galaxies, giving a somewhat efflorescent texture to the whole arrangement.

The emphasis in this chapter is on mapping the nearby Universe. 'Nearby', on a cosmic scale, implies the surrounding Universe out to a redshift approaching 10,000 km/s, that is to about 150 Mpc (500 million light years). It is to this distance that we have reasonable coverage over the whole sky. Beyond that, we have only incomplete or fragmentary coverage (as will be explored in the following chapter).

This chapter attempts a synthesis of the findings of the various redshift surveys. It incorporates the early structures found by Einasto, Gregory, Thompson and others in the Virgo–Coma region, afterwards confirmed and mapped in far more detail by the Center for Astrophysics surveys. It involves the Perseus–Pisces region pioneered by Giovanelli and Haynes, and the Center for Astrophysics. In the southern skies, it involves structures that this author has long known and their confirmation by the Southern Sky Redshift Surveys. Findings from global surveys, such as those of IRAS galaxies are incorporated.

The survey of nearby structures might well be disrupted by the obscuring band of the Milky Way – the classic Zone of Avoidance of galaxies – or, more accurately, the Zone of Obscuration. Fortunately, as indicated in the preceding chapter, a number of special surveys have sought to narrow the zone. Their success in tracing continuous structures from one side of the Galactic plane to the other enables us to give an almost complete account.

Later in this chapter, an Atlas based on 'catalogued' redshift data will be presented. (The justification for catalogued surveys has been discussed extensively in the previous chapter (Section 3.9)). It is appropriate here because the emphasis is on mapping, with a largely qualitative description of the local large-scale structures. By contrast, the strength of the statistically controlled surveys lies in the extraction of the values of various quantitative parameters. This is to be the topic of Chapter 6.

4.2 DEFINING LARGE-SCALE STRUCTURES

In spite of large-scale structures being the topic of this book, no formal definition of them has yet been presented, though they are obvious in character. Nevertheless, a simple definition would prove useful.

Very few (if any) galaxies exist in isolation. Virtually every galaxy has a neighbour. The experience of the author suggests that galaxies and immediate neighbours lie within 200 km/s of one another, in redshift space. For example, when this criterion was applied to very nearby galaxies (redshifts less than 2,000 km/s) in making the maps shown later in this chapter, less than 1 per cent of them were seen to be isolated. If the sampling was absolutely complete, then it is likely that even these isolated galaxies would have been found to possess neighbours.

We therefore propose a simply working definition: *Large-scale structures are formed by condensations of galaxies with galaxy–galaxy separations less than 200 km/s in redshift space.*

For many of the maps that accompany this chapter, such large-scale structures are shown as stippled regions, formed simply by expanding spheres of radius of 100 km/s around each galaxy, but rejecting cases of isolated galaxies. (In fact, we have used cylindrical cells rather than spheres, as explained later.) Such large-scale structures form three-dimensional shapes, appropriately described as walls or filaments, that account for less than 10 per cent of the volume of redshift space (and, by reference to Figures 3.1–3.3, somewhat less of normal three-dimensional space). As already indicated, large-scale structures are not in themselves isolated but interconnect as a labyrinth.

It is probably worth mentioning in passing that clusters of galaxies never exist in isolation, but are always contained within large-scale structures.

4.3 AN OVERVIEW OF NEARBY LARGE-SCALE STRUCTURES

Figure 4.1 is provided as a general reference map that shows the most prominent features. It is simplified by projecting the structures into two dimensions: the classic Supergalactic-plane of Gerard de Vaucouleurs (as described in Chapter 1; see also 3.3 and 4.4.1 below).

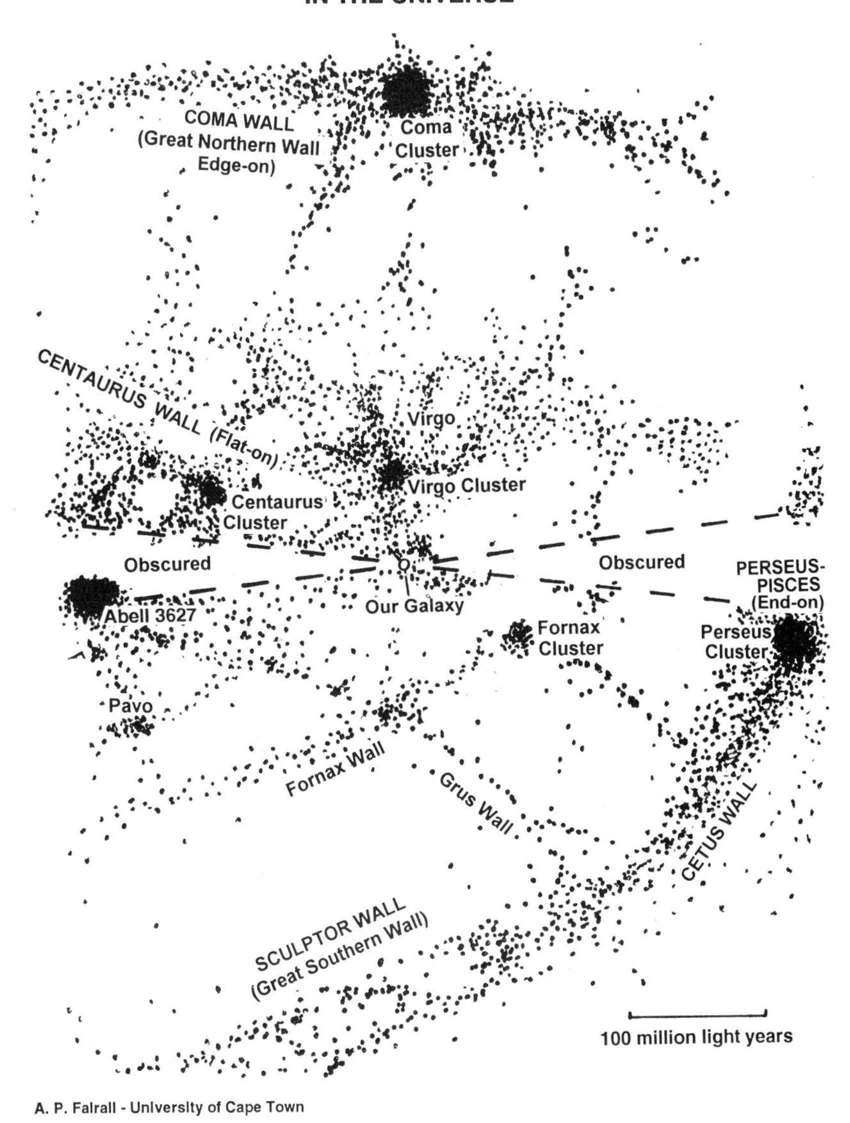

Fig. 4.1. A schematic representation of the principal large-scale structures and features within 7,000 km/s (or equivalent) of the position of our Galaxy. (See text.)

Most of the major nearby condensations of galaxies tend toward this plane. Note that this is a schematic representation, accurate in its representation of the large-scale features, but the individual data points are not real.

The largest features shown by the map are 'great walls' of galaxies. These are concentrations of galaxies within slab-like volumes in redshift space, and have dimensions typically several thousand km/s long, a few thousand km/s wide, and only a thousand km/s thick. Most are straight walls, but some show gentle curvature.

Four such features occur in the map. At the top is the Coma Great Wall, the first such feature to be widely recognised as such. In the map it is depicted in cross-section. The Perseus–Pisces 'Supercluster' is probably a wall-like formation, and is interconnected with the Sculptor Wall which is shown in cross-section at the bottom of the map. By contrast, the irregular structure within the Centaurus Wall is seen flat-on. An appendage of this structure is the Local Supercluster where our own Galaxy is situated.

Clusters of galaxies occur within the great walls. The three most dominant clusters in the map – Coma, Perseus and ACO 3627 – dominate three of the great walls just mentioned. Lesser clusters – Centaurus, Virgo and Fornax – are also shown.

The map also suggests the interconnections that exist between the great walls, and also portrays the voids that separate the structures as though they were seen in cross-section. Later in this chapter, we shall discuss the character of these voids. For now, we shall describe the gross concentrations of galaxies, shown in the map.

4.4 INDIVIDUAL LARGE-SCALE STRUCTURES

This section concerns itself with large-scale structures – regions of overdensity – that form recognisable entities in themselves. We shall review them in sequence of increasing redshift. Aside from the diagrams shown in this section, the features concerned can be identified in the general map (Figure 4.1) and in the accompanying Atlas of Nearby Large-Scale Structures, which appears between pages 80 and 81.

Almost all the features are named after their foreground constellations; the star patterns that gave rise to the constellation names are obviously very much in the foreground, but the system is convenient. Many have been long established (e.g. Virgo Cluster, Coma Cluster), but some are of more recent nomenclature (e.g. Hercules Supercluster, Cetus Wall) put forward by various investigators (including the author). Some of the features are sufficiently large that they span many constellations: if so only a central constellation is used; thus Centaurus Wall replaces the Centaurus–Telescopium–Pavo–Indus concentration. Perhaps it should be emphasised that these names are by no means officially recognised by the appropriate commission of the International Astronomical Union (which ratifies all astronomical nomenclature) but are used in this book as a convenience.

4.4.1 The Virgo Supercluster

As detailed in Chapter 1 (Section 1.2), Sir John Herschel was the first to recognise large-scale structures in the form of the Virgo Supercluster. His description is still accurate today: a roughly spherical system, centred on Virgo, with branches or protuberances running outward from the denser core, our Galaxy forming an element of one such protuberance, i.e. being involved within the outlying members of the system.

The Virgo Cluster is the nearest and most easily studied cluster of galaxies, at a redshift of about 1,100 km/s, a distance of approximately of 15 Mpc, or 50 million light years. The cluster itself contains about 80 elliptical or S0 galaxies, 120 spiral and about 900 catalogued dwarf galaxies. Furthermore, if the distribution is similar to that of our Local Group of galaxies, then there will also be numerous dwarf galaxies of too low a surface brightness to detect, though their collective mass is probably not significant; similarly, their gaseous neutral hydrogen may not be detectable by radio telescopes (though this is a matter of some controversy).

When a cluster is 'gravitationally relaxed', it is possible to determine its mass by the 'virial theorem' (as discussed later in Section 7.4). There is some question as to whether this is appropriate for the Virgo Cluster, in that its spiral galaxies show a much higher velocity of dispersion (900 km/s) within the cluster than do the elliptical and SO galaxies (550 km/s). If it is relaxed, then its mass is at least 5×10^{14} solar masses (one solar mass is the mass of our Sun, i.e. 2×10^{30} kg). At least a tenth of this mass lies within the giant elliptical M87, the dominant central galaxy. Impressive though its membership may seem, the Virgo Cluster still does not rate as a particularly rich cluster; rich clusters would have a mass perhaps an order of magnitude larger.

The segregation between the dynamics of the spiral and the elliptical galaxies in the Virgo Cluster is also apparent in their spatial distribution. The elliptical galaxies form an inner core, sometimes described as having a bow-tie shape, as seen in the sky. In the centre is M87, also known for its nuclear activity. The spiral galaxies form something of a halo. This is quite acceptable to those who believe that the interactions that must occasionally occur in denser environments cause the spirals to evolve to ellipticals. However, the segregation is not quite so symmetrical and there seems to be systematic offsets in the redshifts; even M87 differs from the mean redshift by 200 km/s. All this suggests that the cluster is still far from being gravitationally relaxed.

There is no edge or sharp cutoff; instead, the Virgo Cluster is the central condensation of the more extensive Virgo Supercluster. Brent Tully's *Atlas of Nearby Galaxies* and its associated papers provide probably the best mappings of this supercluster (such as that shown in Figure 4.2). The distribution of galaxies has an irregular though somewhat flattened shape. Herschel's 'protuberances', now seen in redshift space, radiate from the Virgo core as finger-like appendages. Often described as the 'Local Supercluster', this structure is large enough to include our own Galaxy and the Local Group of galaxies; it extends past our Galaxy, though we seem to lie towards its edge. Another prominent appendage leads towards the Ursa Major Cluster, a much weaker conglomeration of spiral galaxies.

Clusters of galaxies, such as Virgo, tend to have reasonably symmetrical shapes, because the 'crossing time' for a galaxy to fall from one side of the cluster and then climb out on the opposite side is usually a small fraction of the age of the Universe; for Virgo it is considered to be about 0.08. This ought to have been sufficient for the Virgo Cluster to have become gravitationally 'relaxed'. By contrast, the crossing time of the extended Virgo Supercluster is many times the age of the Universe, and the structure has an understandably irregular or 'unrelaxed' structure.

Gerard de Vaucouleurs, who for many years championed the existence of our Local Supercluster, referred to it as the 'Supergalaxy'. Since the structure was somewhat flattened, de Vaucouleurs proposed 'Supergalactic coordinates' based on the plane of flattening (as introduced in Section 3.3).

The use of Supergalacticcoordinates is now well established in the literature. Furthermore, as suggested above in Figure 4.1, many neighbouring structures to the Virgo Supercluster also lie close to the extended Supergalactic plane.

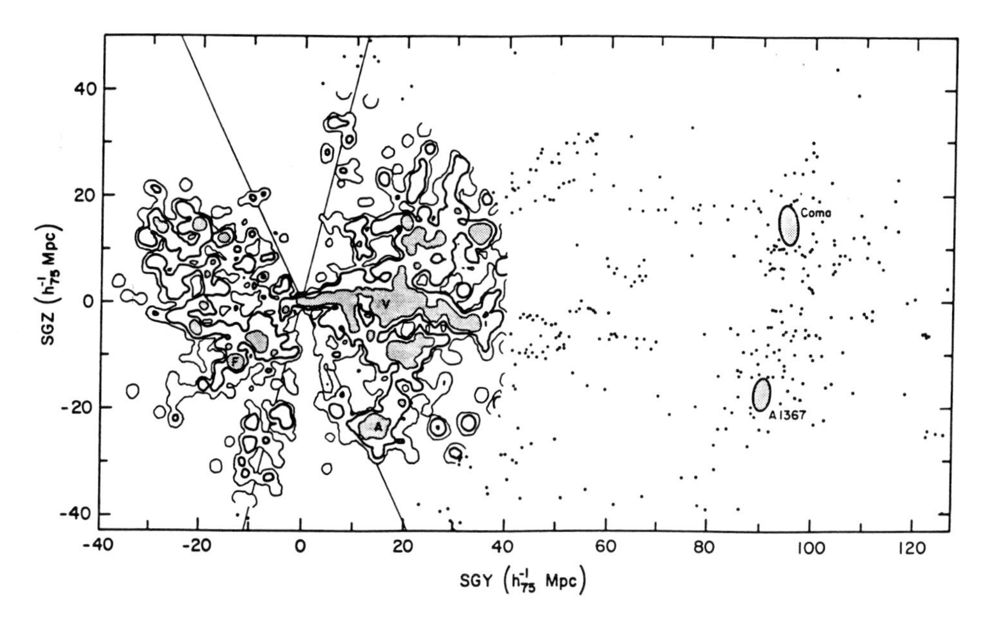

Fig. 4.2. Brent Tully has used data from his *Atlas of Nearby Galaxies* to extract a surface density map of the Virgo Supercluster. Clusters shown are Virgo (V), Antlia (A) and Fornax (F); the positions of the more distant Coma and Abell 1367 clusters are indicated. Rectangular supergalactic coordinates are used. (Reproduced with permission from B. Tully and the *Astrophysical Journal* (**303**, 25, 1986).)

The Virgo Supercluster dominates nearby redshift space in the northern skies; out to 2,000 km/s, as can clearly be seen in the accompanying Atlas of Nearby Large-Scale Structures. However, the extension of its Supergalactic plane to neighbouring structures is an indication that it may not be an independent entity. Although it is surrounded by a number of voids, it forms an appendage to a much larger structure, in the general direction of the constellation Centaurus.

4.4.2 The Centaurus Wall

In the 1970s and early 1980s, the distribution of galaxies in the southern sky was revealed by the UK and ESO Sky Surveys (Section 3.8). A concentrated band of galaxies, roughly coincident with the Supergalactic plane, was seen to cut across the southern Milky Way. Even the earliest redshift data (e.g. those plotted by my student Hartmut Winkler (Figure 1.11)) suggested it was a continuous large-scale structure, variously referred to as 'Hydra–Centaurus–Telescopium–Pavo–Indus' or combinations of these. Today, many more redshifts are available, and the feature is conspicuous in sky plots such as that shown in Figure 4.3. It is a structure very much larger than the Local Supercluster. Its dimensions, in redshift space, are approximately 7,000 by 5,000 by 1,000 km/s. Such a large slab-like structure is usually termed a 'Great Wall'. The term 'Centaurus Great Wall' is appropriate, though not yet in common use. Our Galaxy and the Virgo Supercluster are a part of it; the Virgo Southern Extension – one of the protuberances stretching out from the Virgo Cluster core – merges with the Centaurus Wall.

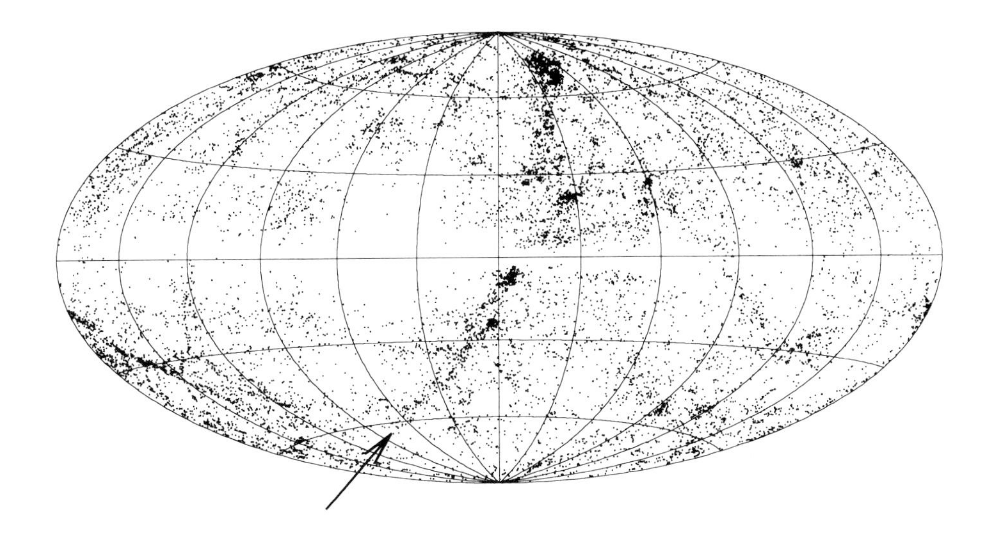

Fig. 4.3. The Centaurus Wall is the conspicuous central feature in this all-sky equal-area projection for galaxies with redshifts less than 6,000 km/s. The Galactic plane (obscuration by the Milky Way) runs along the central horizontal axis. (Diagram by Patrick Woudt, University of Cape Town, 1997.)

We lie close to the plane of the Centaurus Wall and see the structure edge-on. The plane of the Centaurus Wall is tilted about 15 degrees to the Supergalactic plane, with a pole at approximately RA = 5 hrs 45 min, Dec. = –30°. Our Galaxy lies near its edge, such that we view the Centaurus Wall as a band of galaxies that stretches over an angle of more than 180 degrees, but does not quite seem to encircle us.

Georges Paturel and co-workers have identified a 'hypergalactic' plane – roughly coincident with the Centaurus Wall plane – that they propose could be extended over the full 360 degrees to include even more distant structures. In 1988, they gave its pole at 5 hrs, –28.5°, but have since adjusted it to 6 hrs 25 min, –24°.

The importance of the Centaurus region also lies in the fact that a general movement or streaming of galaxies – our own Galaxy included – takes place towards it. Alan Dressler dubbed it the 'Great Attractor', and the name has stuck. Although briefly mentioned in the opening chapter, this streaming will be discussed in detail in Chapter 7.

A great wall contains considerable sub-structure within its flattened form. There are numerous openings or hollow passages permeating the wall. While some researchers see the galaxies concentrated into a multitude of internal filaments, the author sees an internal labyrinth of cellular structure, within tiny voids ranging from perhaps only 200 to 1,000 km/s in diameter. Certainly, such tiny voids are apparent in the nearby Centaurus Wall. They give the wall a thickness of about 1,000 km/s and make its outer boundary somewhat irregular.

Many of the great wall structures have a central massive cluster of galaxies. The Centaurus Wall has Abell 3627, only recently recognised for what it is, as (perhaps by Murphy's law) it lies in that short portion of the wall obscured by the foreground Milky Way.

Abell 3627 is heavily obscured but can just be seen through the dust clouds of our own Galaxy. It is one of the three nearest rich clusters of galaxies (the others are the Perseus and Coma Clusters), each of which is many times more massive than the Virgo Cluster, described above. Like the Coma Cluster, its core contains mainly elliptical and SO galaxies, which predominate in such dense environments. A trio of massive ellipticals marks the centre of the cluster. One of these is a well-known active galaxy, PKS 1610-60. Like Coma, spiral galaxies tend to occur in a halo (including a Seyfert galaxy).

Abell 3627 is not the only cluster within the Centaurus Great Wall. The less massive Centaurus Cluster and a number of Abell/ACO clusters are more clearly visible as they lie clear of the plane of the Milky Way. The Centaurus Cluster, like Virgo, is one of the best known in the sky. Like Virgo, its elliptical and spiral galaxies tend to segregate towards separate concentrations (as seen in the sky). The redshift histogram of the cluster also shows two pronounced peaks – a main one at 3,000 km/s and a secondary one at 4,500 km/s – which have been the cause of much debate. Though one group of investigators has ascribed this to the presence of two dynamical units, it would seem more likely that the true peak is the 3,000 km/s one and the 4,500 km peak results from contamination by an extended background structure within the Centaurus Wall.

The presence of such massive clusters might be expected to have a disruptive effect on their surroundings, such that the texture within the wall would show a change in character in the neighbourhood of the clusters. Within the Centaurus Wall, this does not seem to be the case. Moreover, there is no clue as to why and where such clusters ought to be situated. There is a general misconception that massive clusters occur where walls intersect. While this is vaguely true for the Centaurus Cluster (which is situated close to where the Hydra Wall intersects, as discussed below), it is apparently not true for Abell 3627. The surrounding structures to this massive cluster within the wall – small voids and bubbly texture – seem quite normal, though some of the immediate surrounding structures are totally hidden by the obscuration of the Milky Way.

In the accompanying Atlas of Nearby Large-Scale Structures the Centaurus Wall appears in all the southern sky maps out to 6,000 km/s.

4.4.3 The Fornax Wall

While the Centaurus Wall is seen edge-on, the nearby Fornax Wall is seen flat-on. Being so close, its galaxies are spread over a large fraction of the southern sky. In the accompanying Atlas, it dominates the southern 1,000–1,999 km/s redshift shell, yet the preceding and subsequent shells have relatively few galaxies over the same general area (due to the Volans Void in front of it, and the Eridanus Void behind). The Fornax Wall was first identified by Gerard de Vaucouleurs as the 'southern supercluster' (see also Figure 1.4)

The wall is named for the Fornax Cluster – a nearby cluster at a redshift of 1,500 km/s, containing mainly spiral galaxies (several times less massive than the Virgo Cluster) that lies within it. The wall has dimensions of 7,000 (at least) × 3,000 × 700 km/s. The galaxies of the Fornax Wall are spread quite thinly and in an irregular fashion. The wall passes within 1000 km/s of our Galaxy (near the Fornax Cluster), and it intersects the Centaurus Wall at around 1,500 km/s.

Atlas Section

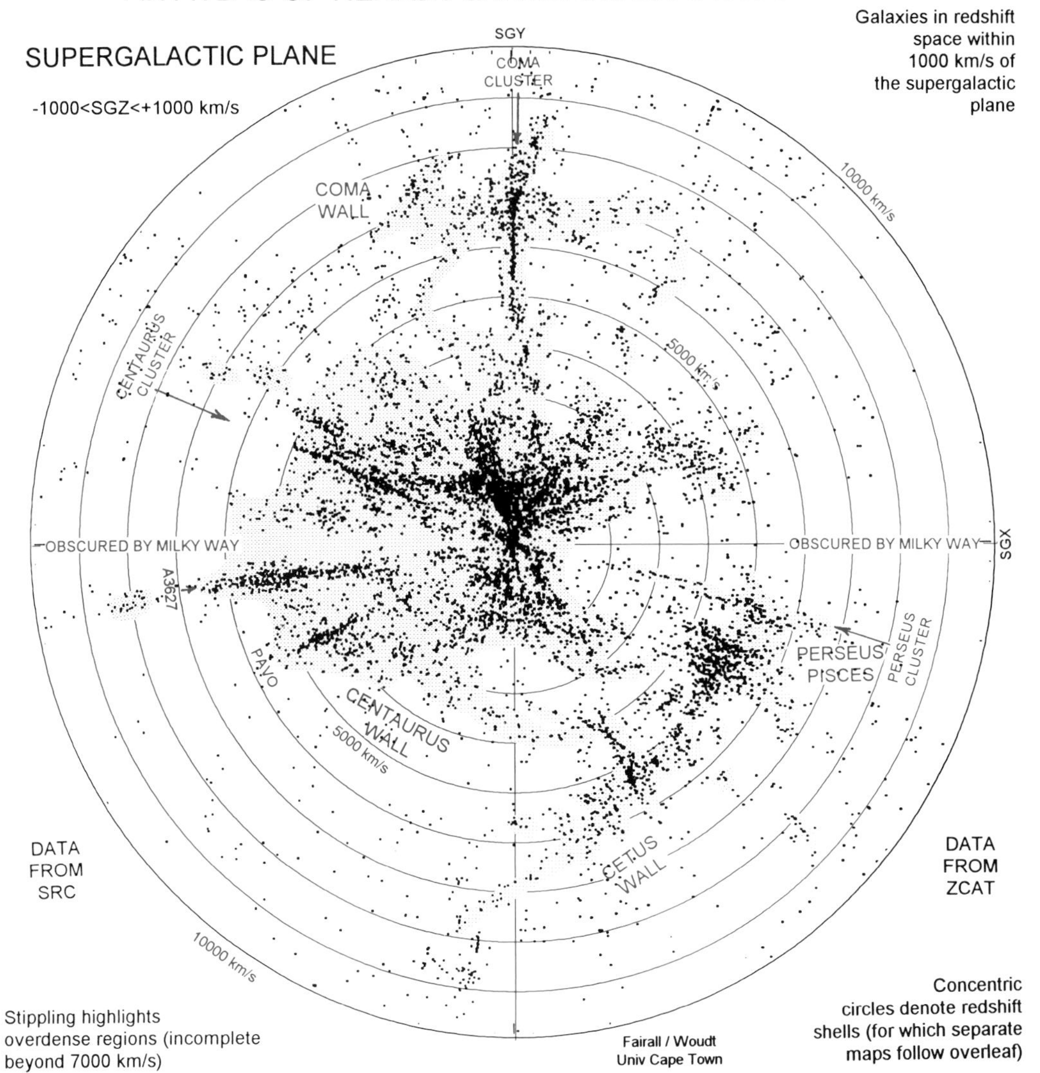

This 26-page Atlas (see Section 4.7) provides maps of large-scale structures in redshift space out to 10,000 km/s (the topic of Chapter 4). The majority of the most prominent concentrations are situated close to the supergalactic plane (defined in Section 3.3) and show up in cross-section in the map above. The Virgo supercluster is the concentration immediately above centre, within the more extensive Centaurus Wall, seen here flat-on. The stippling is based roughly on that used in subsequent maps; it suggests the mapping is only complete to 7,000 km/s.

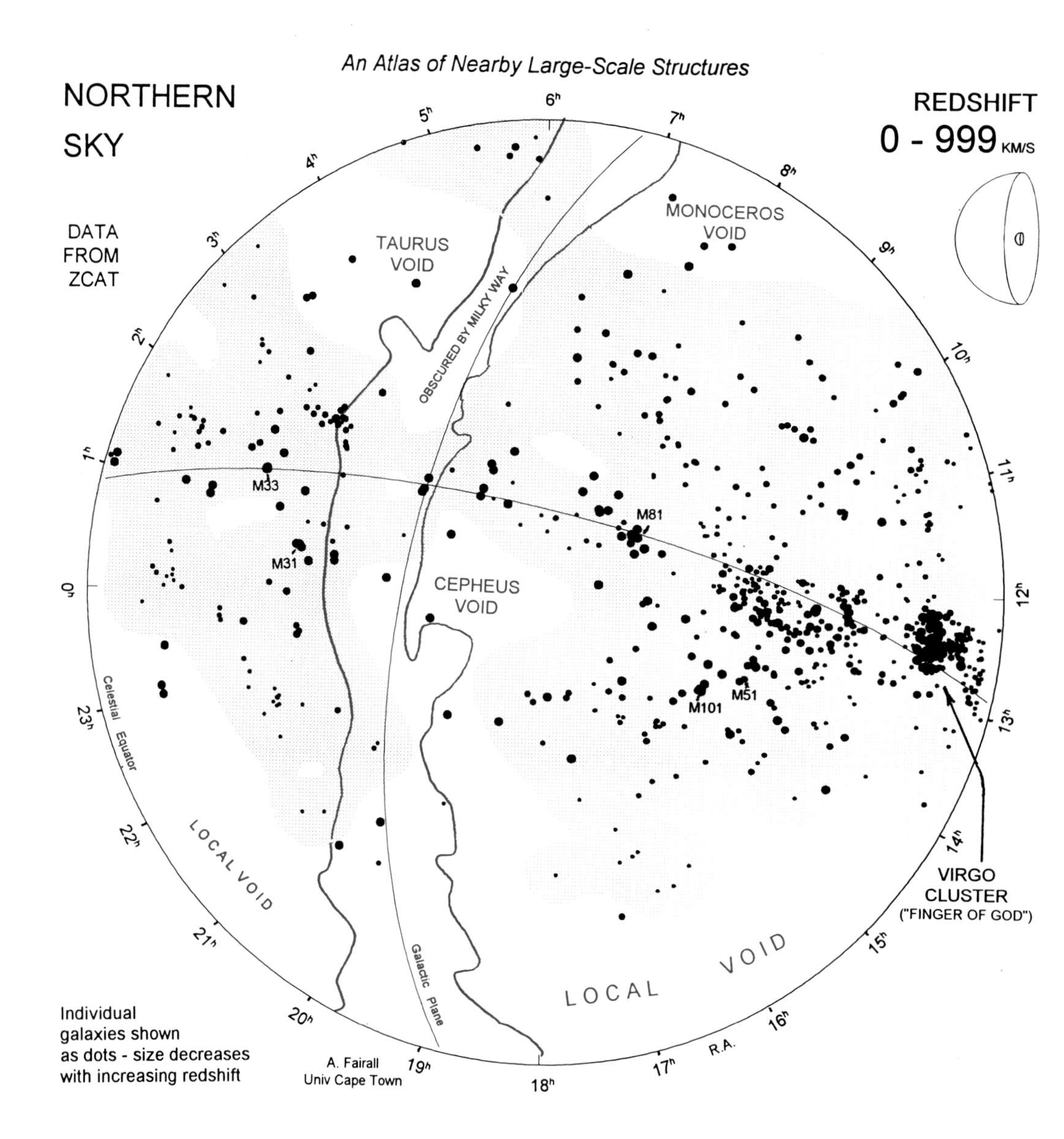

These two maps are the first of ten pairs of all-sky maps, centred on the north and south celestial poles and divided at the celestial equator. The first pair looks at only those galaxies with a redshift out to 999 km/s. In terms of nearby structures, this is our very immediate neighbourhood – only one thousandth of the volume covered by the Atlas. In the foreground are members of our Local Group of galaxies, especially M31 (the Andromeda Galaxy) and M33 (spiral galaxy in Triangulum), with other nearby groups being centred around M81 (the Ursa Major Group) and NGC 5128 (the Cen-

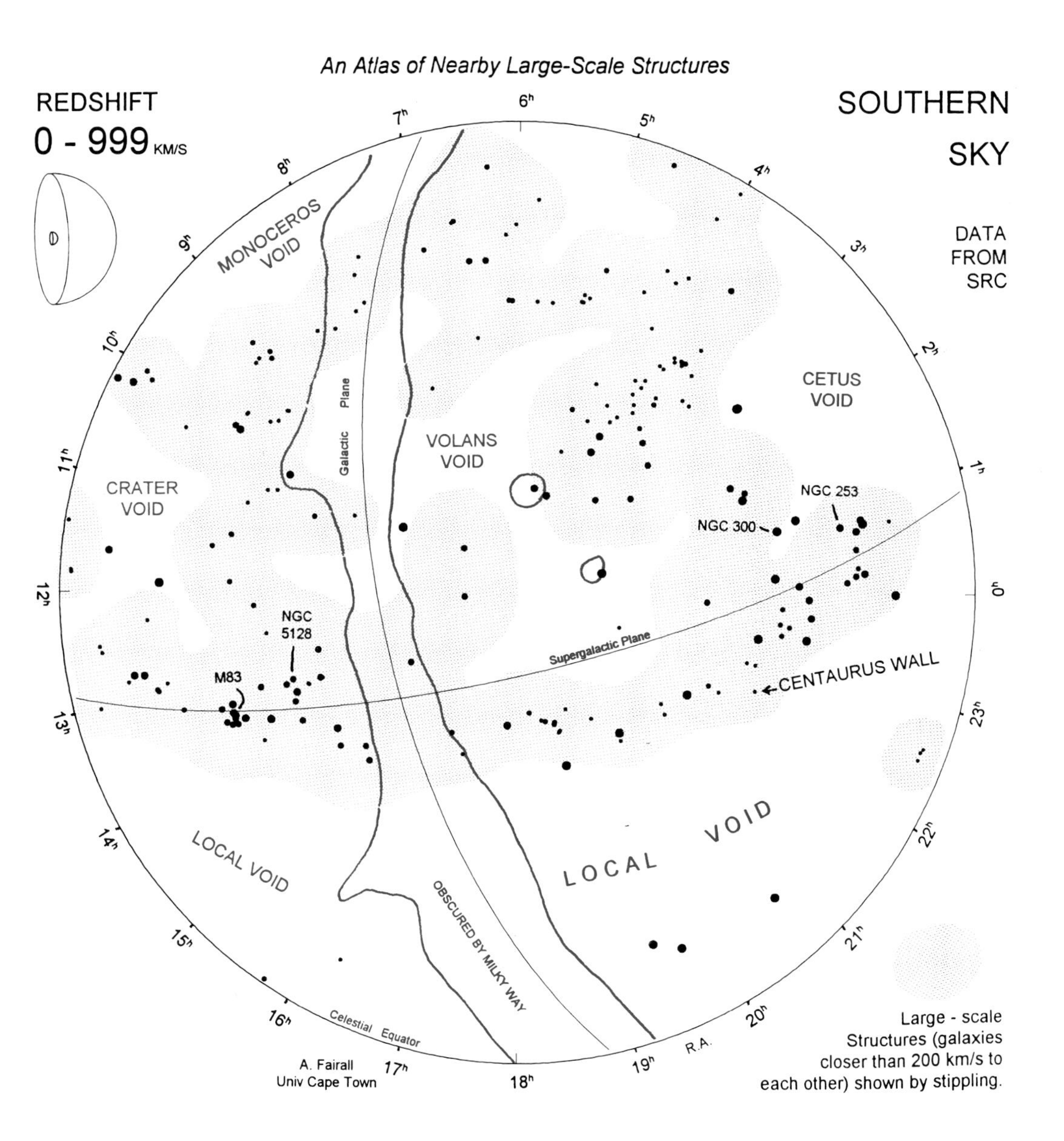

taurus Group). Beyond that the distribution is totally dominated by the Virgo cluster, from which John Herschel's 'protuberances' extend over much of the sky. The stippled large-scale structures (percolations at 200 km/s; see Section 4.7.2) are only those beyond redshift 400 km/s; closer than that the structures fill the entire sky, as our Galaxy is contained within them. Notice the remarkably empty Local Void that covers a third of the sky. Other much smaller voids are indicated.

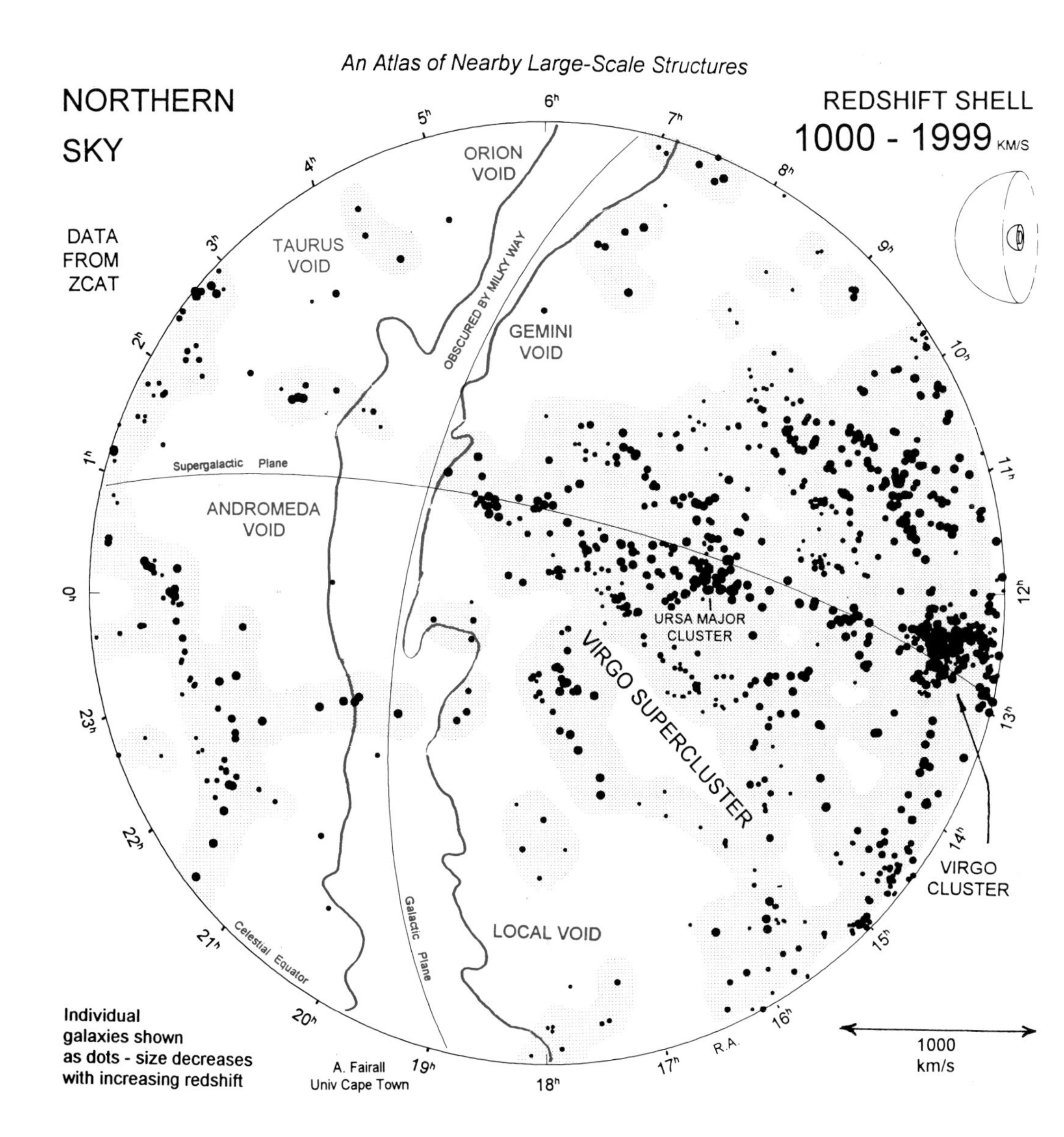

The core of the Virgo cluster forms the centre of the extended Virgo supercluster, which includes the Ursa Major cluster and a southern extension into the Centaurus Wall. The nearby Fornax Wall surrounds the Fornax cluster. It is seen flat-on, and almost all of it is contained in this map (whereas the Centaurus and Hydra Walls, both more or less edge-on, are still seen in subsequent shells). The Local Void still covers much of the southern sky, while the Orion Void cuts off structures at the top of the maps. In all these maps, the band of the Milky Way obscures, or partially obscures, the view.

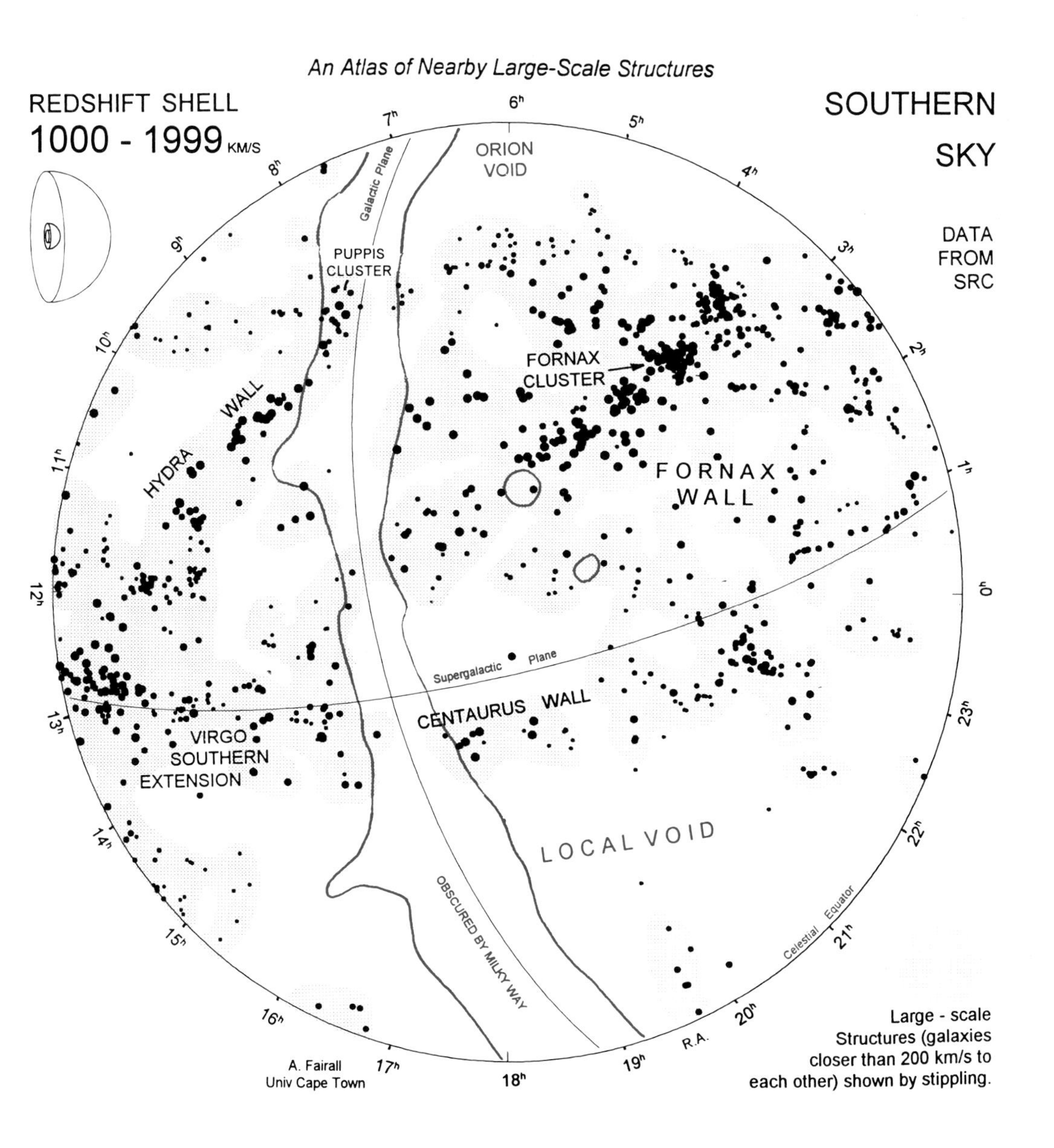

Usually the Milky Way within a few degrees of the Galactic plane would be completely opaque, while the remainder (within the boundary indicated in the maps) would suffer heavy obscuration. In the southern sky, the two Magellanic Clouds also cause slight obscuration. It should be appreciated that the volume contained in this shell is seven times larger than that of the first pair of maps, but still only 0.7 per cent of the total volume covered by the Atlas.

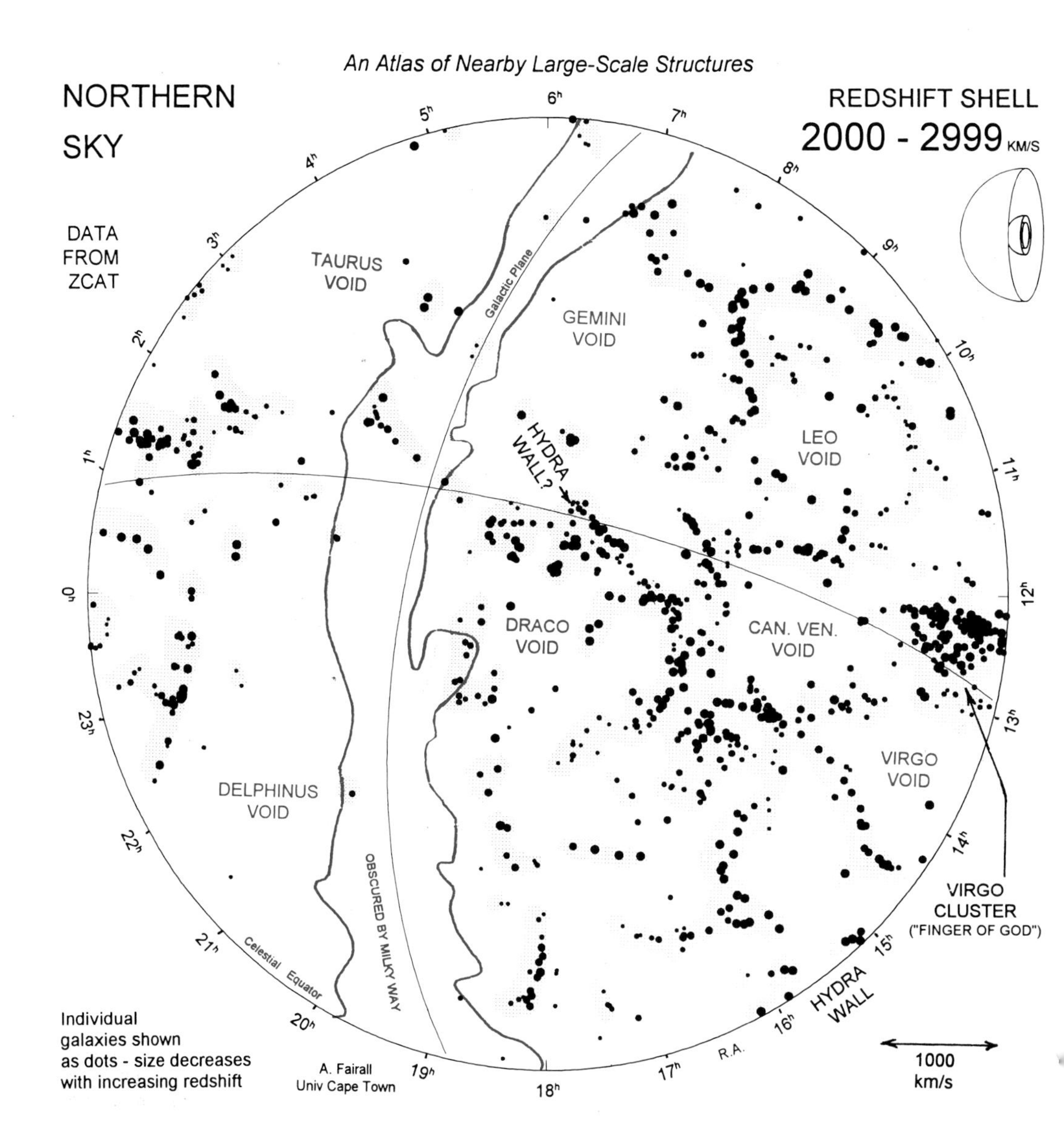

The 'finger of God' of the Virgo cluster is still dominant; similarly the plane of the Centaurus Wall, which is seen edge-on, stretching over more than 180 degrees. The Hydra Wall is a ribbon-like structure that stretches right across the southern sky, with a suggestion that it continues even on towards the north celestial pole (the centre of the left-hand map), in which case it almost encircles us. The two walls intersect almost orthogonally, the Centaurus cluster being close to, but not exactly on, the point of intersection. The Hydra Wall is named for the Hydra cluster that it contains. Be-

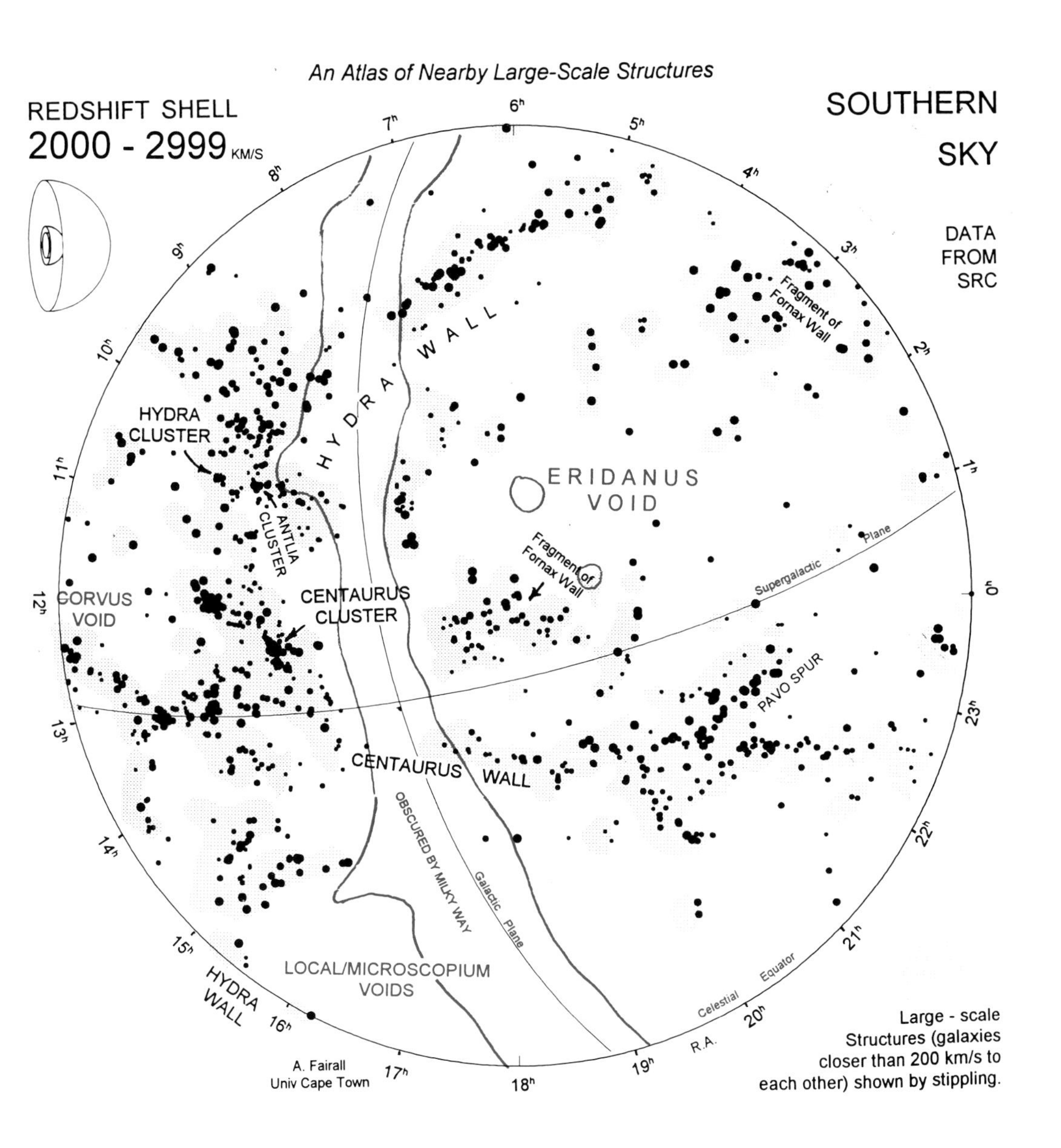

tween the two walls in the southern sky is the Eridanus Void, immediately beyond the flat-on Fornax Wall that dominated the same portion of sky in the previous shell. The Local Void, which leads into the more distant Microscopium Void also accounts for a major part of the sky. By contrast, the northern sky – beyond the Virgo supercluster (previous shell) – shows a network of curved filamentary structures that contain a number of small voids. The volume contained in this shell is 2.7 times larger than that in the previous shell (1.9 per cent of our total volume).

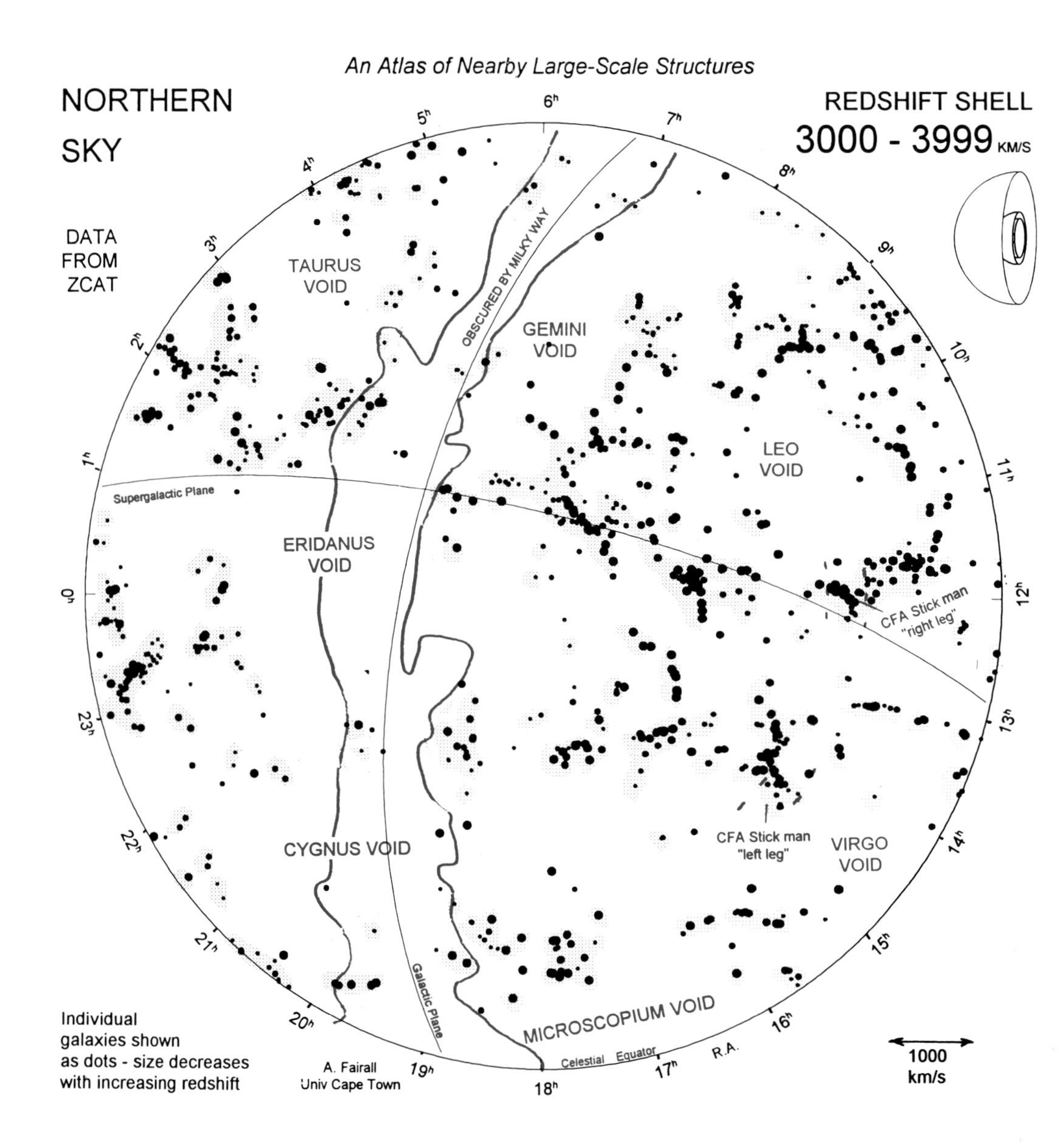

The Hydra and Centaurus clusters are the most dominant features, though they appear as condensations within more extended walls. However, outside the vicinity of these clusters this shell displays a delicate filigree of structures. Curiously, the Centaurus Wall seems to be much shorter, as if it were fading out; yet this is not so, since it is 'revived' in subsequent shells. Notice that the tubular Eridanus Void occurs in both northern and southern skies (see the 0h Right Ascension slice). We also see here the start of the Microscopium Void – an enormous and almost empty void that contin-

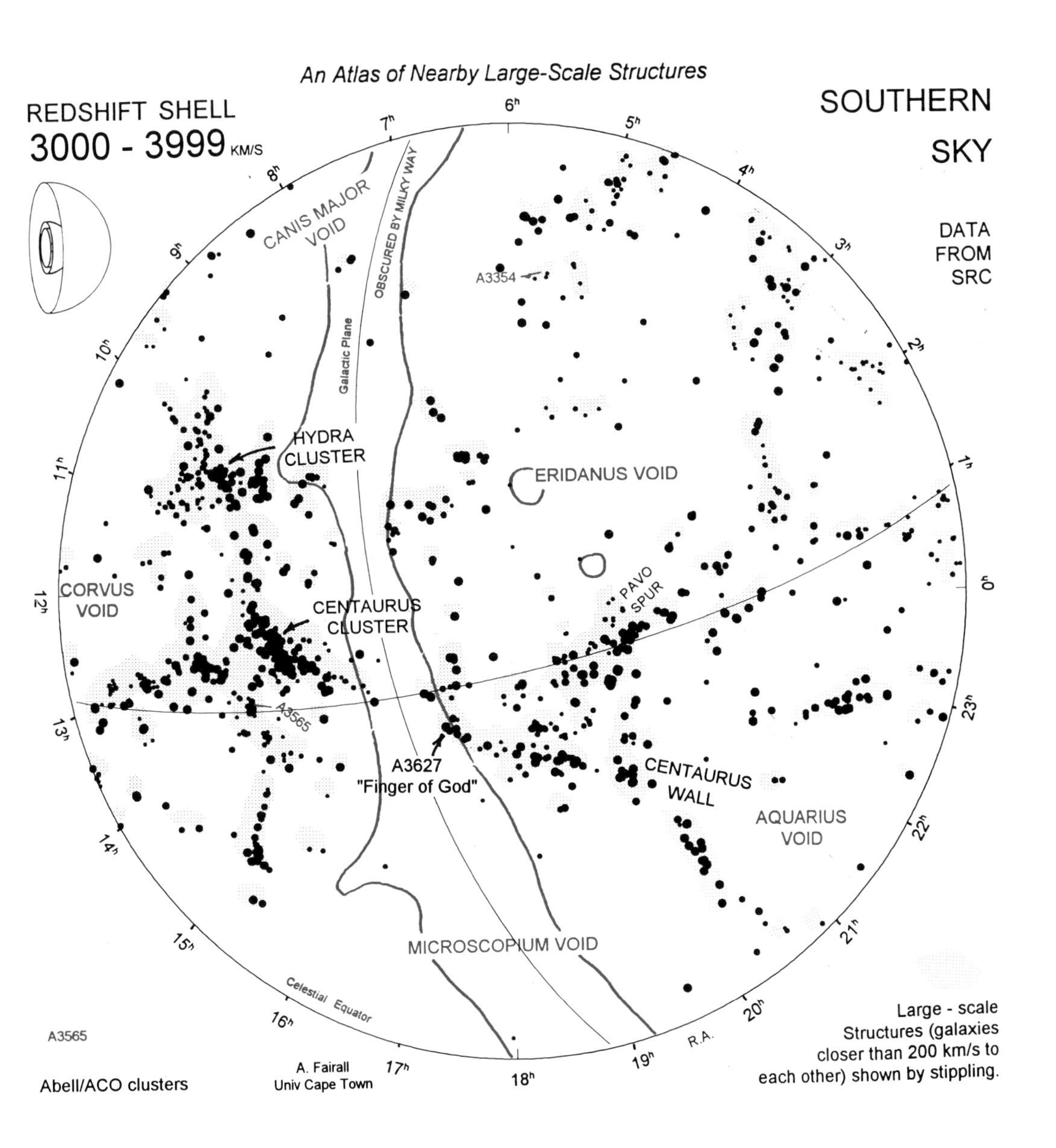

ues in subsequent shells. Since some readers will be familiar with the 'stick man' figure that occurred in the well-known Center for Astrophysics 'Slice of the Universe', we have identified the structures responsible – both here and in subsequent shells. The 'stick man' figure also appears later in the selected northern Declination slice. The volume represented here is 3.7 per cent of our total volume – twice that displayed in the previous shell.

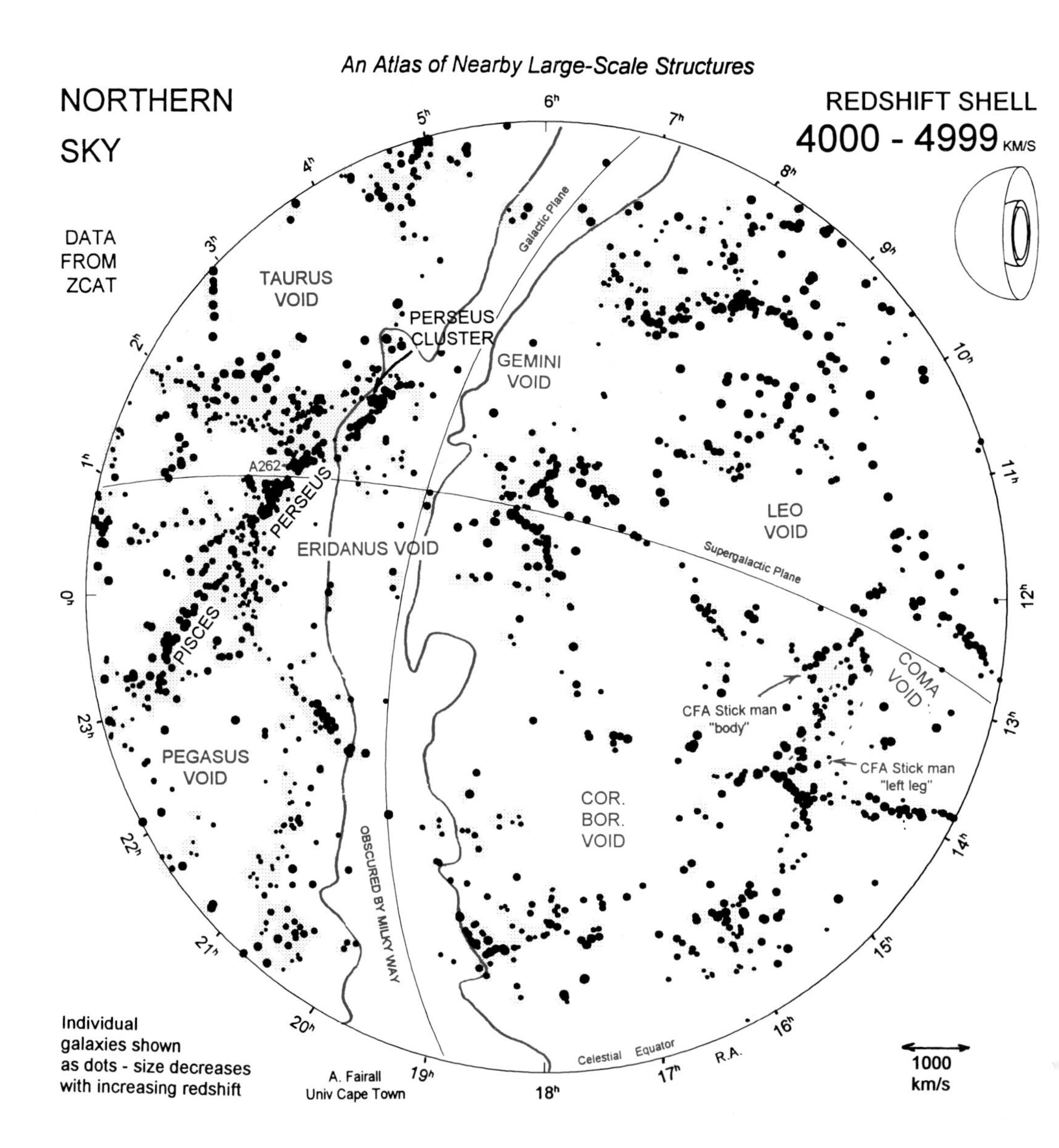

The northern sky is dominated by the beginnings of the massive Perseus–Pisces region (a ribbon seen edge-on). Included towards one end of it is the Perseus cluster, which may seem under-represented in the map due to the overlapping of points. The sprinkling of points to the upper left of Perseus–Pisces is the beginning of the Cetus Wall that runs southward from here. The southern sky is dominated by the Centaurus Wall, now seen centred about the massive cluster A3627. The cluster is, however, heavily obscured by the Milky Way, and is as yet only mildly represented in the

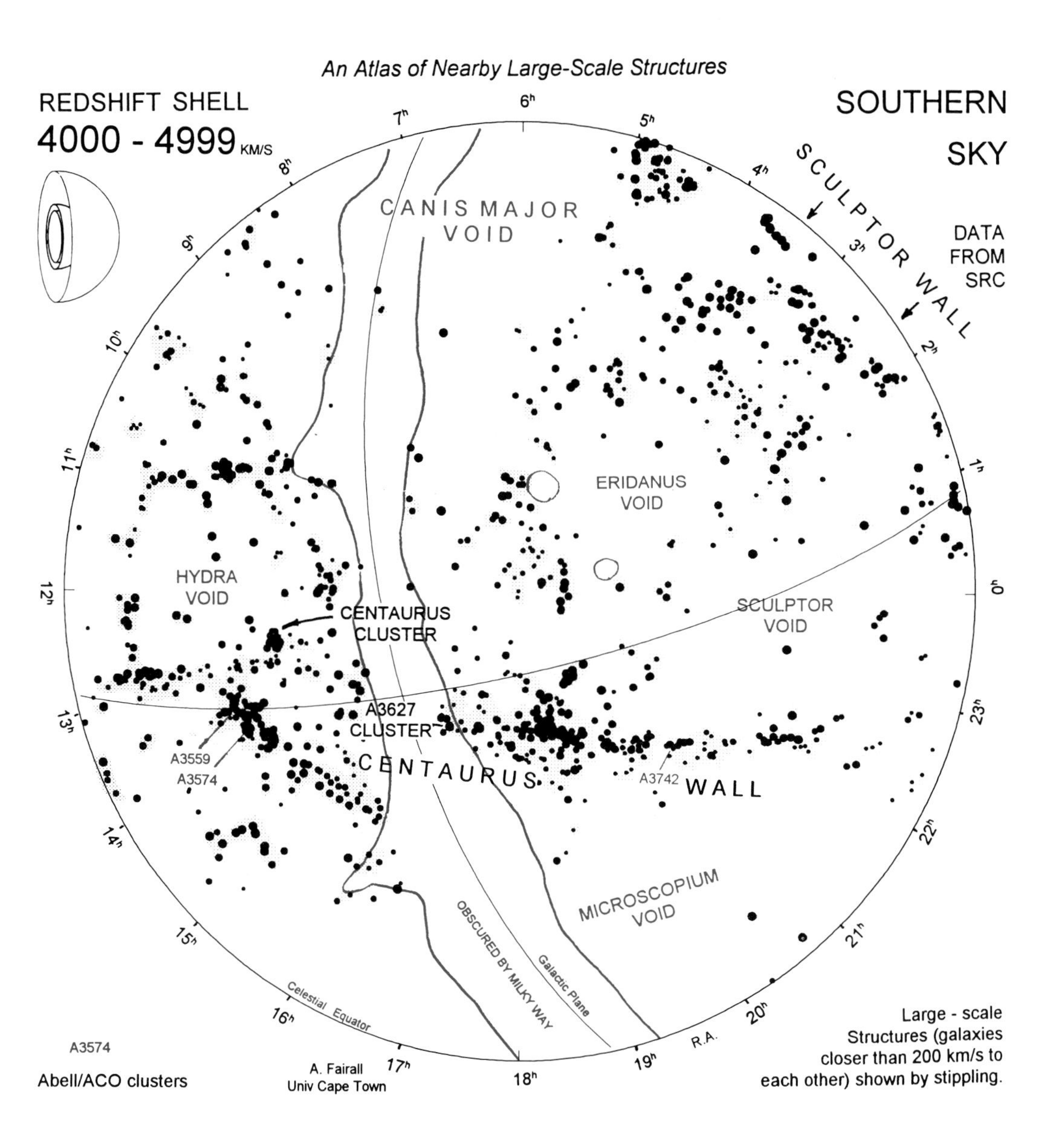

database used to create these maps. (A later map – Selected slice in the Galactic plane – will show it far better.) Other clusters are seen in the Centaurus Wall, as well as the further component of the Centaurus cluster. A portion of the Sculptor Wall (seen neither flat-on nor edge-on) is apparent. Elsewhere the distribution of galaxies exhibits the frothy-like texture, resulting in a labyrinth of interconnected voids. The tubular Eridanus Void is again seen in both northern and southern maps. The enormous Microscopium Void is obvious.

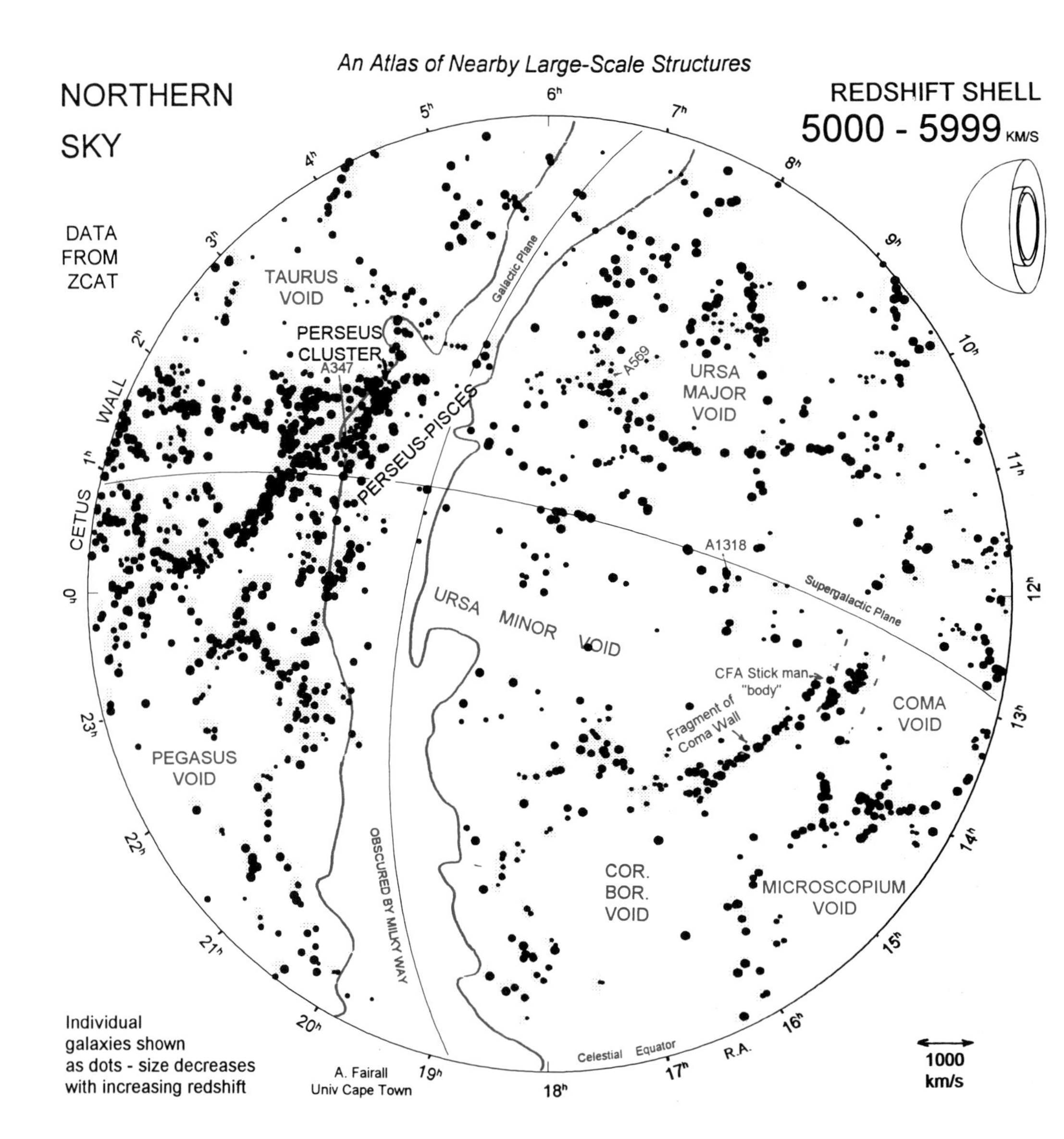

The massive Perseus–Pisces region is seen interlaced with the Cetus Wall, which dominates both the left-hand portion of the northern map and its continuation as the right-hand portion of the southern map (where it interconnects with the Sculptor Wall). Like Perseus–Pisces itself, the Cetus Wall is a ribbon-like structure, but broader and seen flat-on. Consequently, one can see clearly the distribution of its galaxies – very far from uniform, with considerable substructure – filaments and probably embedded voids. The Centaurus Wall – which is present in all five of the preceding redshift

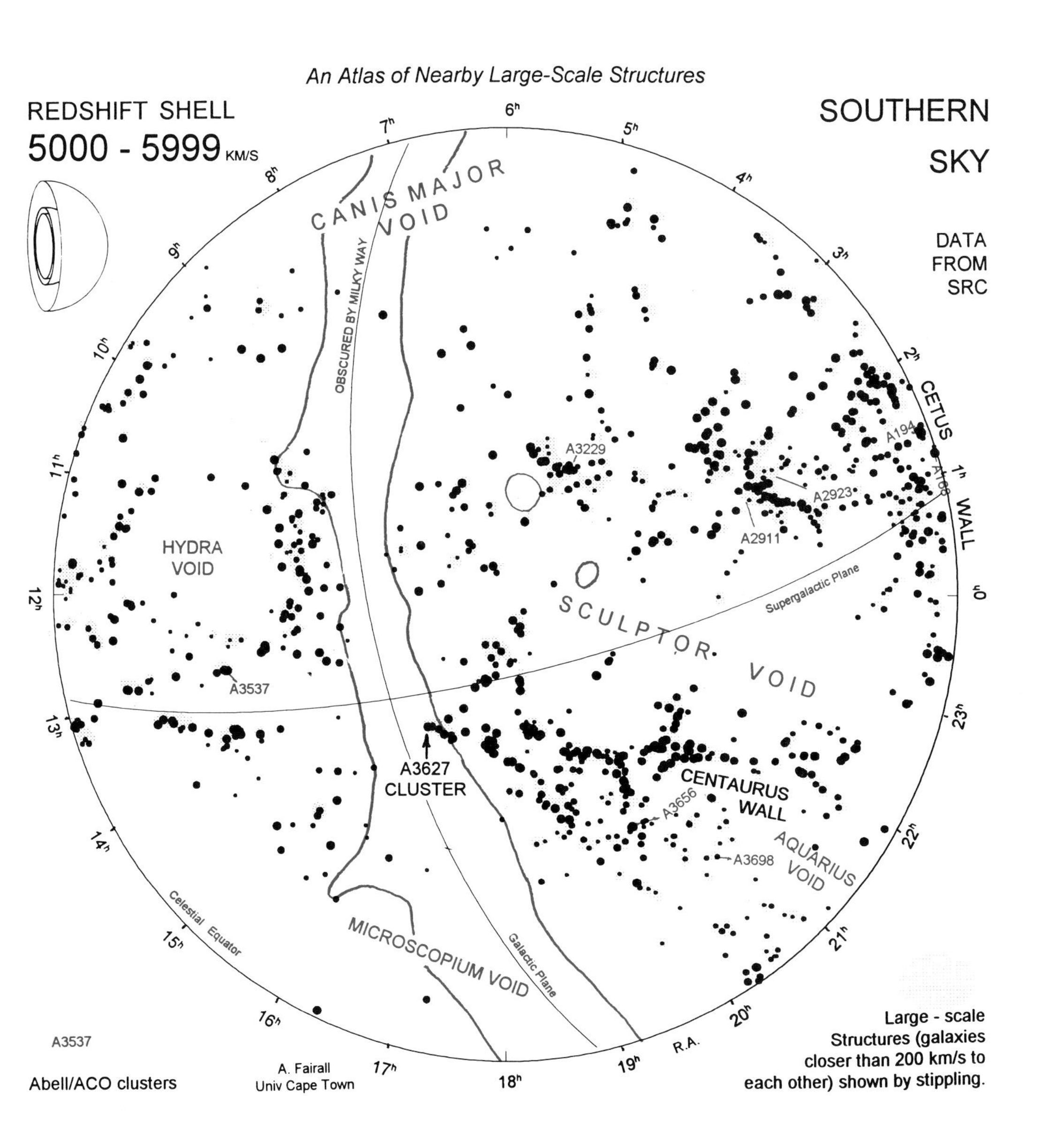

shells – is still seen. However, inspection of the dot sizes shows that it fizzles out beyond 5,500 km/s, though clearly it is interconnected with a structure beginning to wrap around the far side of the Microscopium Void. Elsewhere the labyrinth of weaker structures and voids continues. It is interesting to note that the 'body' of the Center for Astrophysics 'stick man' which elsewhere appears to be a major connecting structure (though part of it is the 'finger of God' of the Coma cluster) is not very prominent here when viewed in cross-section.

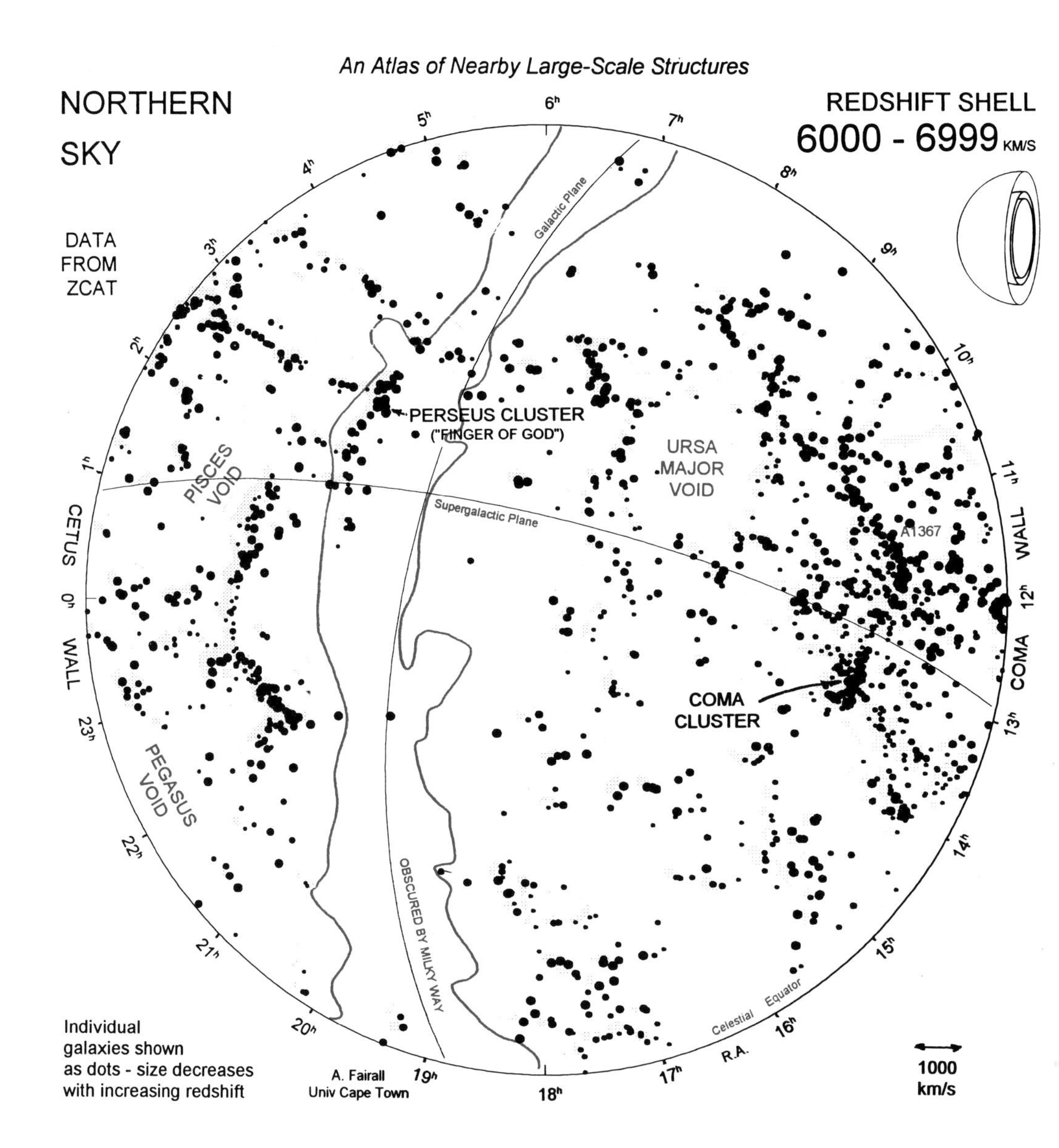

Centred on the rich Coma cluster, the Coma Wall (the original 'Great Wall') makes its appearance – and will also appear in the following shell. The Coma Wall is seen towards flat-on, such that its redshift increases downwards in the diagrams. The central filament that interconnects the Coma and Abell 1367 clusters is seen to extend far beyond the latter. As with the Fornax and Cetus Walls (also seen more or less flat-on), the distribution of galaxies within the Coma Wall reveals considerable substructure; the central filament seeming like a backbone. By contrast, the Perseus–Pisces region

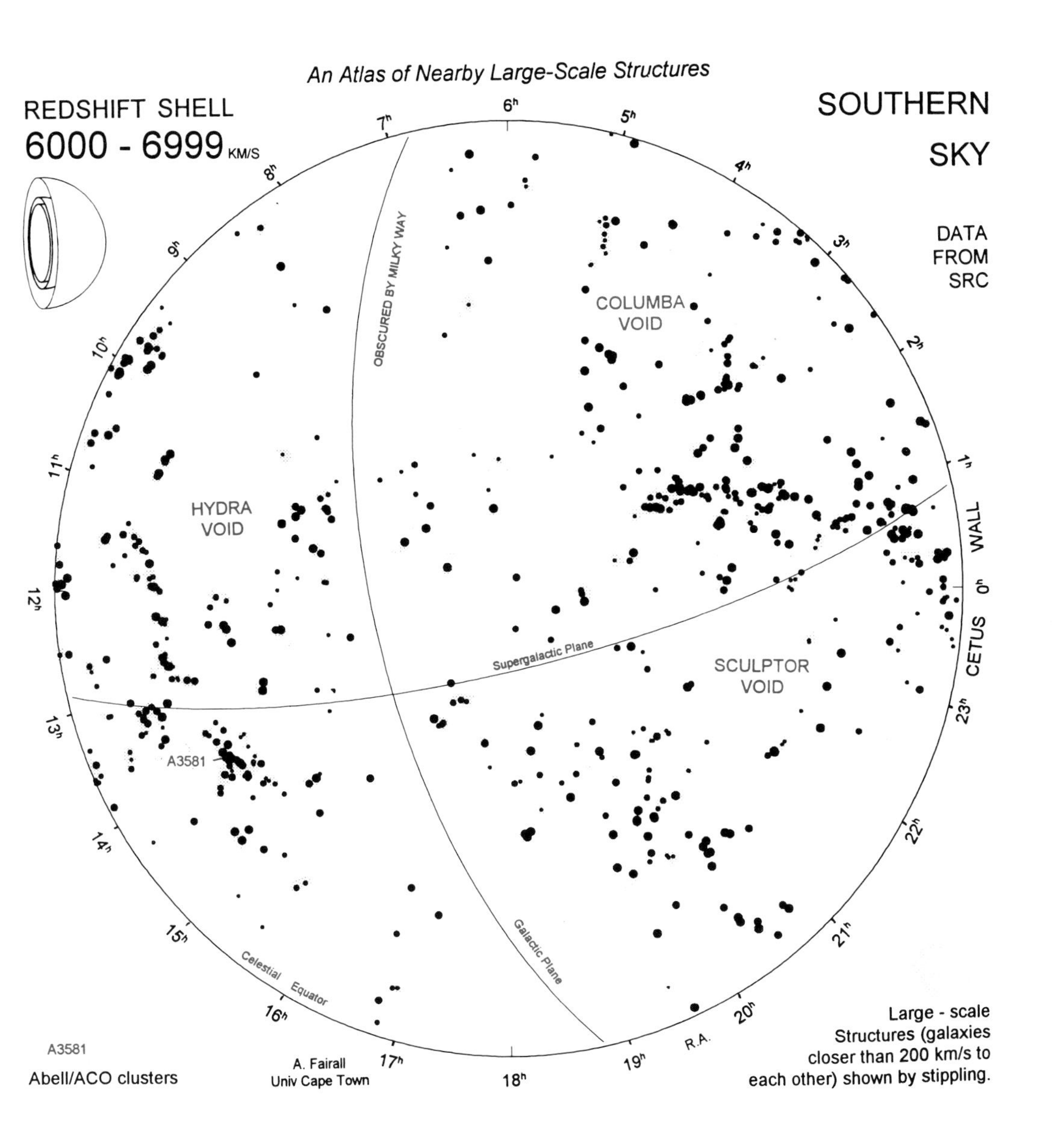

is thinning out, though the very elongated 'finger of God' of the Perseus cluster persists. Fragments of the Cetus Wall are still seen. There is an absence of major structures in the southern skies. The Centaurus Wall has completely vanished and only the Cetus Wall is prominent. Although a number of voids are labelled, incompleteness of data is now such that much of the sky cannot be charted in this fashion.

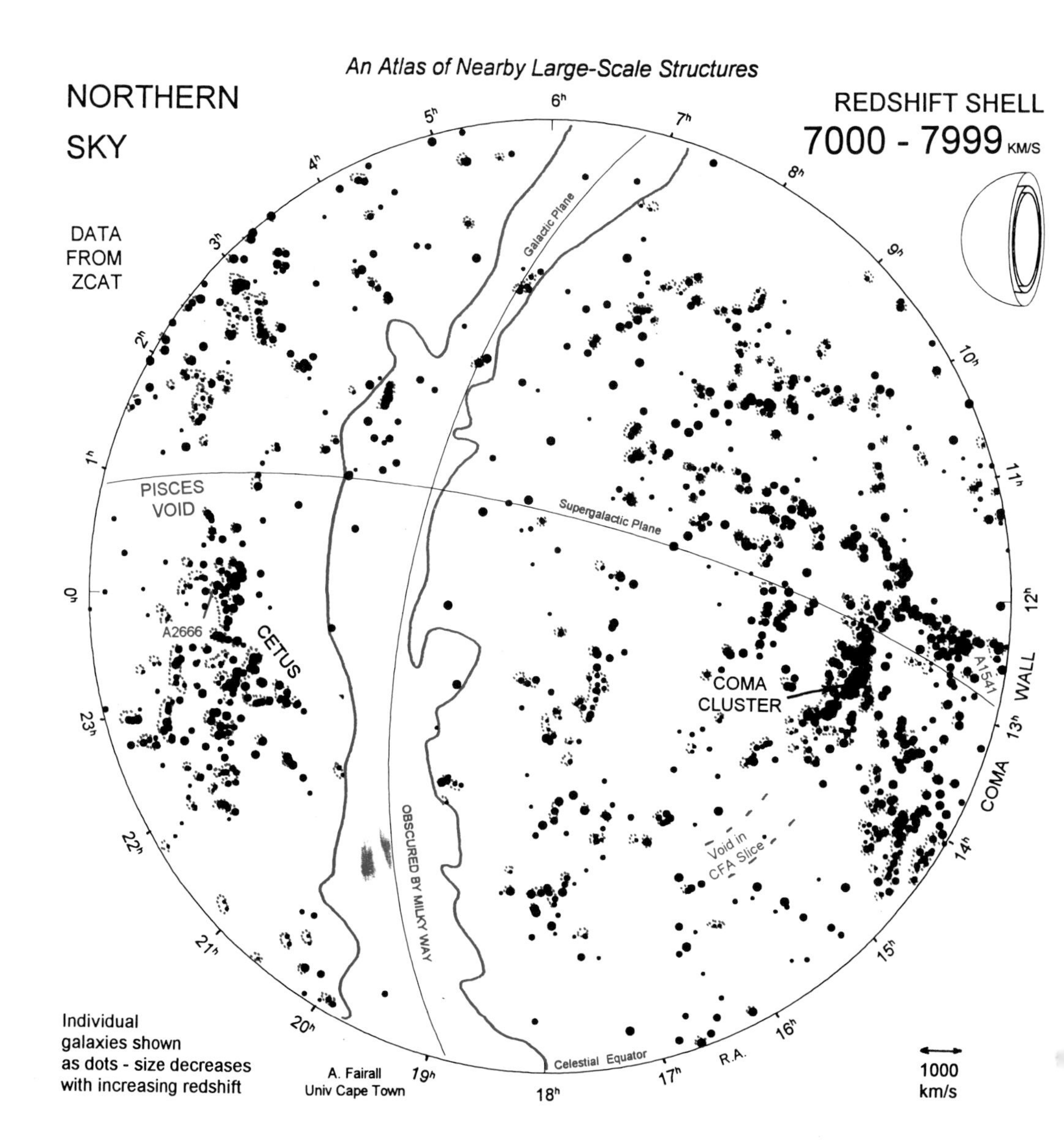

The central portion of the Coma Wall dominates the northern sky. Again notice the massive filament – now interconnected with a perpendicular filament running southwards. Voids intrude into the wall; these are larger than one might realise – perhaps 1,500 to 2,000 km/s across. (The transverse scale, showing 1,000 km/s, always appears with each pair of maps, and can be seen to diminish with increasing redshift). The condensation labelled Cetus – on the other side of the obscuration of the Milky Way – is associated with the Cetus–Perseus–Pisces complex, but may even be a further

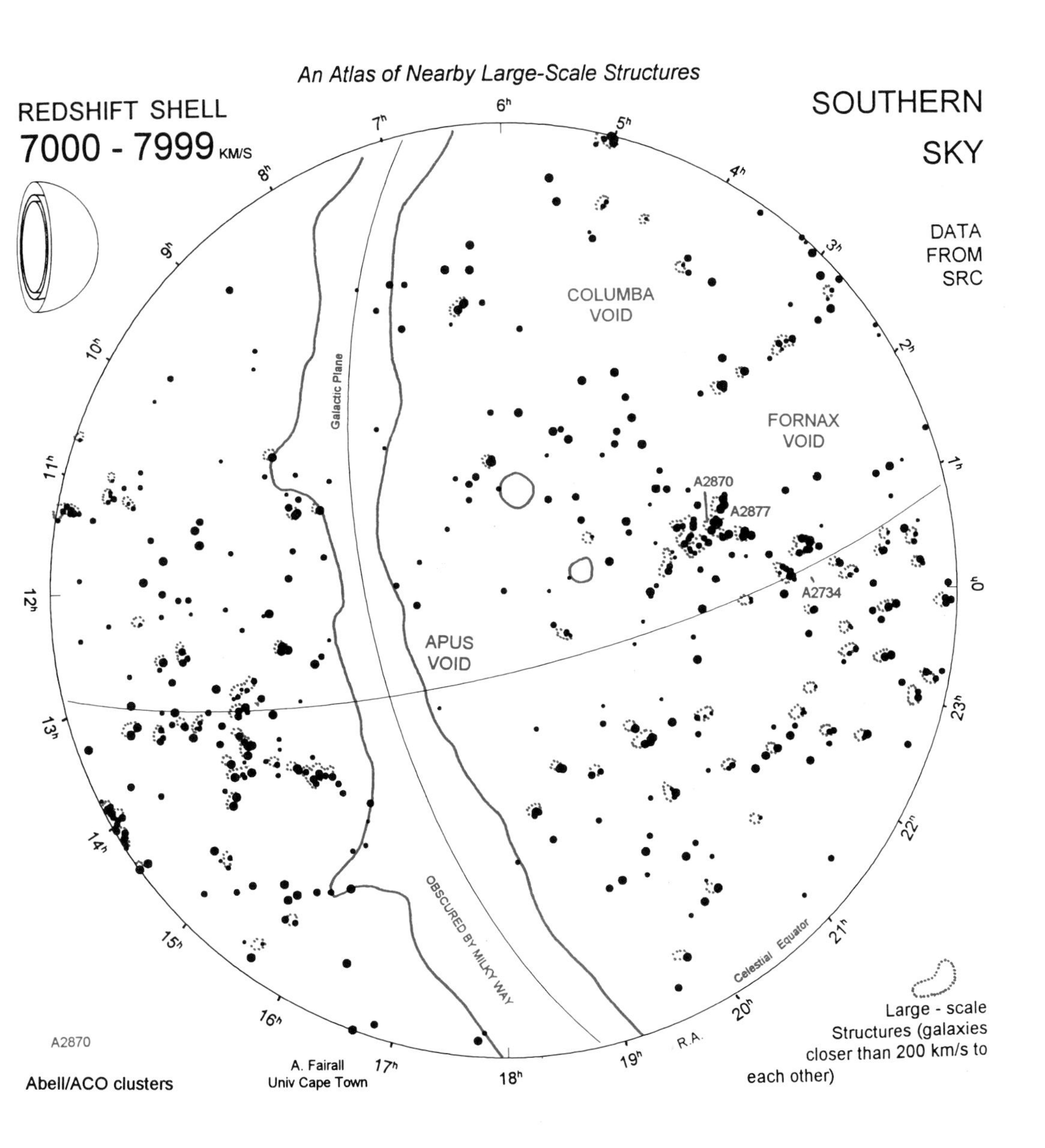

wall-like structure roughly at right angles to both Cetus and Perseus–Pisces (see also the 0h Right Ascension slice later). The data in the southern sky is now much thinner and obviously incomplete due to the greater redshift. While tentative indications are apparent, the absence of substantial structures (that have been so apparent at lower redshifts) shows that we have reached the limit of mapping such features. This is much as suggested in the opening map that showed the distribution of observed galaxies in the supergalactic plane.

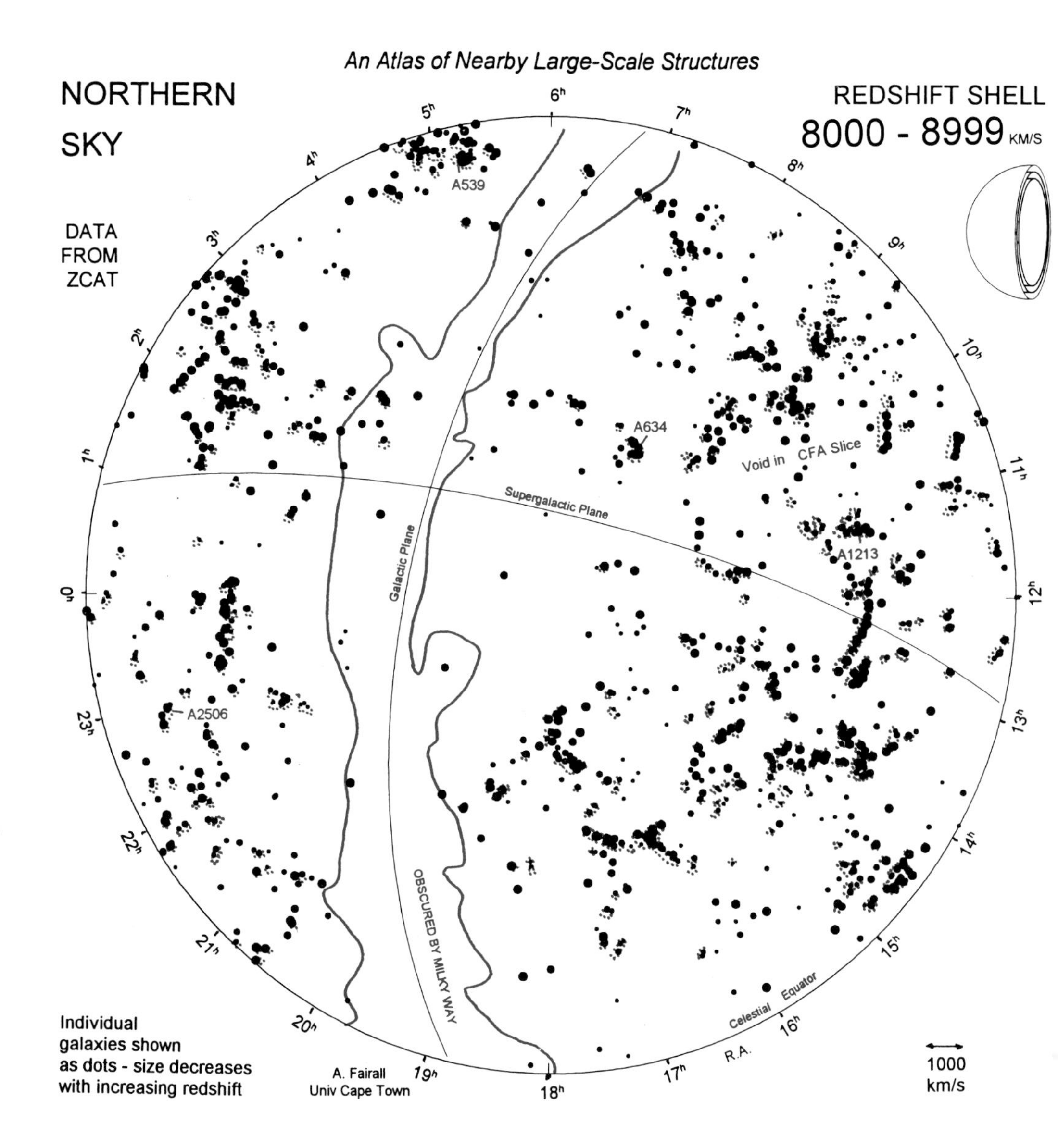

The redshift data is again very incomplete – more so in the underworked southern sky. However the numerous filaments and condensations provide indications of large-scale structures. It must be borne in mind that the volume shown on the map is so much greater than that seen in earlier redshift shells; this shell alone accounts for 22 per cent of the total volume surveyed by the Atlas. The features seen are therefore comparable in size to the Virgo supercluster and other significant over-densities seen earlier. Those towards the lower right of the northern sky map are part of the Coma

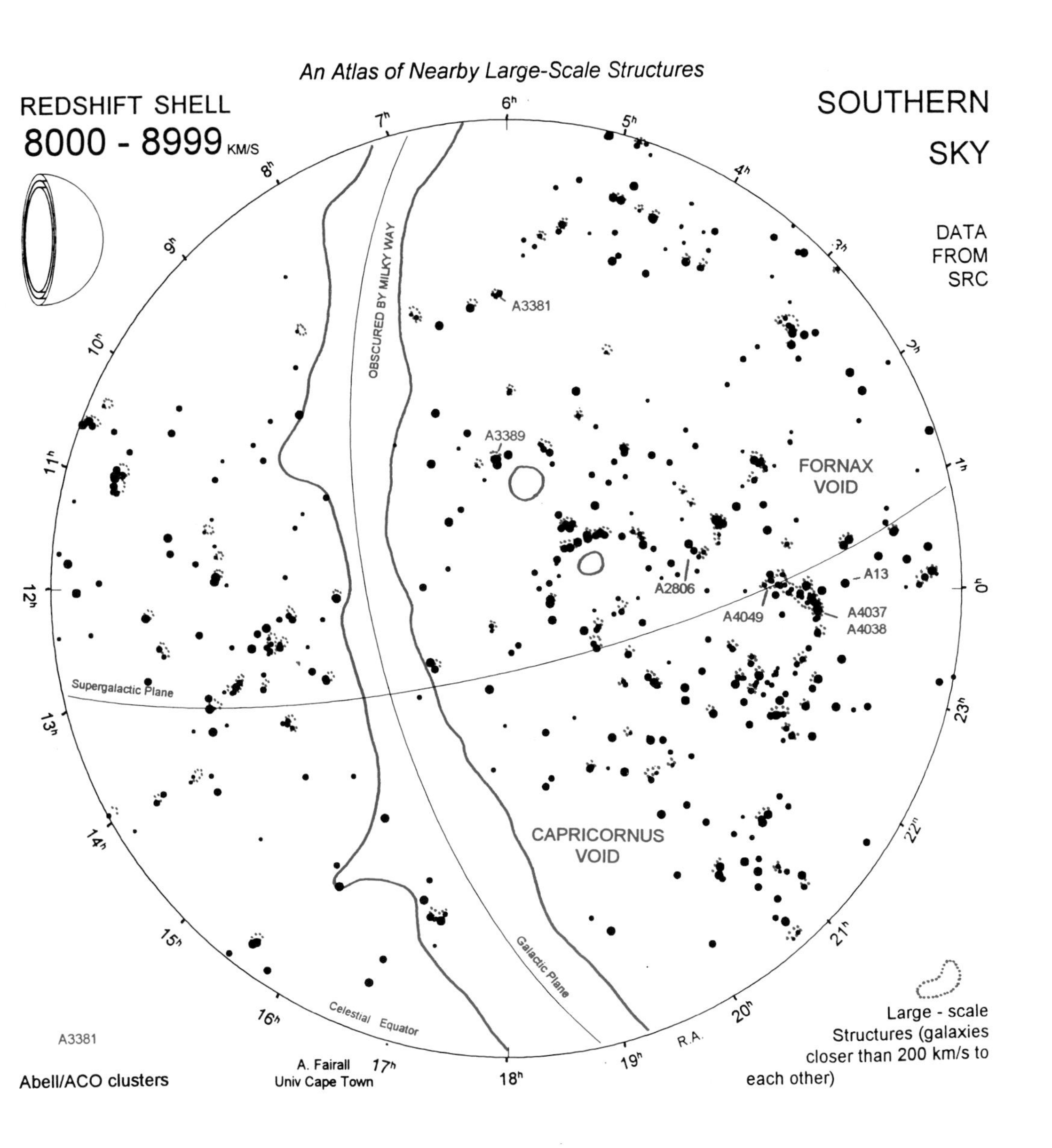

Wall. While the mapping is so incomplete, the occurrence of Abell clusters is now more frequent. Since such clusters are associated with higher density structures, it can be appreciated that the mapping of such clusters in redshift space gives a good indication of the presence of large-scale structures. The apparent emptiness towards the bottom left of the southern sky map may be quite false. There are already indications of a major structure at this redshift (data not yet published) stretching down from the northern sky.

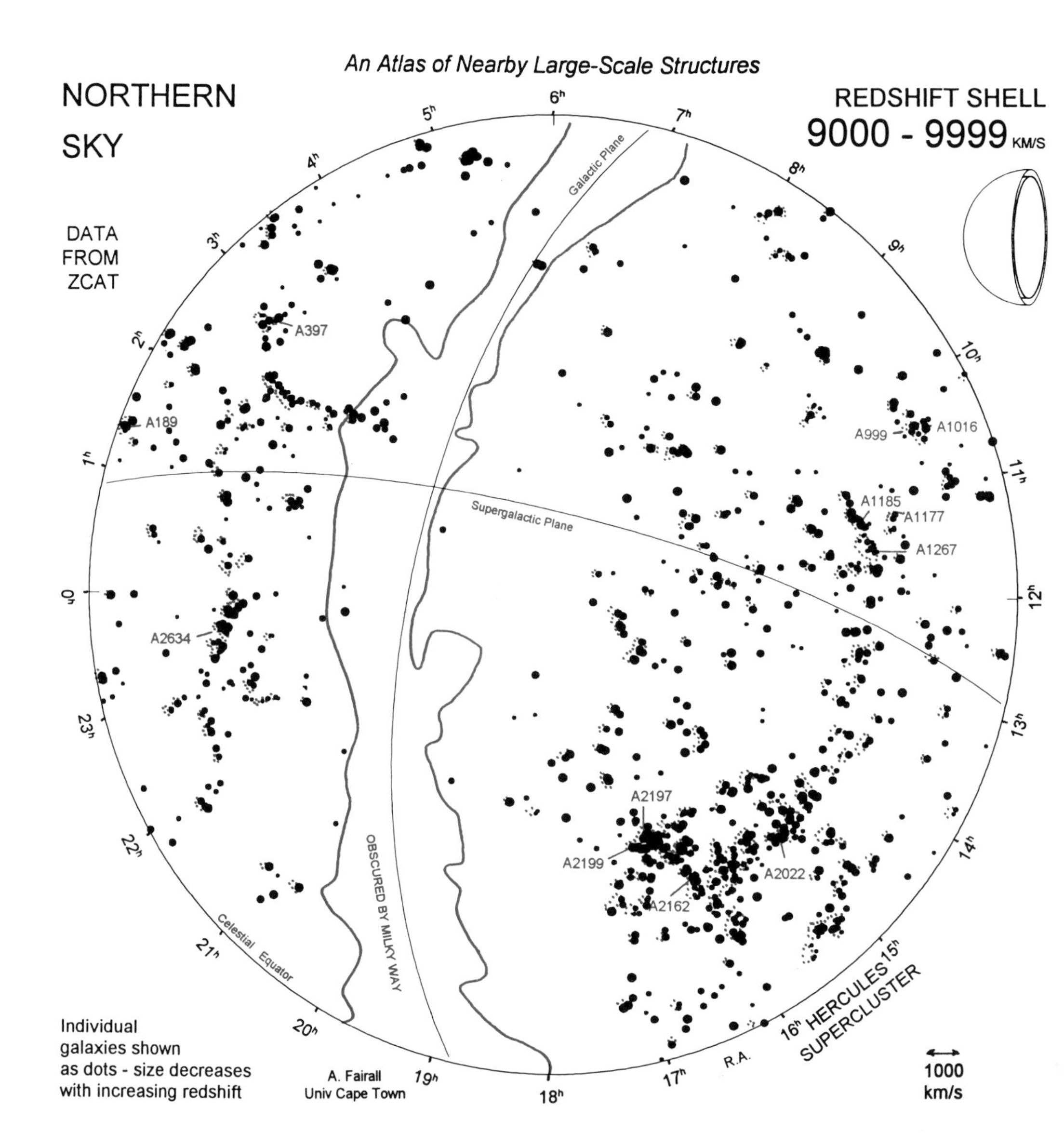

Although very incomplete, one of the justifications for carrying the maps to 10,000 km/s is that they reach the Hercules supercluster, probably the most significant overdensity at this redshift. This (nearer) Hercules supercluster also forms the end of the Coma Wall. Note that it is far removed from the supergalactic plane, and so did not feature in the opening map. In the southern sky, the Sculptor Wall runs almost in a radial direction. However, extensive intersecting structure spreads over most of the southern Galactic hemisphere (to the right of the Milky Way). The volume shown is 27 per

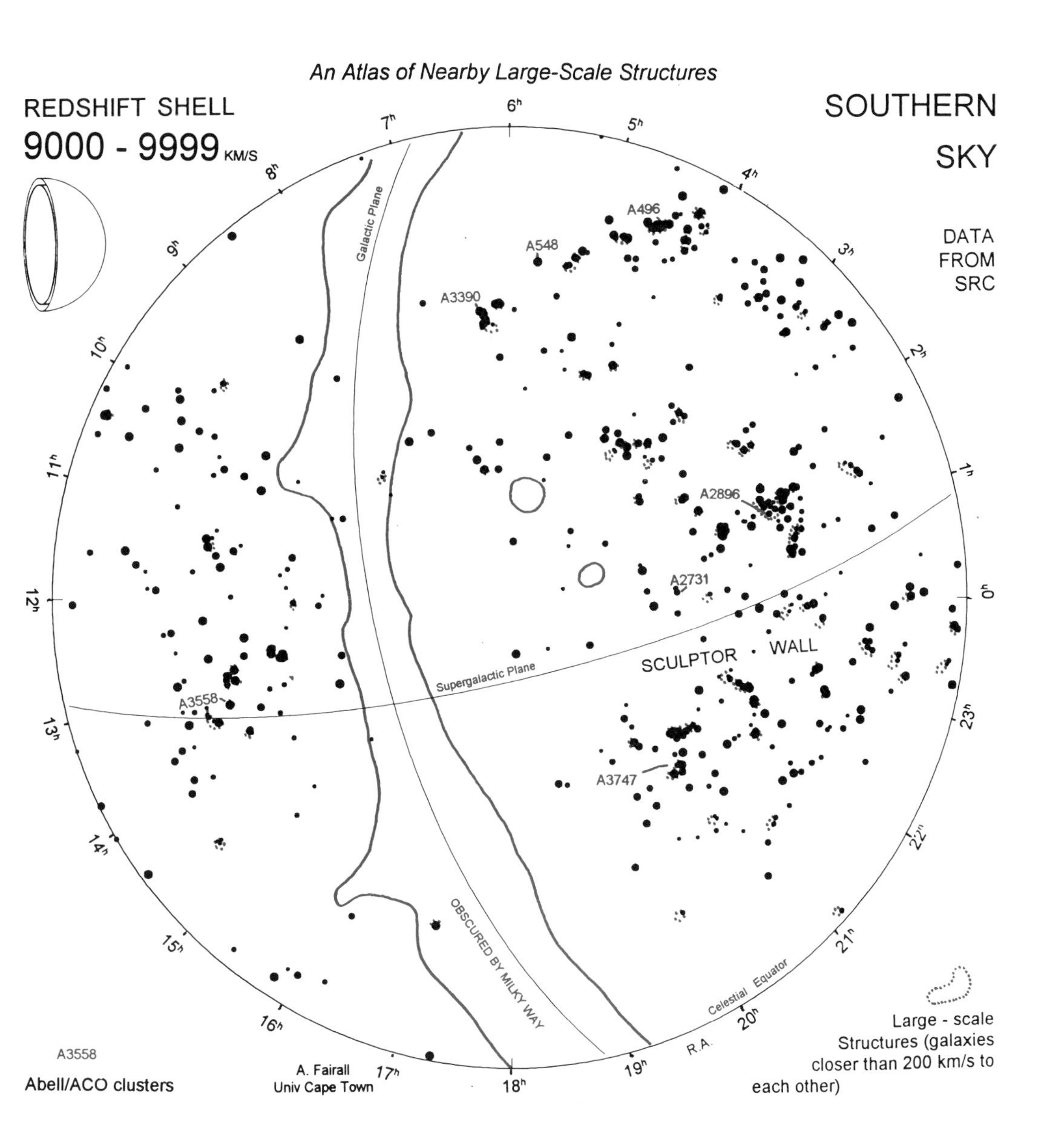

cent of the Atlas volume, and a significant number of Abell clusters are apparent. Their distribution here again demonstrates their association with large-scale structures, such that at higher redshifts, when the accompanying structures cannot be discerned in redshift space, the clusters nevertheless serve as beacons. Already deeper redshift surveys have probed way beyond these maps, but only in limited regions of the sky, and not on the global basis attempted in these maps.

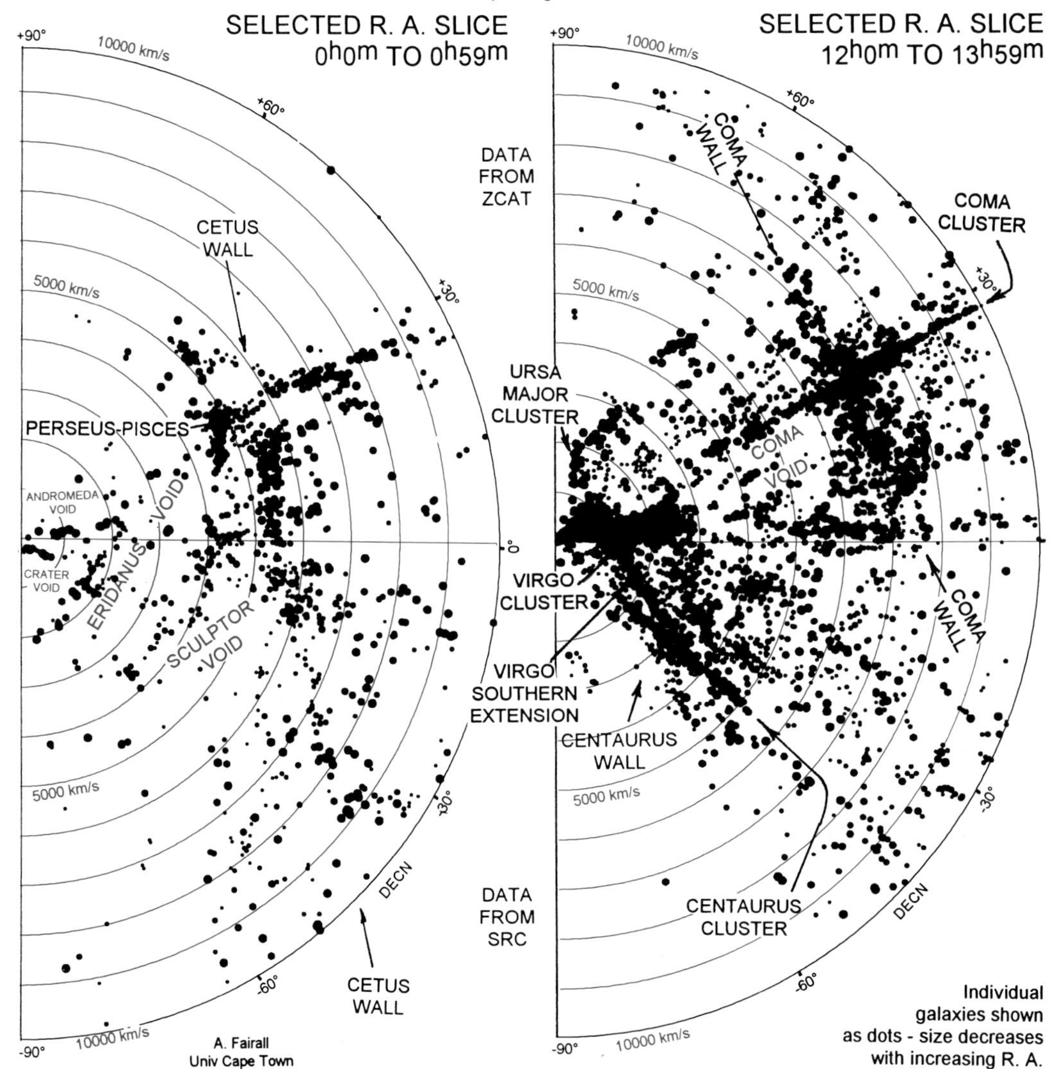

Four slices in Right Ascension are presented here, as a supplement to the main set of redshift-shell maps. The ranges have been selected to show up particular large-scale features. The first slice shows the remarkably straight and narrow Cetus Wall seen almost edge-on (the redshift shells show it flat on), closely associated with Perseus–Pisces. It is not clear whether a real radial feature runs outward from Perseus–Pisces or whether this is just due to selection effects. The tubular Eridanus Void – perhaps formed from the merger of a line of spherical voids – runs north–south. The second slice

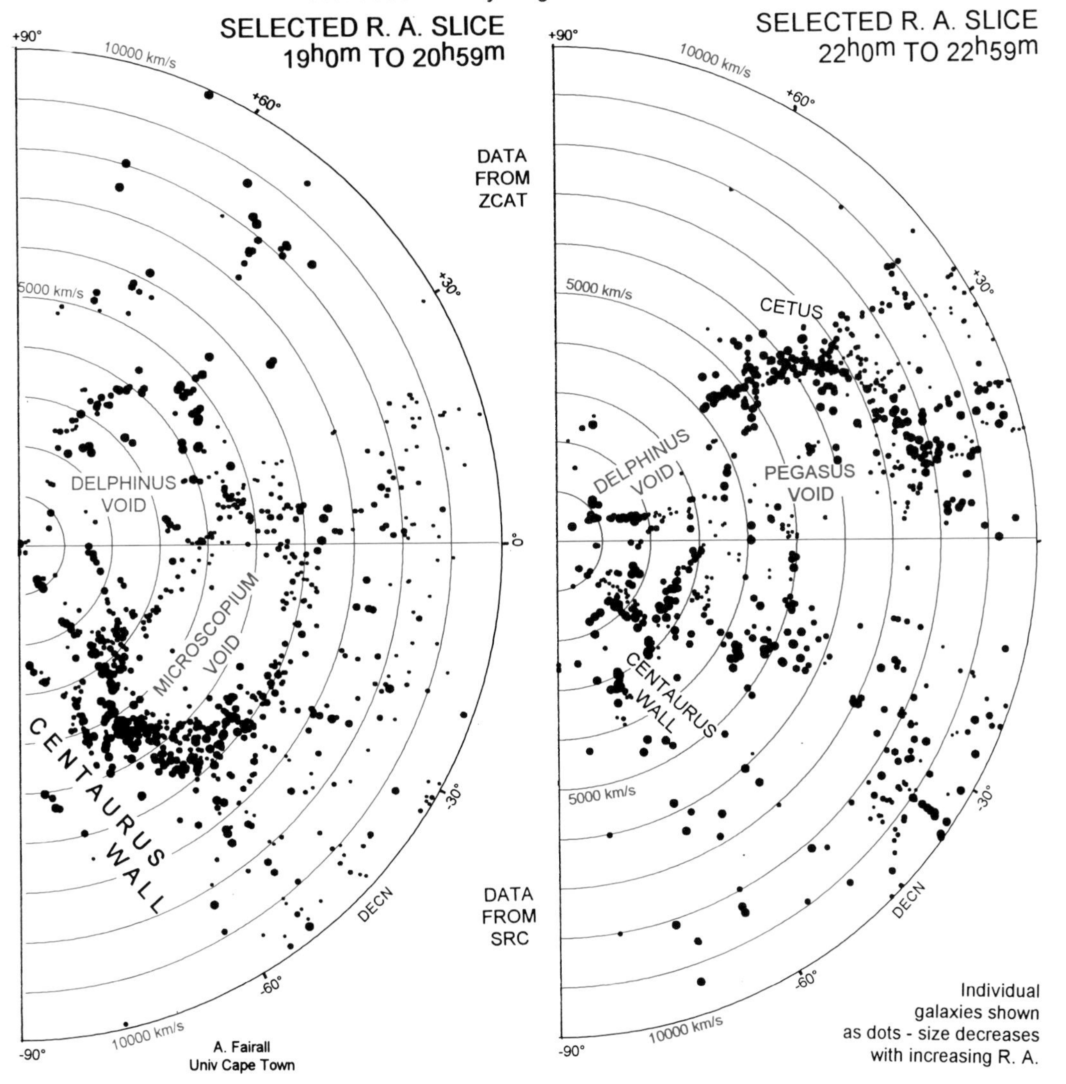

shows the Coma Wall end-on, such that its north–south extent can be assessed. The 'fingers of God' of both the Coma and Centaurus clusters are very conspicuous. The underdense region between Virgo/Centaurus and Coma is also obvious. The third and fourth slices are selected for showing particular voids. The core of the Microscopium Void is remarkably empty. By contrast, a small group of active galaxies inhabit what seems to be the central spot of the Pegasus Void.

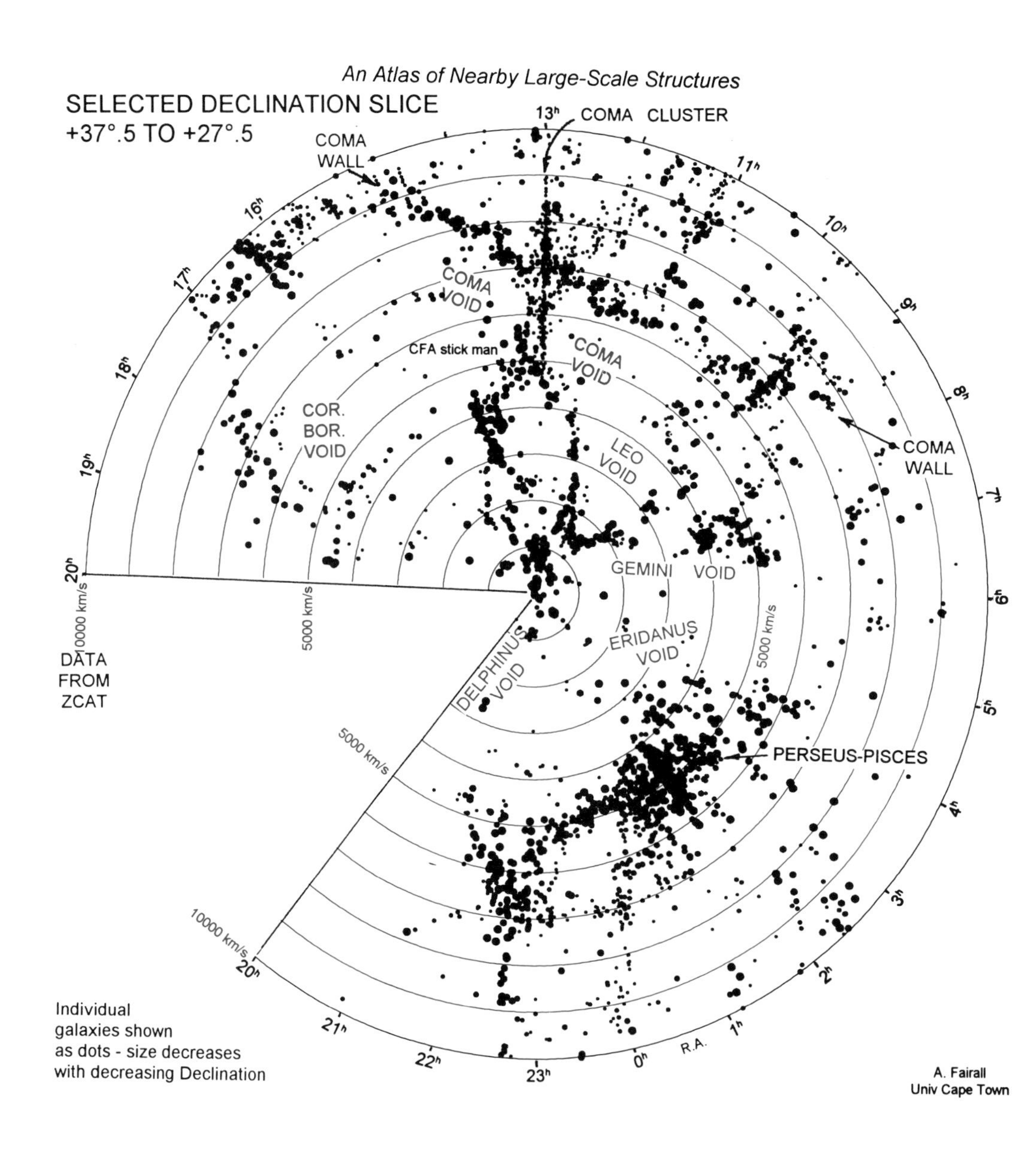

These two selected slices include most of the large-scale structures identified in this Atlas. It should be appreciated that features seen are not coplanar, the plots being roughly conical in nature. The northern slice (left) includes the well-known 'stick man' figure, the 'arms' of which extend as the Coma Wall. That to the left leads to the nearer Hercules supercluster (at 16h Right Ascension). Note that the 'body' is more restricted; the void on either side also runs 'beneath' it in the diagram. The

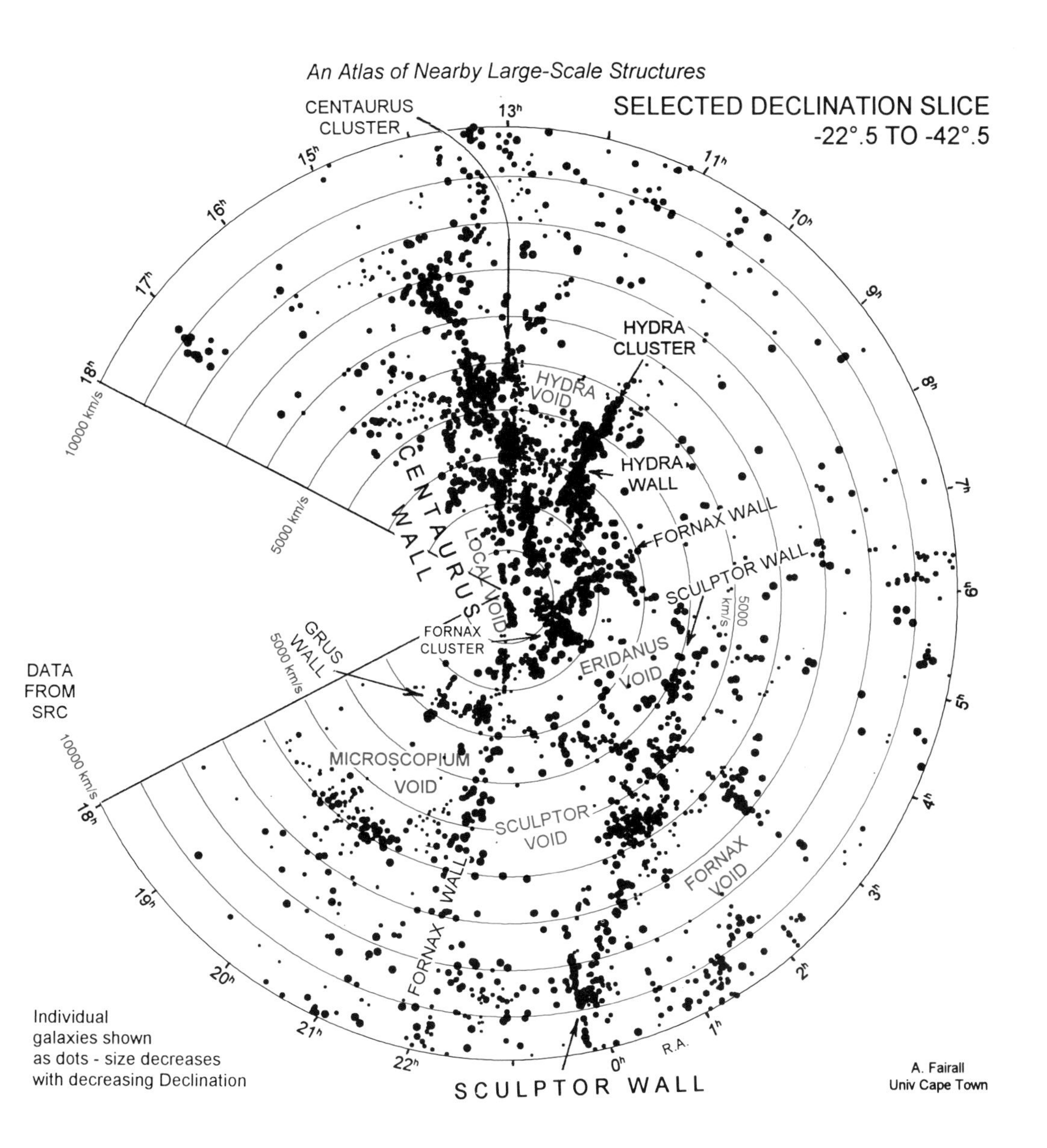

Perseus–Pisces filament is seen, but the Perseus cluster itself is outside the Declination range portrayed. The southern slice shows the Centaurus–Hydra conglomeration; the 'fingers of God' of the Hydra and Centaurus clusters being barely discernible in such an overdense region. The lower portion of the slice reveals an almost rectilinear array of structures. The Fornax and Sculptor Walls run virtually parallel to one another, while the Grus Wall cuts across orthogonally.

A SEGMENT OF THE GALACTIC PLANE

Galaxies with latitude
less than 10°

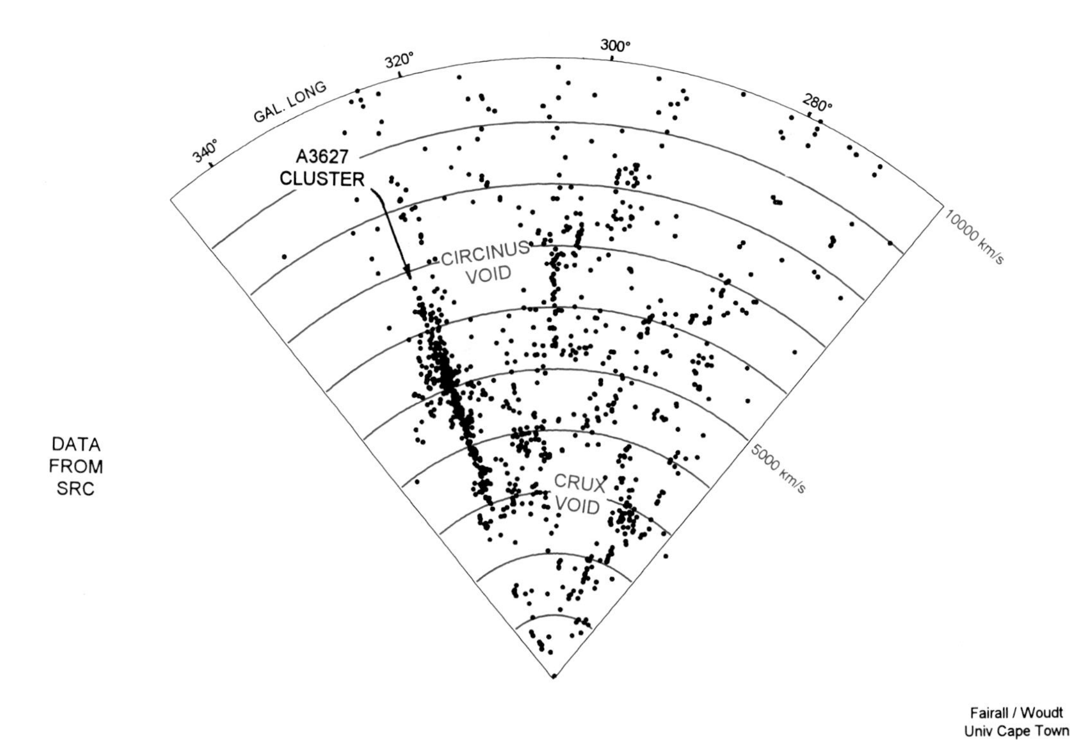

The least mapped portion of the extragalactic sky is that obscured by the Milky Way. However, thanks to a number of special searches for obscured galaxies, with follow up redshift observations, the 'Zone of Avoidance' has been narrowed such that large-scale structures close to the Galactic plane can now be discerned. Our final map stands as testimony. It shows (from a collaborative effort involving the author) only galaxies within 10 degrees of the plane. It also includes extended data for the cluster ACO 3627, which is seen as a dominant cluster comparable to the other two rich clusters (Coma and Perseus) within the Atlas volume. The familiar labyrinth of weaker structures and voids is also obvious, much as it is away from the Galactic plane.

4.4.4 The Hydra Wall

The Hydra Wall is similarly named for the Hydra Cluster. It runs nearly perpendicular to the Supergalacticplane, and is therefore best seen as a very elongated structure in the southern Atlas redshift shells 1,000–1,999 and particularly 2,000–2,999 km/s. It passes behind the Milky Way in the Puppis region – where it includes the Puppis Cluster – and intersects and crosses the Centaurus Wall in the vicinity of the Centaurus Cluster. In shape it is a narrow 'great wall', about 1,500 by 500 km/s in cross section, that might be better termed a 'great ribbon', though there are considerable variations in width and thickness along its length. The Hydra Wall spans some 6,500 km/s across the southern sky, but may extend even half that distance again as an alignment of filamentary structures in the northern sky.

Some years ago, when 'superclusters' rather than 'walls' were being recognised, there was considered to be a Hydra–Centaurus Supercluster surrounding the Hydra and Centaurus Clusters. However, the interconnection between the two clusters was relatively weak, and much of the Centaurus Wall was hidden behind the Milky Way and ignored. The Atlas view clearly shows two separate features that intercept orthogonally.

4.4.5 The Perseus–Pisces region

The sky distribution of galaxies reveals a distinct 'filament' of galaxies in the constellations of Perseus and Pisces. When examined in redshift space (see Figure 4.4), its member

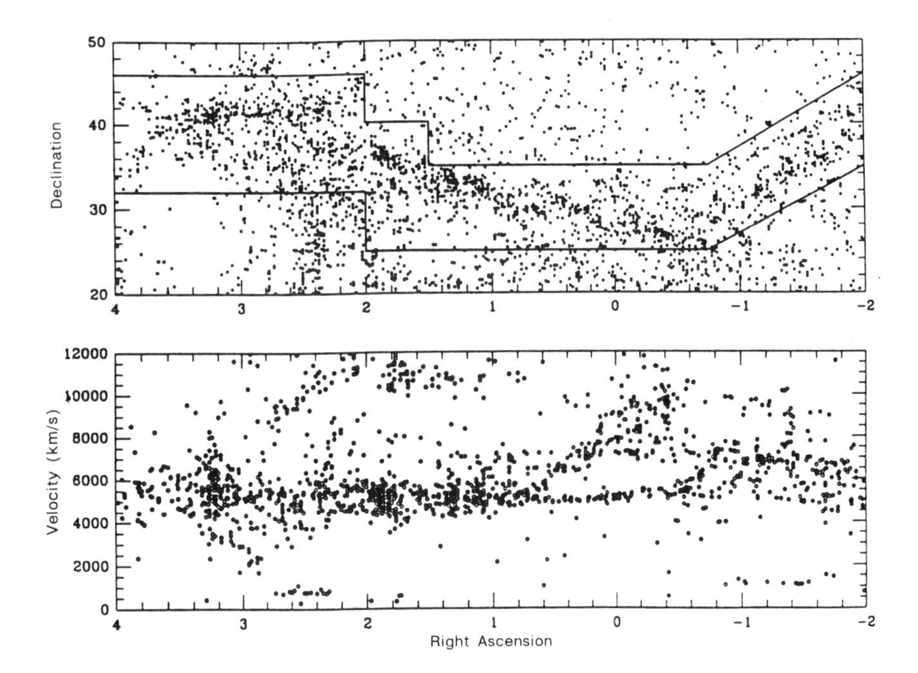

Fig. 4.4. The conspicuous filament of galaxies in Perseus–Pisces (top diagram) show redshifts mainly in the 4,000–6,000 km/s range (bottom). (Reproduced with permission from R. Giovanelli and the *Astronomical Journal*, (**105**, 1251, 1993).)

galaxies are found to have redshifts in the 4,000–6,000 km/s range. The shape of this concentration is therefore something of another long narrow great wall seen edge-on, its depth being 2,000 km/s, its thickness 500 km/s and its length about 5,000 km/s. As with the other walls, the distribution of galaxies within the wall is far from uniform, and there are indications of substructure (see lower panel of Figure 4.4).

Embedded within this structure, close to where it passes behind the Milky Way, is the Perseus Cluster. Though, as a cluster, it does not seem as compact as the Coma Cluster, its 'finger of God' in redshift space stretches from 2,500 km/s to 8,000 km/s (that is centred at 5,300 km/s, with a dispersion as high as 1,300 km/s)

The Perseus–Pisces filament has been extensively studied, particularly through the efforts of Riccardo Giovanelli and Martha Haynes.

4.4.6 The Cetus Wall

Yet another major wall stretches southwards from the Perseus–Pisces region. Running at approximately right angles to the Perseus–Pisces ridge, the Cetus Wall measures 7,000 × 4,000 × 500 km/s, and is seen close to flat-on. Aside from appearing in the usual redshift shell in the Atlas, a special slice plot in Right Ascension reveals the feature.

4.4.7 The Coma Wall (Great Wall)

The first great wall to be recognised is centred on the famous Coma Cluster, long recognised as the nearest of the rich clusters (until the recent discovery of Abell 3627). Like other such clusters, it is rich in elliptical and SO galaxies with a symmetrical profile. Its dynamics were the first to suggest (to Zwicky in 1937) the existence of substantial amounts of non-luminous matter, The application of the virial theorem to measured velocities gives a total mass for the cluster of approximately 10^{15} solar masses. In redshift space, the peculiar velocities within the cluster ($\sigma = 900$ km/s) add and subtract to the cosmological velocity of recession of the cluster, so stretching the cluster radially to form the axis of the 'stick man' figure in the CfA 'Slice of the Universe' (shown earlier in Fig 1.12).

An update on this famous region is shown as Figure 4.5, using the CfA2 data. The 'arms' of the stick man figure spread out across to our line of sight (at cz = 7,000 km/s). They are really a cross-section through the Coma Great Wall; the same structure is repeated in adjacent slices. That it was only found in 1989 is because the wall is seen almost flat-on, so its component galaxies are spread thinly over a substantial fraction of the sky. The character of the wall is also defined by there being voids both in front and behind. There are, however, interconnections to adjacent large structures. The one end of the wall ends on the Hercules Supercluster. Filamentary condensations also serve as links to the Centaurus Wall, and probably the Sculptor Wall.

The Coma Wall has impressive dimensions: some 12,000 km/s (east–west) by 5,000 km/s (north–south), with a thickness of about 1,000 km/s. It really is a 'Great' wall.

As seen earlier in Figure 4.1, the Coma Great Wall together with Perseus–Pisces seems to form encircling formations about the position of our Galaxy. Plotting in hypergalactic coordinates, G. Paturel and H. di Nella have gone so far as to propose a complete encircling ring of structures, centred on our position. However, such an apparent structure is nevertheless offset towards the northern sky; the ring is not closed in the south. Nevertheless, the question of formations concentric with our Galaxy's position will be raised again in Chapter 5.

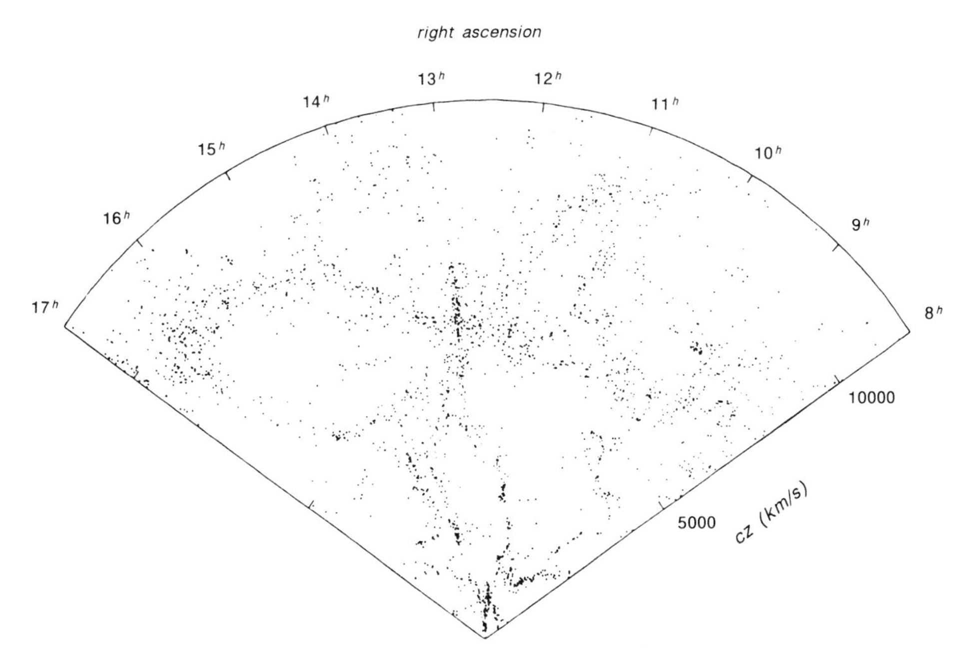

Fig. 4.5. Cone diagram for the Declination range 26.5° to 44.5° from the Center for Astrophysics (CfA2) survey. Over 2,300 galaxies are shown. The (Coma) Great Wall runs from the Hercules Supercluster (left) through the Coma Cluster (centre) and beyond. (Reproduced with permission from M. J. Geller and the *Astrophysical Journal* (**353**, 51, 1990).)

4.4.8 The Sculptor Wall

The first great wall to be recognised in the skies of the southern Galactic hemisphere was the Sculptor Wall, sometimes referred to as the 'Great Southern Wall'. First recognised by Luiz da Costa, it is very much the equivalent of the Coma Great Wall in the north.

The Sculptor Wall is somewhat different in that it does not contain any rich cluster. Only weak clusters occur within it, and can barely be discerned, in redshift space, since their 'fingers of God' are superposed on the general distribution of the other galaxies in the wall. The Sculptor Wall is an extremely long wall – perhaps 20,000 km/s long – and approximately 3,000 × 1,000 km/s in cross-section. It seems to be traced from a point in the northern Galactic hemisphere (in Cancer, at 5,500 km/s redshift); then it runs behind the Milky Way as a dominant feature in the southern Galactic hemisphere, passing closest to our Galaxy at a redshift of around 4,000 km/s. Since it runs at an angle to our line of sight, it shows best in an appropriately orientated redshift slice (rather than a redshift shell); in this case the special Southern Declination slice in the Atlas. The Sculptor Wall is so long, however, that only half of it is included within our 10,000 km/s limit. The remainder bends slightly to continue radially to a point some 20,000 km/s out (in the direction RA 0 hrs, Dec. –35°).

In the nearer section, the Sculptor Wall and its neighbours display a remarkable rectilinear pattern. As shown in the Atlas Southern Declination slice, it runs almost parallel to the Fornax Wall – and the orientation of the planes of these two walls are within 15 degrees. Running almost perpendicular between the two is a somewhat anaemic 'Grus Wall'.

4.5 VIEWING IN THREE DIMENSIONS

The two-dimensional pages of this book cannot do justice to the three-dimensional distribution of galaxies. To identify some structures, or to examine the internal details, it is often necessary to view the data in three dimensions. This author has experimented with various techniques, the most successful of which involves wearing 3-D spectacles. These are not the usual (red-and-green) version but an ingenious system employing chromostereoscopy, whereby colour translates into depth. Software (developed by my collaborator Wayne Paverd) allows individual large-scale structures to be displayed from any angle or distance. The author has made conference presentations of these views, with the audiences equipped with the spectacles. A more spectacular rendition is to create 3-D images over a hemispherical planetarium dome, and the success of projections of the southern sky, done in this manner, inspired the production of the redshift shells presented here in the accompanying Atlas. Furthermore, on the basis of the 3-D examinations, some further remarks on the character of large-scale structures follow.

Great walls are the most dominant elements of large-scale structure. However, these slablike – or thick ribbon-like – concentrations of galaxies are not without their own internal substructures. Great walls are not perfectly coplanar; instead they have a thickness of about 500–1,000 km/s in redshift space. Within their volumes, the galaxies are not distributed uniformly, nor even randomly. At first sight, numerous 'holes' seem to permeate the walls – places where the galaxies seem to open apart, beyond what one might expect of a random arrangement. Inspections, in three dimensions, suggest that the holes are in fact small voids, partly surrounded by galaxies. This could be an extrapolation of the cellular structure on the larger scale (a few thousand km/s) extended down to a much smaller scale (a few hundred or less km/s). Alternatively, there could be a finer texture of small voids throughout. (Further discussion follows in Sections 4.6 and 6.1).

In great walls like Coma or Sculptor, the number of galaxies surrounding each of these small voids is so low, and because there is still the tendency towards filamentary structures, an alternative view might be to see a network of filaments embedded within the walls. However, if we do lie within the Centaurus Great Wall, then we have a chance to view such substructure up very close. We do then see filaments of galaxies surrounding small voids, so the author's opinion would be that small voids do exist within the great walls.

4.6 VOIDS

The visible concentration of galaxies towards great walls is, of course, only apparent because the intervening spaces have so few galaxies. Such 'voids' may therefore seem to arise as a consequence of the 'design' of the cosmic texture – they are simply the spaces between walls.

However, if smaller completely empty voids do permeate the larger structures, as though their growth had pushed the galaxies aside, then voids are more than just spaces. Moreover, voids are normally neither flattened nor elongated (they have similar diameters in all three dimensions) no matter what their size. Some are very close to being spherical, those surrounded by walls tend to be more cubical, and a few of the larger ones are very

Table 4.1. Approximate coordinates of voids in the Atlas

	Equatorial				Galactic		Supergalactic		
Designation	Size km/s	RA hr	Dec deg	cz km/s	l deg	b deg	SGX km/s	SGY km/s	SGZ km/s
Cetus	500	2.0	−20	700	192	−72	100	-600	-200
Cepheus	500	23.0	+65	800	112	5	700	0	300
Monoceros	1,000	4.0	+5	800	185	−34	400	−400	−500
Volans	~700	7.0	−70	800	281	−25	−600	−300	−500
Local Void	3,000	18.0	−10	1,500	18	6	−700	0	1,300
Andromeda	1,500	1.0	+35	1,500	126	−28	1,300	−700	200
Crater	500	11.5	−15	1,500	276	43	−800	1,100	−600
Orion	1,500	6.5	+5	1,500	206	−2	500	−100	−1,400
Can Ven	1,000	13.5	+40	2,500	92	75	500	2,300	700
Draco	1,000	18.0	−80	2,500	314	−25	−2,300	−1,000	−300
Delphinus	3,000	20.0	+20	2,500	59	−6	500	−500	2,400
Eridanus	~4,000	1.0	0	2,500	129	−62	1,100	−2,200	−100
Corvus	~2,500	12.0	−15	3,000	286	46	−1,800	2,300	−800
Gemini	~2,500	6.0	+40	3,000	172	9	2,400	600	−1,600
Virgo	2,500	13.5	+10	3,000	334	70	−1,000	2,800	600
Cygnus	2,500	20.5	+25	3,500	67	−9	1,200	−900	3,200
Leo	3,000	10.5	+30	3,500	200	60	900	3,200	−1,200
Taurus	~4,000	3.5	+20	4,000	167	−29	3,100	−1,700	−1,900
Aquarius	3,000	22.0	0	4,500	60	−41	700	−3,300	3,000
Microscopium	3,500	18.0	−20	4,500	10	1	−2,700	300	3,500
Canis Major	5,000	6.5	−20	5,000	229	−13	−100	−600	−5,000
Coma	3,000	13.5	+15	5,000	343	74	−1,200	4,700	1,100
Hydra	3,000	12.0	−30	5,000	291	31	−3,800	2,800	−1,600
Cor Bor	4,000	17.0	+35	5,200	58	37	800	2,600	4,400
Pegasus	3,000	22.0	+15	5,500	74	−31	2,100	−3,300	3,900
Sculptor	6,000	0.5	−35	5,500	330	−81	−800	−5,400	−400
Ursa Minor	5,000	19.0	+85	5,500	117	27	4,600	2,300	1,900
Ursa Major	3,000	9.0	+50	6,000	169	41	3,900	4,200	−1,900
Columba	~5,000	4.0	−30	6,500	228	−48	−100	−4,300	−4,800
Pisces	3,000	1.0	+15	7,000	127	−48	4,600	−5,200	300
Apus	~5,000	20.0	−75	7,500	320	−31	−6,400	−3,900	−200
Fornax	~5,000	2.0	−25	8,000	208	−74	700	−7,400	−2,900
Capricornus	~5,000	19.0	−35	8,500	2	−17	−5,800	−3,200	5,400

close to being completely empty of galaxies. Other larger voids apparently have a completely empty core, with smaller numbers of galaxies lying inward from the surrounding walls. More usually, the interiors of larger voids show anaemic traces of substructure, not dissimilar to the character of that within the great walls.

Such is the preponderance of voids that many early interpretations put forward the idea that the growth of voids created the fabric of large-scale structure, like the small spherical bubbles that grow and often merge within a loaf of bread when it rises. Curiously, a bath sponge is a better analogy to the voids found in large-scale structure. Its bubbles are also close to spherical, yet are interconnected so that they can be filled with bath water or emptied by squeezing. Such interconnections characterise the nature of the voids within the large-scale structures. In Chapter 6, we shall see how such spongelike topology can be assessed in a quantitative fashion.

However, the earlier ideas of the growth of voids shaping the fabric of the cosmos preceded the recognition of the great wall structures. Whilst great wall-like structures might occasionally arise within sponge-like topologies from a chance arrangement of surrounding voids, a labyrinth of interconnecting wall-like structures that sometimes shows rectilinear formations could certainly not be formed in this way. However, our concern at the moment is to convey the character of the voids, rather than their interpretation – which follows in later chapters.

Table 4.1 lists the larger voids with redshifts of less than 10,000 km/s. Like those used for the walls, the names given to the voids are simply derived from the foreground constellations. This system avoids the inconsistencies that arise from individual numbering systems; the author has advocated its use in the southern skies, and a number of other researchers have adopted the same labels. Most of the names have been used in the literature, but some of the names are used here for the first time, as a convenience and not a prescription.

Large as some of the voids in Table 4.1 are, none are larger than 5,000 km/s in diameter. Whilst one might have to look further afield to find rarer larger voids, it may also be that this is the upper limit in size.

An ongoing concern about voids was whether they might not be filled with faint dwarf galaxies, such that the whole labyrinth structure was more a case of galaxy segregation than separation. There have now been a number of studies, though involving relatively small numbers. All of these conclude that, in general, the dwarfs follow similar distributions to the giant galaxies; they do not fill in the voids. This is not to exclude some degree of segregation, about which more will be said later (in Chapter 6).

We shall not attempt to describe all the voids in detail. However, a few of the voids in the table are worthy of special note.

4.6.1 The Local Void

The Local Void, noted by Tully and others, is not a particularly large void (though it is more than 2,000 km/s across), but it lies so close that it covers a third of the sky. Its core seems completely empty, and may truly be empty, considering the very large number of surveys that have looked straight through it. However, a portion of the core is obscured by the central bulge of our Galaxy. As with probably all voids, the Local Void is not isolated; in particular it is interconnected with the larger and more distant Microscopium Void. The Local Void is obvious in the first two redshift shells displayed in the accompanying Atlas of Nearby Large-Scale Structures.

4.6.2 The Sculptor Void

This is as close to a 'text-book' void as possible. Its core appears to be completely empty out to a radius of 1,500 km/s, beyond which are found inlying galaxies from the surrounding walls. It is bounded on either side by the the Fornax and Sculptor Walls, as best seen in the Atlas Southern Declination slice. Though it looks to be completely enclosed, there are nevertheless openings to adjacent voids.

4.6.3 The Eridanus Void

The Eridanus Void is appropriately named for the long and rambling central foreground constellation. While voids are not normally similarly shaped, it is likely that it has formed by merging of a chain of spherical voids into a tubular form running approximately north–south, as seen in one of the Atlas selected Right Ascension slices. It is a prominent feature in both the southern sky and the northern sky, where it lies in front of the Perseus–Pisces ridge.

4.6.4 The Microscopium Void

Although bisected by the Milky Way, this void is remarkable for its general spherical shape, size and emptiness. It is the cleanest void within our survey area – see the Atlas Right Ascension slice.

4.6.5 The Pegasus Void

Seen in north–south cross-section (again in the Atlas Right Ascension slices), this void shows a circular profile, but with a small clump of galaxies, including at least one active galaxy, at its centre.

4.7 AN ATLAS OF NEARBY LARGE-SCALE STRUCTURES

Incorporated within this chapter is an Atlas, consisting of a set of maps of the distribution of galaxies out to the 10,000 km/s redshift limit, and the large-scale structures they form.

This Atlas comprises a main set of redshift 'shells' 1,000 km/s thick, each an all-sky plot of the galaxies within the redshift range indicated. The plots show all of the features described in this chapter. However, a few of the key features are revealed by redshift 'slices', rather than shells, and for that reason the main set of maps is supplemented by a number of selected slices in Right Ascension and Declination and Galactic latitude. All the maps are plots in redshift space, using heliocentric velocities of recession and 1950 epoch positions.

4.7.1 The data

The data used for the Atlas is Huchra's ZCAT (obtained courtesy of Harold Corwin and the NASA Extragalactic Database) for sky north of the celestial equator, and this author's SRC (*Southern Redshift Catalogue*) for the southern sky. The Atlas therefore uses catalogue data, rather than statistically-controlled data, for the reasons debated in Section 3.9.

Though the database for the northern sky has a greater numbers of galaxies – over 14,000, compared with less than 9,000 in the south – there is nevertheless reasonable balance, since the Virgo, Perseus and Coma Clusters all lie in the north. Only beyond 8,000 km/s is it noticeable that there is more data available for the northern sky than there is for the southern sky.

To enable some three-dimensional assessment to be made within each redshift shell, the dots representing the galaxies diminish in size between the lower and upper redshift limits.

4.7.2 The large-scale structures

A crucial statement of the Atlas is its representation of large-scale structures – the entities shown by stippled areas. These are based on the working definition stated earlier that large-scale structures are formed by galaxies having separations of less than 200 km/s in redshift space. Thus a sphere of radius 100 km/s can be expanded about each galaxy, and impinging spheres used to define large-scale structures.

For the maps, some licence has been taken by using cylindrical elements instead of spheres. The data have been subdivided into fifty redshift shells of thickness 200 km/s. For each of the (one hundred resulting) plots, circles of a radius corresponding to 100 km/s have been drawn about each galaxy. Overlapping circles then define large-scale structures. Each plot is then superposed in sequence on its neighbouring plots, where again overlapping circles establish large-scale structures. Also, since the plots represent either the northern or southern sky, the peripheries of each plot are compared with the three possible adjoining neighbouring plots from the opposite hemisphere for overlapping circles. (Though somewhat labour-intensive, such a procedure was rewarding in the measure of acquaintance gained with the structures.)

Single galaxies, having no neighbour within overlapping circles, are rejected as being no part of any large-scale structure. The impressive finding is the very low number of such isolated galaxies – less than 1 per cent of all galaxies at low redshift. Obviously, many of these single galaxies still lie close to large-scale structures, but not close enough to meet the 200 km/s criterion.

The shapes of the large-scale structures so established include the 100-km/s circles – like a 'top-hat' filter being applied to the distribution. Consequently, a line of galaxies, for example, is given an artificial width. However, the shapes created are more easily visually assimilated than, say, lines of interconnected dots. They also allow a perception of distance since their angular sizes, particularly widths, diminish with increasing redshift. As one pages through the maps, the large-scale stuctures are clearly seen to become more distant.

With increasing redshift, there is obviously increasing incompleteness in the data. The number of isolated galaxies increases dramatically beyond 7,000 km/s. It was very tempting to modify the 200 km/s criterion to accommodate the greater distance. However, this would then create the false impression that angular sizes did not diminish with distance, and that the mapping could continue to the same standards. Instead, the 200 km/s criterion has been retained, and the structures are seen to fragment – as a measure of incompleteness of data and the limitations in mapping.

4.7.3 Clusters of galaxies

Well-known clusters of galaxies are identified in the maps, as are all Abell clusters that lie within the 10,000 km/s redshift limit. A list of the clusters is also provided as Table 4.2.

Table 4.2. Approximate coordinates of clusters in the Atlas

Designation	Equatorial			Galactic		Supergalactic		
	RA hr	Dec deg	cz km/s	l deg	b deg	SGX km/s	SGY km/s	SGZ km/s
Virgo	12.4	12	1,150	281	74	–300	1,100	–100
Fornax	3.7	–36	1,400	237	–53	–100	–1,000	–1,000
A3565	13.6	–34	3,250	314	28	–2,900	1,500	0
Centaurus	12.8	–41	3,300	303	22	–3,000	1,300	–600
Hydra	10.6	–27	3,400	270	27	–2,000	1,800	–2,100
A3354	5.5	–29	4,000	233	–29	–300	–1,500	–3,700
A3559	13.5	–29	4,200	314	33	–3,500	2,300	0
A3574	13.8	–30	4,250	318	31	–3,600	2,200	300
Centaurus	12.8	–41	4,550	303	22	–4,100	1,800	–9,00
A3627	16.1	–60	4,700	325	–6	–4,600	–600	600
A262	1.8	+36	4,850	136	–25	4,400	–2,100	–100
A3742	21.1	–47	4,850	353	–43	–2,900	–3,500	1,700
A3537	13.0	–32	5,000	306	31	–4,200	2,700	–600
A2923	1.5	–31	5,100	237	–80	–100	–4,900	–1,400
A3229	4.2	–63	5,100	275	–42	–2,800	–3,100	–2,900
Perseus	3.0	40	5,300	149	–16	5,000	–1,300	–1,100
A194	1.4	–2	5,350	143	–63	2,400	–4,700	–800
A168	1.2	0	5,400	135	–62	2,500	–4,800	–400
A426	3.3	+41	5,500	151	–13	5,200	–1,100	–1,400
A347	2.4	+42	5,600	141	–17	5,300	–1,600	–500
A3656	20.0	–39	5,650	2	–30	–3,500	–3,200	3,100
A569	7.1	+49	5,900	168	23	4,700	2,600	–2,500
A1318	11.3	+55	5,900	148	58	3,100	5,000	0
A3698	20.5	–25	5,950	19	–33	–2,300	–3,700	4,100
A2911	1.4	–38	6,000	271	–77	–900	–5,700	–1,600
A3581	14.1	–27	6,400	323	33	–5,300	3,400	900
A1367	11.7	+20	6,450	235	73	–200	6,300	–1,200
A779	9.3	+34	6,800	191	45	2,900	5,200	–3,300
Coma	12.9	+28	6,950	62	89	100	6,900	900
A400	2.9	+6	7,200	170	–45	4,300	–4,800	–3,300
A2877	1.1	–46	7,250	294	–71	–2,200	–6,700	–1,700
A1541	12.4	+9	7,350	284	71	–2,000	7,100	–600
A2870	1.1	–47	7,500	294	–70	–2,300	–6,900	–1,800
A2734	0.1	–29	7,800	21	–80	–600	–7,800	300
A2666	23.8	+27	7,950	107	–34	5,700	–4,800	2,900
A3389	6.4	–65	8,000	275	–27	–5,200	–3,100	–5,200
A634	8.2	+58	8,000	160	34	6,200	4,600	–2,000

Table 4.2. (continued)

Designation	Equatorial			Galactic		Supergalactic		
	RA hr	Dec deg	cz km/s	l deg	b deg	SGX km/s	SGY km/s	SGZ km/s
A1213	11.2	+30	8,050	200	69	1,400	7,700	−1,800
A13	0.2	−19	8,100	76	−78	800	−8,000	600
A2806	0.6	−56	8,100	307	−61	−3,800	−7,000	−1,500
A3381	6.1	−33	8,450	240	−23	−1,600	−2,400	−7,900
A4038	23.8	−28	8,500	27	−76	−700	−8,400	900
A4049	23.8	−29	8,500	23	−76	−800	−8,400	900
A539	5.2	+6	8,650	196	−18	4,300	−1,900	−7,300
A2506	22.9	+13	8,650	85	−41	4,000	−6,200	4,500
A4037	23.7	−29	8,750	23	−75	−900	−8,600	1,100
A2199	16.4	+40	9,050	63	44	1,800	5,600	6900
A2197	16.4	+41	9,150	65	44	2,000	5,600	6,900
A3747	21.1	−44	9,150	357	−43	−5,200	−6,700	3,600
A548	5.8	−26	9,300	231	−24	−500	−2,900	−8,800
A2731	0.1	−57	9,350	315	−59	−4,700	−8,000	−1,100
A2634	23.6	+27	9,350	104	−33	6,500	−5,500	3,800
A1171	11.1	+22	9,450	220	66	600	9,000	−2,800
A2896	0.3	−37	9,550	335	−78	−1800	−9,400	−400
A999	10.3	+13	9,550	228	52	100	8,100	−5,000
A189	1.4	+1	9,600	14	−60	4,700	−8,300	−1,200
A1016	10.4	+11	9,600	232	52	−300	8,200	−5,000
A1267	11.4	+27	9,600	209	71	1,100	9,300	−1,900
A3558	13.4	−31	9,600	312	31	−8,100	5,100	−300
A2162	16.2	+30	9,600	49	46	200	6,100	7,400
A397	2.9	+16	9,750	162	−37	7,100	−5,500	−3,800
A1185	11.1	+29	9,750	203	67	1,700	9,300	−2,400
A496	4.5	−13	9,850	209	−37	2,500	−5,000	−8,100
A3390	6.4	−37	9,900	245	−21	−2,800	−2,500	−9,200

The Atlas shows redshift space in which clusters are distorted into radial 'fingers of God' (explained in Section 1.4). In some cases, such extended fingers, rather than the cluster centre, are labelled.

4.7.4 Comparison with other surveys

The Atlas represents a more detailed mapping of nearby structures than found elsewhere, but is obviously not the only representation. Aside from earlier all-sky mappings of nearby structures carried out by the author and colleagues, various other maps – global or otherwise – appear in the literature. Figures 4.6 to 4.9 show some key examples. Though obvi-

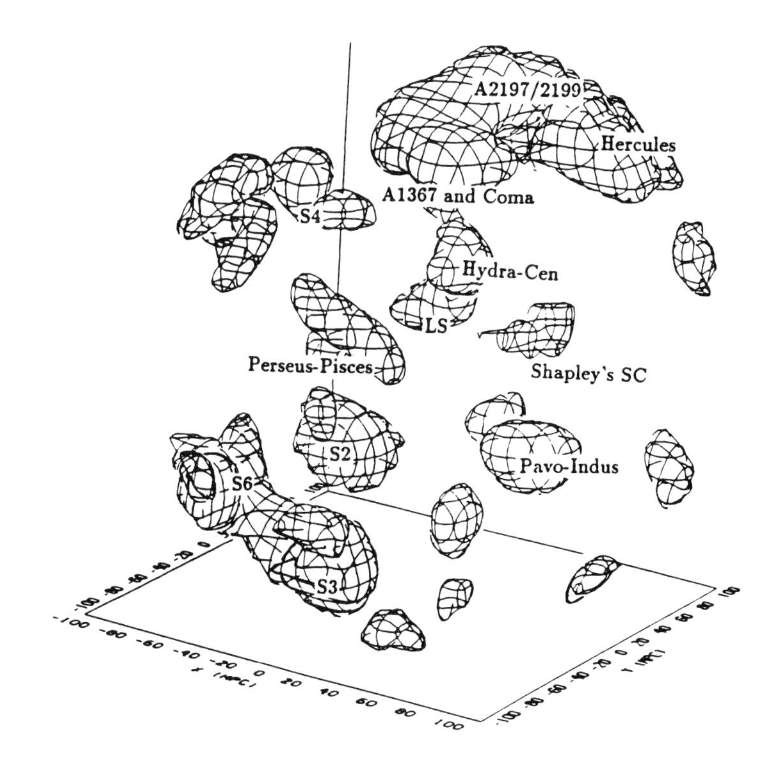

Fig. 4.6. A three-dimensional representation of the high-density regions, within 10,000 km/s, identified by IRAS galaxies in the QDOT survey. (Reproduced with permission from W. Saunders and the *Monthly Notices of the Royal Astronomical Society* (**256**, 477, 1992).)

ously different in character, they map all of (or a good portion of) that volume of redshift space covered by the Atlas. There is agreement between the features shown and those in the Atlas. Finally, Table 4.3 on page 95 provides cross-identifications between voids listed earlier and some of those in the literature.

4.8 FURTHER READING

Specialised

Global coverage

Fairall, A.P. *et al.*, Large-scale Structure in the Universe: Plots from the updated *Catalogue of Radial Velocities of Galaxies* and the *Southern Redshift Catalogue*, *Mon. Not. R. astr. Soc.*, **247**, 21P (1990).

Fairall, A.P. *et al.*, The Topology of Nearby Large-Scale Structures, [in] *Observational Cosmology*, ASP Conference Series, **51** (*Ed.* G. Chincarini, A. Iovino, T. Maccacaro, D, Maccagni), p.148 (1993).

Hudson, M.J., Optical galaxies within 8000 km s^{-1} – I. The density field., *Mon. Not. R. astr. Soc.*, **265**, 43 (1993).

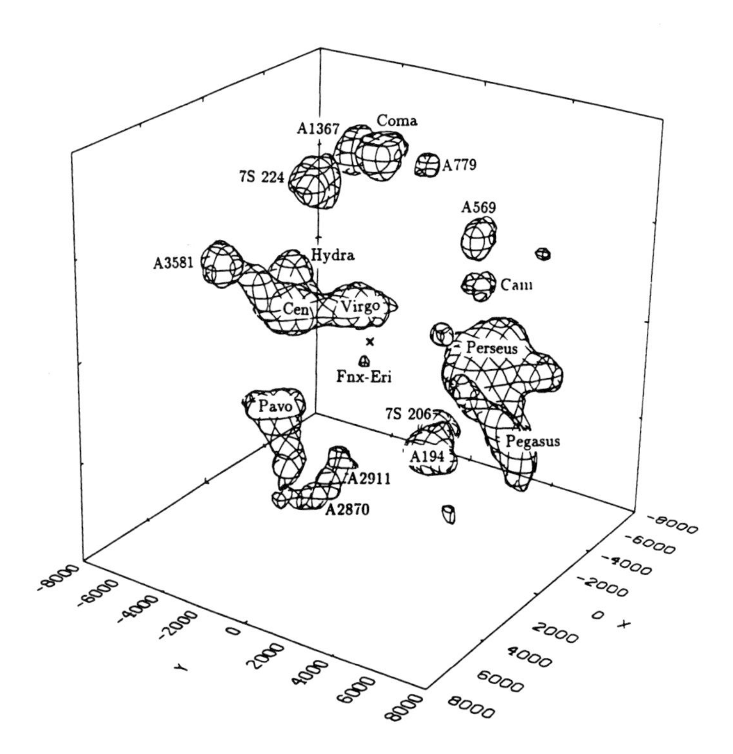

Fig. 4.7. A plot similar to the previous diagram that shows high-density regions (optical catalogues) within 8,000 km/s, as identified by M. Hudson. The density contours are set slightly different from the QDOT diagram, but the features match, except that they are seen from a different direction. (Reproduced with permission of M. Hudson and the *Monthly Notices of the Royal Astronomical Society* (**265**, 43, 1993).)

Kolatt, T. *et al.*, Large-scale mass distribution behind the Galactic plane, *Mon. Not. R. astr. Soc.*, **275**, 797 (1995).

Moore, B. *et al.*, The topology of the QDOT IRAS redshift survey, *Mon. Not. R. astr. Soc.*, **256**, 477.

Rowan-Robinson, M. *et al.*, The QMW IRAS galaxy catalogue: a highly complete and reliable IRAS 60-mm galaxy catalogue, *Mon. Not. R. astr. Soc.*, 253, 485 (1991).

Saunders, W. *et al.*, The density field of the local Universe, *Nature*, **349**, 32 (1991).

Shaver, P.A., Radio Surveys and Large Scale Structure, *Aust. J. Phys.*, **44**, 759 (1991).

Local Group and Virgo Supercluster

Bureau, M. *et al.*, A new *I*-band Tully–Fisher relation for the Fornax cluster: implication for the Fornax distance and Local Supercluster velocity field, *Astrophys. J.*, **463**, 60 (1996).

Guthrie, N.G. and Napier, W.M., Redshift periodicity in the Local Supercluster, *Astron. Astrophys.*, **310**, 353 (1996).

Han, C. *et al.*, The Orientation of the spin vectors of galaxies in the Ursa Major filament, *Astrophys. J.*, **445**, 46 (1995).

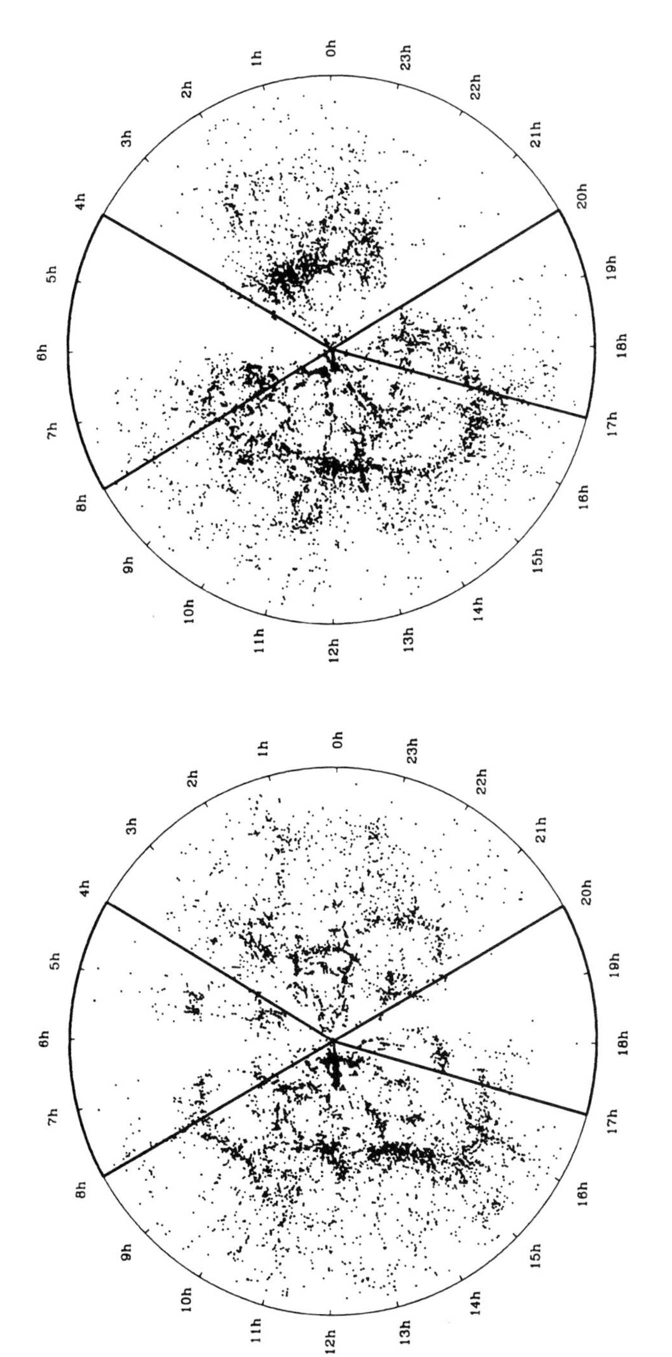

Fig. 4.8 Galaxies in the Center for Astrophysics surveys, declination ranges 0° to 20° (left) and 20° to 45° (right), out to 15,000 km/s. The (Coma) Great Wall is the dominant feature in the left-hand sector and Perseus–Pisces is to the right. (Reproduced with permission from J. Huchra and the *Astronomical Journal* (**112**, 1803, 1996).)

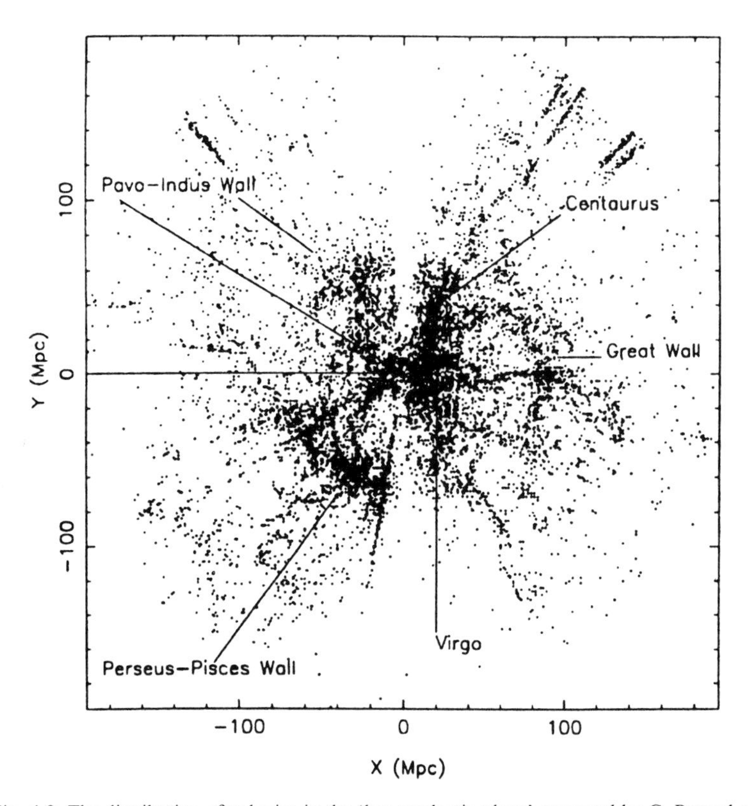

Fig. 4.9. The distribution of galaxies in the 'hypergalactic plane' proposed by G. Paturel and H. di Nella. Major features seem almost to form an encircling ring. Those labelled 'Centaurus' and 'Pavo–Indus Wall' are described in this chapter as the 'Centaurus Wall' (seen here flat-on and bisected by the obscuration of the Milky Way). The continuity between the Virgo Supercluster and Centaurus is very clear. (Reproduced with permission from G. Paturel.)

Huchtmeier, W.K. *et al.*, Two new possible members of the IC 342–Maffei 1/2 group of galaxies, *Astron. Astrophys.*, **293**, L33 (1995).

Karachentsev, I., The local group in comparison with other nearby groups of galaxies, **305**, 33 (1996).

Tully, R.B. *et al.*, The Ursa Major Cluster of Galaxies. I. Cluster Definition and Photometric Data, *Astron. J.*, **112**, 2471 (1996).

Tully, R.B., Unscrambling the Local Supercluster, *Sky and Telescope*, June 1982, p.550. (Popular level)

Vallée, J.P., Further analysis of a 'Complete Sample' in the Virgo supercluster of galaxies, *Mon. Not. R. astr. Soc.*, **264**, 665 (1993).

Van den Bergh, S., The Outer Fringes of the Local Group, *Astron. J.*, **107,** 1328 (1994).

Centaurus Wall

Fairall, A.P. and Paverd, W.R., Large-Scale Structure in the Southern Sky to 0.1c: Visualisations of Cellular Structure within Great Walls and our Position within a possible

Table 4.3. Some cross references for Atlas void designations

Cetus KF2
Monoceros KF(14)
Local Void KF(11)
Andromeda KF26
Orion KF(16/17)
Can Ven V1
Draco KF(19)
Delphinus KF(15/25/35)
Eridanus W(iii/vii) F5 SSRS:1 KF30/33/37/48/55
Corvus KF50
Gemini KF28
Virgo KF24 L(A6)
Cygnus KF(36/54)
Leo KF27
Taurus KF3/51
Microscopium F2 SSRS:2 KF31/(49)
Canis Major KF(100)
Coma V5/6 KF63 L(C5)
Hydra Wii
Cor Bor KF(62)
Pegasus KF40/(71)
Sculptor Wi F1 SSRS:3 KF(58)/89 SSRS2
Ursa Minor KF67/77
Ursa Major KF43/(52)/73
Columba Wvi SSRS:4 KF(91) SSRS2
Pisces KF72
Apus F4 KF(29)
Fornax KF101
Capricornus KF(108)

W=Winkler 1983, *Mon. Not. Astr. Soc. Sthn. Africa*, **42**, 74
F=Fairall 1984, *Publ. Dept. Astr. Univ. Cape Town*, No.6
V=Vettolani *et al*, 1985, *Astron. Astrophys.* **144**, 506
SSRS=da Costa *et al*, 1988, *Astrophys. J.*, **327,** 544
KF=Kauffmann & Fairall, 1990, *Mon. Not. R. astr. Soc.,* **248**, 313
L=Lindner *et al.* 1995, *Astron. Astrophys.*, **301,** 329
SSRS2=El-Ad *et al*, 1996, *Astrophysics. J.*, **462**, L13

'Centaurus Great Wall', [in] *Wide-Field Spectroscopy and the Distant Universe* (The 35th Herstmonceux Conference), *Ed.* S.J. Maddow and A. Aragon-Salamanca, p.121, World Scientific, 1995.

Lahav, O. *et al.*, The spatial distribution of X-ray clusters of galaxies, *Mon. Not. R. astr. Soc.*, **238**, 881 (1989). (See Figure 2).

Paturel, G. *et al.*, New determination of the pole of a 'hypergalactic' large-scale system, *Astron. Astrophys.*, **189**, 1 (1988).

Coma region

Colless, M. and Dunn, A.M., Structure and Dynamics of the Coma Cluster, *Astrophys. J.*, **458**, 435 (1996).

Hoffman *et al.*, The Large-Scale Distribution of Late-Type Galaxies between Virgo and the Great Wall, *Astrophys. J.*, **441**, 28 (1995).

Lindner, U. *et al.*, The distribution of galaxies in voids, *Astron. Astrophys.*, **314**, 1 (1996).

Longo, M.J., Evidence for Structure in the distribution of active galactic nuclei with $z < 0.05$, *Astrophys. J.*, **372**, L59 (1991).

Pustil'nik, S.A. *et al.*, The Spatial Distribution of Blue Compact Galaxies in the Second Byurakan Survey, *Astrophys. J.*, **443**,, 499 (1995).

Ramella, M. *et al.*, The Distribution of Galaxies within the 'Great Wall', *Astrophys. J.*, **384**, 396,(1992).

Thorstensen, J.R. *et al.*, Redshifts for fainter galaxies in the first CfA Slice. III. To the Zwicky Catalog limit, *Astron. J.*, **109**, 2368 (1994).

Thuan *et al.*, The Spatial Distribution of Dwarf Galaxies in the CfA Slice of the Universe, *Astrophys. J.*, **315**, L96 (1987).

Perseus–Pisces region

Batsuki, D.J. and Burns, J.O., A Possible 300 Megaparsec filament of clusters of galaxies in Perseus–Pegasus, *Astrophys. J.*, **299**, 5 (1985).

Chamaraux, P. *et al.*, A connection between the Perseus–Pisces supercluster and the A569 cloud?, *Astron. Astrophys.*, **229**, 340 (1990).

Focardi, P. *et al.*, The north extension of the Perseus Supercluster, *Astron. Astrophys.*, **136**, 178 (1984).

Giovanelli, R. *et al.*, A 21 cm survey of the Pisces–Perseus supercluster. II. The declination zone +21.5 to +27.5 degrees, *Astron. J.*, **92**, 250 (1986).

Giovanelli, R. *et al.*, Morphological segregation in the Pisces–Perseus Supercluster, *Astrophys. J.*, **300**, 77 (1986).

Hauschildt, M., The Perseus Supercluster at low Galactic Latitudes, *Astron. Astrophys.*, 184, 43 (1987).

Wegner, G. *et al.*, A survey of the Pisces–Perseus Supercluster V. The Declination Strip +33.5° to +39.5° and the main supercluster ridge, *Astron. J.*, **105**, 4, 1260 (1993).

Willick, J.A. *et al.*, A Redshift Survey toward a proposed void of galaxies suggested by the distribution of Abell clusters, *Astrophys. J.*, **355**, 393 (1990).

Southern regions

Fairall, A.P., *Southern Redshift Catalogue*. The maps in the accompanying Atlas of Nearby Large-Scale Structures include updated versions of plots made from this catalogue. Earlier sets of plots appear in *Mon. Not. Astr. Soc. Sthn. Africa,* **42,** 74 (1983), and *Publ. Dept. Astron. Univ. Cape Town,* Numbers 6 (1984), 10 (1988) and 11 (1991).

di Nella, H. *et al.*, Are the Perseus–Pisces chain and the Pavo–Indus Wall connected?, *Mon. Not. R. astr. Soc.,* **283**, 367 (1996).

Pellegrini, P.S. *et al.*, Distribution of Galaxies in the Southern Galactic Gap, *Astron. J.*, **99**, 751 (1990).

Ramella, M. *et al.*, The redshift-space neighbourhoods of 13 SSRS groups of galaxies, *Astron. Astrophys.,* **312**, 745 (1996).

Zone of Avoidance

Chamaraux, P. *et al.*, A search for IRAS galaxies behind the Taurus molecular clouds, *Astron. Astrophys.,* **299**, 347 (1995).

Hau, G.K.T. *et al.*, Visual search for galaxies near the northern crossing of the Supergalactic Plane by the Milky Way, *Mon. Not. R. astr. Soc.,* **277**, 125 (1995).

Kraan-Korteweg, R.C. *et al.*, Extragalactic Large-scale structures behind the Southern Milky Way, *Astron. Astrophys.,* **297**, 617 (1995).

Lahav, O. *et al.*, The Puppis cluster of galaxies behind the Galactic plane and the origin of the 'Local Anomaly', *Mon. Not. R. astr. Soc.,* **262**, 711 (1993).

Lu, N.Y. *et al.*, Large-Scale Structures in the Zone of Avoidance: The Galactic Anticenter region, *Astrophys. J.,* **449**, 527 (1995).

Marzke, R.O. *et al.*, Large-Scale Structure at Low Galactic Latitude, *Astron. J.*, **112**, 1803 (1996).

Saito, M. *et al.*, A search for galaxies behind the Milky Way between $l = 230°$ and 250°, *Publ. Astron. Soc. Japan*, **43**, 449 (1991).

Seeberger, R., From Galactic to Extragalactic Structures: Galaxies in the 'Zone of Avoidance' between 90° and 110°, *Publ. astr. Soc. Pacific,* **107**, 301 (1995).

Takata, T. *et al.*, A Systematic Search for IRAS galaxies behind the Milky Way, *Astrophys. J.,* **457**, 693 (1996).

Takata, T. *et al.*, Search and redshift survey for IRAS galaxies behind the Northern Milky Way, *Astron. and Astrophys. Suppl. Ser.*, **104**, 529 (1994).

Visvanathan, N. and Yamada, T., Redshift distribution of galaxies in the Southern Milky Way Region, *Astrophys. J. Suppl.*, **107**, 521 (1996).

Weinberger, R. *et al.*, Penetrating the 'zone of avoidance'. I. A compilation of optically identified extragalactic objects within $|b| < 5°$, *Astron. and Astrophys. Suppl. Ser.*, **110**, 269 (1995).

Yamada, T. *et al.*, A Search for IRAS galaxies behind the Southern Milky Way, *Astrophys. J. Suppl.*, **89**, 57 (1993).

Yamada, T. *et al.*, Connection of large-scale structures of the galaxy distribution behind the southern Milky Way, *Mon. Not. R. astr. Soc.,* **262**, 79 (1993).

Void catalogues

El-Ad, H. *et al*, Automated Detection of Voids in Redshift Surveys, *Astrophys. J.*, 462, L13 (1996).

Kauffmann, G. and Fairall, A.P., Voids in the distribution of galaxies: an assessment of their significance and derivation of a void spectrum, *Mon. Not. R. astr. Soc.*, **248**, 313 (1990).

Lindner, U. *et al.*, The Structure of Supervoids, *Astron. Astrophys.*, **301**, 329 (1995).

Vettolani, G. *et al.*, The Distribution of Voids, *Astron. Astrophys.*, **144**, 506 (1985).

5

Mapping the cosmos - deep surveys

5.1 LOOKING DEEPER INTO THE COSMOS

The previous chapter has given us an overview of 'nearby' features – those large-scale structures within 10,000 km/s, for which reasonable all-sky coverage is available. Now we shall look further and consider the findings of surveys that have penetrated much deeper into redshift space. We shall find quite incomplete sky coverage and a wide variety of styles of redshift survey.

Greater depth can only be achieved by limiting the sky coverage. Any survey must have a limit in the total number of galaxies it can cover. Hence a compromise has to be made between sky coverage and the depth of the survey. For example, the sky coverage could be limited to just a few small fields, if one wanted to sample to maximum depth (such 'pencil-beam' surveys are described below). However, the best compromise between survey volume and depth of coverage is a thin 'slice' in Declination. The survey covers only a very narrow spread in Declination, but stretches over a wide range of Right Ascension as the rotation of the Earth makes it accessible for observation. Such thin slice volumes emulate the highly successful Center for Astrophysics 'Slice of the Universe' that pioneered the surveying of the nearby Universe. The product is of course a cross-section through the large-scale structures, but from which general characteristics are apparent and quantitative parameters can be extracted. A further alternative is to make use of limited 'markers' of large-scale structure, such as clusters.

The quest to define the fabric of the cosmos on ever-larger scales has led to a 'Pandora's Box' of redshift surveys of one sort or another; far too many to describe individually, though a representative selection has been mentioned in Chapter 3. In any case, the purpose of this book is to outline the approaches and methods, and to provide only a general overview of findings.

We shall therefore look at some well known but relatively distant large-scale structures. We will then discuss some of the avenues that have been followed in probing to much greater depths and concentrate on a few of the major and most representative surveys.

5.2 SOME INDIVIDUAL SUPERCLUSTERS AND VOIDS

The previous chapter dealt with large-scale structures within a redshift of 10,000 km/s. There are, however, a number of major structures, particularly in the 10,000–20,000 km/s range that have been known for a considerable time. In a sense, they have been more visible than nearby structures, and were recognised much earlier. This is because their member galaxies are confined to a smaller region of the sky, of size similar to a single constellation; by contrast, a number of the nearer structures escaped early recognition since their member galaxies were so scattered over such a large portion of the sky. For instance, even as far back as the 1930s, Harlow Shapley recognised 'metagalaxies' in Centaurus and in Horologium, while galaxy 'clouds' have long been known in Hercules, Corona and Serpens–Virgo (see Chapter 1).

Surveys such as QDOT have penetrated beyond 10,000 km/s. Figure 5.1 shows overdense regions, immediately beyond those shown in the previous chapter. They include the well-known concentrations and other features, briefly reviewed below. Where necessary, we shall use the system of expressing directions or positions in redshift space in parentheses: (Right Ascension, Declination, velocity).

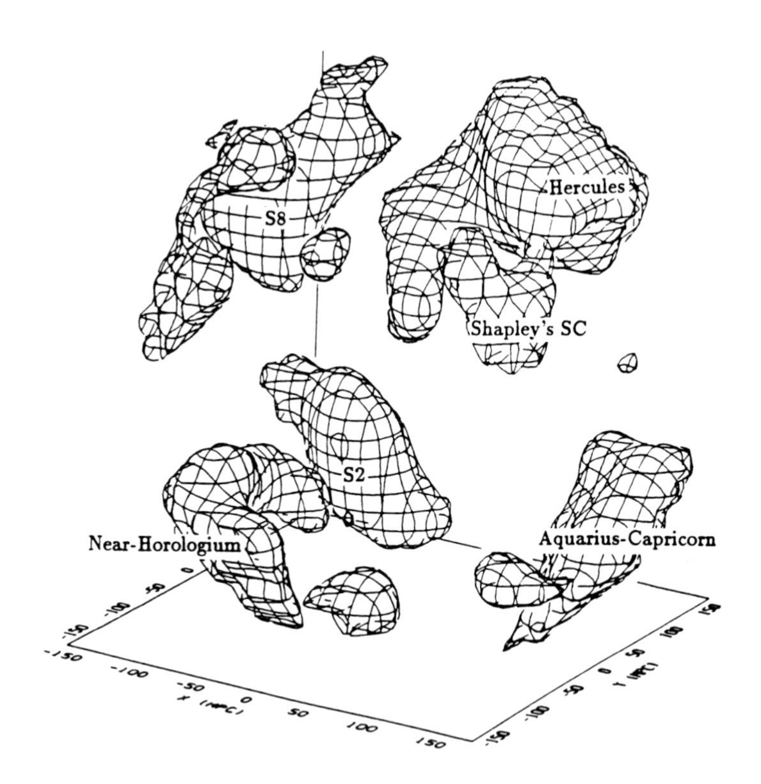

Fig. 5.1. Overdense regions out to 15,000 km/s (more distant than those of Figure 4.6) as shown by IRAS galaxies in the QDOT survey. (Reproduced with permission from W. Saunders and the *Monthly Notices of the Royal Astronomical Society* (**256**, 477, 1992).)

5.2.1 The Shapley region

Though noted so long ago, Shapley's 'metagalaxy' in Centaurus (13 hrs, –25°) only drew attention in the late 1980s when the search for the 'Great Attractor' – the apparent overdensity responsible for the large-scale motion of our Galaxy and its neighbours – got under way. Much more will be said about this in Chapter 7, but suffice to note here that the Shapley concentration is considered a possible contender. It is apparent as a conglomeration of clusters of galaxies, at a redshift of around 13,000 km/s, generally labelled as the 'Shapley' or 'Alpha' region. It may even be the most massive structure within 20,000 km/s. Figure 5.2 shows something of the concentration of clusters in the region. S. Raychaudhury and colleagues, for example, report on its high density of rich clusters, including six of the 46 brightest X-ray emitting clusters in the sky.

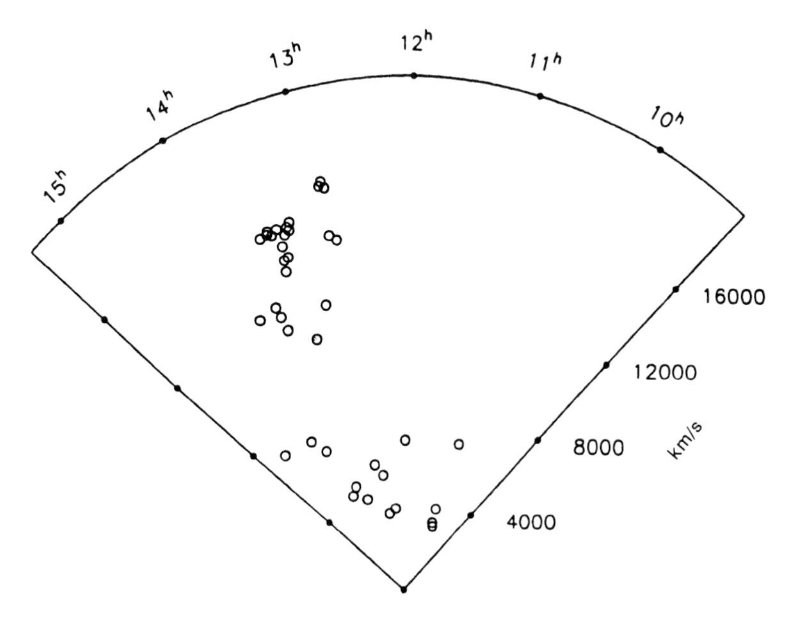

Fig. 5.2. A plot by S. Raychaudhury, showing the marked concentration of clusters at 13,000 km/s, in the direction RA 13 hrs (and Declination −15°), of the Shapley region. (Reproduced with permission from S. Raychaudhury and the *Monthly Notices of the Royal Astronomical Society* (**248**, 101, 1991).)

5.2.2 Horologium region

The Horologium region (3 hrs, −50°) is the other cloud or overdensity originally described by Shapley. A number of redshift surveys were directed towards it in the 1980s, particularly by Guido Chincarini and co-workers. The redshifts reveal that there are two main concentrations at 12,000 and at 17,000 km/s, seen along a common line of sight, both of which include a number of clusters.

5.2.3 Hercules superclusters

Early redshift surveys uncovered two 'superclusters' in Hercules (at 16 hrs, +35°, 9,000 km/s and 16 hrs, +20°, 11,000 km/s). The northern one is connected to the Coma Great Wall (see Section 4.4.7) and appears in the outermost redshift shells in the Atlas section, located between pages 80 and 81.

5.2.4 The Boötes Void

The well-publicised Boötes Void was originally speculated to be a large extensive void 6,000 km/s in depth at a redshift of 15,000 km/s. However, it is more a general under-dense region than the distinct voids displayed in the previous chapter. Soon after its discovery, Vicki Bolzano and Daniel Weedman reported that there are at least 12 emission-line galaxies within it. Since then, at least 12 IRAS galaxies have also been found within the void by A. Dey, M. Strauss and J.Huchra. More recently, A. Szomuru, J. van Gorkom and M. Gregg detected 18 uncatalogued companions to previously known void galaxies, using radio HI observations.

5.2.5 Other features

A number of other superclusters and voids have been reported in the literature, including those following.

A void of diameter 4,000 km/s has been reported by J. Burns and collaborators (at 1.5 hrs, +5°, 11,000 km/s). T. Yamada and M. Saito have found a grouping of three clusters in Monoceros at 10,000 km/s that they believe is an indication of a supercluster. Figure 5.1 – from the QDOT survey of IRAS galaxies – also suggests further overdensities (in Aquarius–Capricorn and 'S8' towards l = 162°, b = 0°). M. Ashby and colleagues report a 'massive' supercluster (at 18 hrs, +66.5°, 26,400 km/s).

Ever more distant features are a candidate supercluster at 46,500 km/s associated with the Abell clusters 3802/3834, and two, both at 56,000 km/s, associated with Abell clusters 2576/2628 – as suggested by D. Batuski and colleagues. Robin Ciardullo reports an investigation of a supercluster (at 14.85 hrs, +22°, 45,900 km/s) associated with Abell 2009. At still higher redshifts, A. Connolly and collaborators report a supercluster or great wall (at 0 hrs 16 min, +16°, 162,000 km/s).

5.3 ABELL CLUSTERS

In 1958, George Abell published a list of 2,700 clusters of galaxies that he had located on the Palomar Sky Survey. Much later, when the British SRC survey appeared, he started extending his work to include the southern sky. Sadly, his early death left the work incomplete, but Harold Corwin and Ron Olowin were to finish the job according to his precepts. The original Abell and the ACO catalogues together therefore give all sky coverage, except of course for the usual Zone of Avoidance caused by the Milky Way.

Abell laid down the various criteria for selecting galaxies, and also graded the clusters into seven scales of estimated distance. Subsequently his most distant clusters have been found to correspond to a redshift of about 75,000 km/s.

Rich clusters represent density peaks in the distribution of visible galaxies, and, presumably peaks in the distribution of mass throughout the Universe. For example, the greatest of all concentrations of Abell clusters coincides with the Shapley region described above. They are therefore seen as cosmological probes that might enable us to map the gross structures of the Universe out to some 1,000 Mpc (3 billion light years).

Some caution has, however, been expressed earlier in Section 3.10. Abell clusters may be subject to selection effects. They are, after all, located by visual scans of the survey photographs, conducted unavoidably over an extended period of time. The survey was carried out one Declination zone at a time, and claims have been made that coverage in some zones has been deeper, while that in other zones has been shallower, in spite of the best of intentions.

Reconciliation with clusters located by computer software from automated scans of the same photographs does not give a one-to-one agreement. In any case, such automated scans have also highlighted field-to-field differences that the human eye could not possibly compensate for, and which must particularly affect the more distant clusters. Nevertheless, in spite of these obvious imperfections, Abell clusters enjoy widespread reference and recognition.

Fritz Zwicky had also tried to catalogue both galaxies and clusters in his monumental work of the 1960s. Like Abell, he adopted grades for distance and richness. However, the large irregular outlines he drew around the peripheries of clusters could not be reproduced by other investigators, so his list of clusters, though initially analysed in the literature, has not had the enduring popularity accorded to Abell's work.

All members of a single cluster are presumably at the same distance from Earth, so that a single redshift measurement – of usually the brightest member – can then do for the entire cluster. This, of course, is the attraction of working with clusters. However, a further caution, also mentioned earlier, applies, since all too often the brightest member may in turn prove to be a foreground galaxy superposed on the cluster, and not a member after all. For reliability, a number of galaxies in the cluster must be observed spectroscopically.

The mapping and analysis of the distribution of Abell clusters in redshift space has given rise to a small industry.

As reported in Chapter 1, Brent Tully made the claim that Abell clusters were congregated into entities as large as 30,000 km/s. Although the reality of such structures was disputed by M. Postman, J. Huchra and M. Geller, who could not detect superclustering on scales more than 7,500 km/s, numerous other researchers have looked to see which 'superclusters of clusters' can be located. A classic work by Neta Bahcall and Raymond Soneira showed significant correlations between clusters (see Section 6.2, which discusses correlation functions).

In 1989, Michael West drew up a catalogue of 48 probable superclusters and analysed their shapes, noting a clear tendency for neighbouring superclusters (3,000 to 6,000 km/s apart) to be aligned with one another.

A major redshift survey of 145 Abell clusters has been carried out by John Huchra and collaborators. Amongst their analyses (reported in Chapters 6 and 7) was the discovery of an Abell 'cluster void' of diameter 20,000 km/s.

A similar major survey of 173 APM clusters (located by an algorithm from Automatic Plate Measuring machine data, many in common with Abell) was undertaken by the Oxford group (Dalton and collaborators), who would claim that their clusters form a more homogeneous sample. Oxford also claim that an analysis of the Huchra *et al* sample shows the clustering of Abell clusters is highly anisotropic, providing evidence of artificial clustering. Other analyses (e.g. Kolatt) have suggested that the distribution of Abell/ACO clusters is purely Gaussian.

Though somewhat thinner on the ground, the spatial distribution of 94 Edinburgh–Durham southern clusters is shown by C. Collins and collaborators (see Figure

5.3). A variation on this approach has been to use a sample of 128 clusters according to a flux limit on their X-ray emission (Figure 5.4).

Independent lists of superclusters of Abell/ACO clusters have also been drawn up by M. Kalinkov and I. Kuneva.

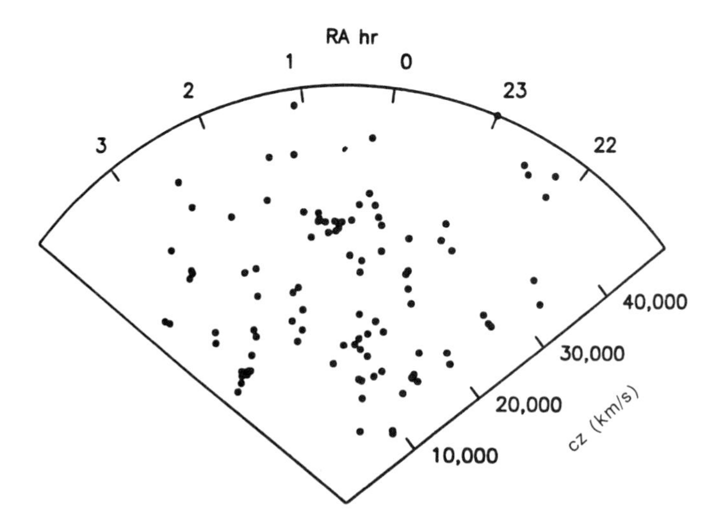

Fig. 5.3. The distribution of Edinburgh–Milan clusters and others in the Declination zone −22.5° to −42.5°. (Reproduced with permission from C. Collins and the *Monthly Notices of the Royal Astronomical Society* (**274**, 1071, 1995).)

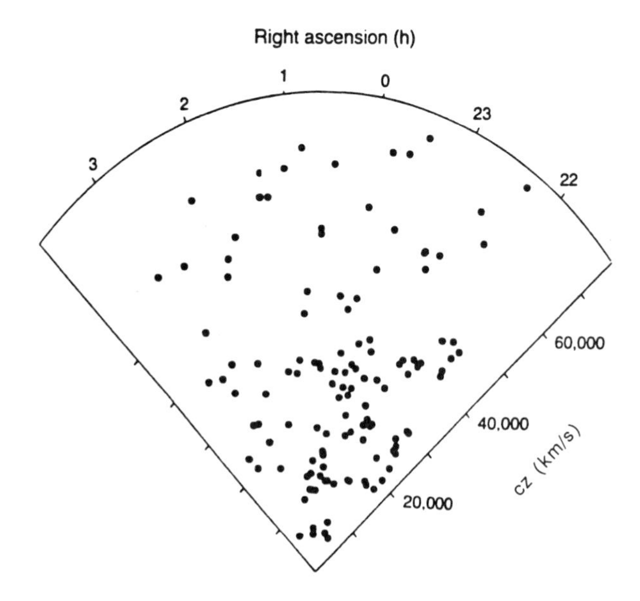

Fig. 5.4. The distribution of 128 X-ray clusters in the Declination zone +2° to −50°. (Reprinted with permission from C. Collins and *Nature* (**372**, 75). Copyright (1994) Macmillan Magazines Ltd.)

The European Southern Observatory has recently completed a very thorough investigation, reporting over 5,600 redshift measurements in the directions of 107 ACO clusters. Using a fixed gap of 1,000 km/s to separate systems and field galaxies, some 220 systems have been identified in the data. These data will allow a clear analysis of ACO clusters; for example, only about 10 per cent of clusters with $R \geq 1$ and $cz < 30,000$ km/s, appear to be the result of the superposition of two systems.

5.4 THE LAS CAMPANAS REDSHIFT SURVEY

Out to a redshift of 50,000 km/s, there is one dominant survey that has so far covered more galaxies than any other – the Las Campanas Redshift Survey (briefly mentioned in Section 3.12.3).

Las Campanas Observatory is a southern observing station of the Carnegie Institution. Its major telescope for many years has been a 2.5-m telescope. Starting in the late 1980s, a collaboration of astronomers – centred around Gus Oemler at Yale, Bob Kirschner at Harvard–Smithsonian and Stephen Shectman at the Carnegie Observatories – began the Las Campanus Redshift Survey, a survey to cut very deep sample slices.

To probe much deeper, the survey needed to observe much fainter galaxies than were catalogued. Consequently, it proceeded in two stages. The first was to make a large number of CCD photographs along the strips of Declination to be surveyed. The positions and magnitudes of galaxies were then extracted from these images. The second stage was to conduct follow-up spectroscopy using a multi-fibre spectrograph. The limited number of fibres and uneven distribution of the galaxies in the sky did, however, mean that not all galaxies down to the desired magnitude limit could be covered (for the 50-fibre system, galaxies in the range $16.0<m<17.3$ were observed; for the 112-fibre system, $15.0<m<17.7$), while galaxies of low surface brightness were excluded. Inevitably, there were further losses due to the nature of some spectra. Consequently the survey does not claim completeness to a fixed magnitude limit, but the variations are documented and can be taken into account in the analyses (cf. M. Hudson's approach in Section 3.9).

The Las Campanas survey has, however, produced more redshifts than any survey had done before: more than 26,000. It demonstrated that redshift observations could be mass-produced, and its success and methodology has led the way to its successor – the Sloan Digital Sky Survey (Section 3.12.5).

The Las Campanas slices (see Figure 5.5) reveal a rich texture of filaments, clusters and voids out to a redshift of 50,000 km/s. It goes four or five times deeper than the material covered in the previous chapter, and strongly suggests 'repetition' of similar structures, rather than an ongoing hierarchy of ever-larger structures. At last, the size of the largest discernible structures (around 15,000 km/s) is but a fraction of the size of the survey volume, strongly suggesting that the scale of homogeneity may have at last been reached (see Section 5.5 below). The Las Campanas researchers were fond of saying that their survey had gone 'beyond bigness'.

The Las Campanas finding – of a repetition of structures of a crude cellular nature of scale around 15,000 km/s – is in accordance with the pencil-beam surveys (described in Section 5.6 below). Figure 5.5 also suggests the occasional occurrence of higher overdensities, such as that (at 4 hrs, 17,000 km/s) which coincides with the Horologium region (Section 5.2.2 above).

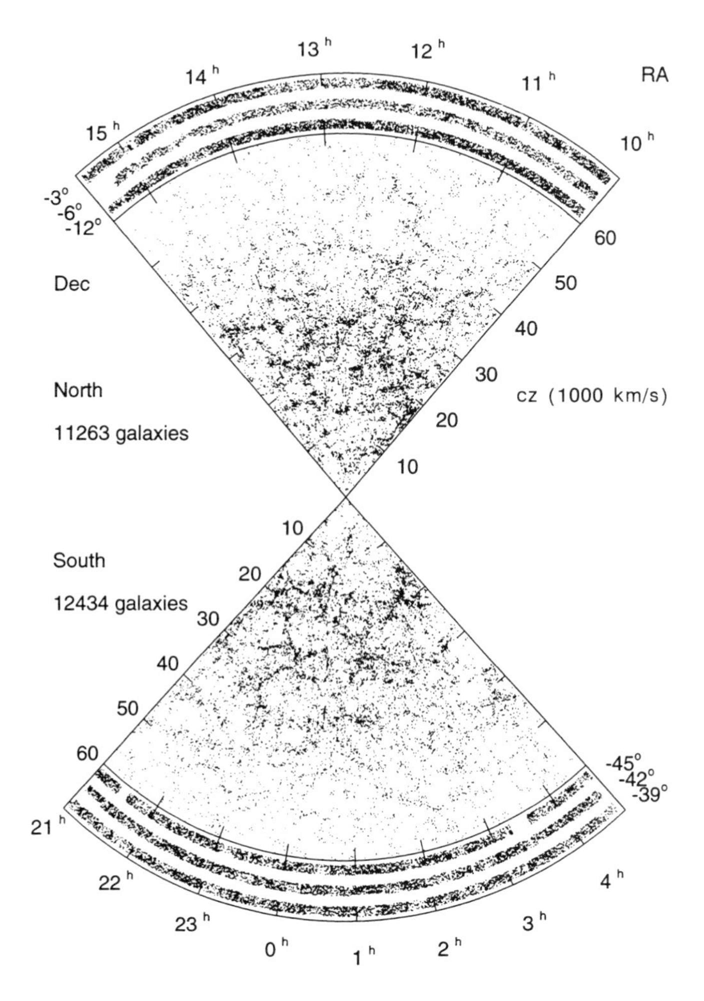

Fig. 5.5. The distribution of galaxies in the Las Campanas Redshift Survey in redshift and Right Ascension, for the northern and southern Galactic caps. The distribution of the same galaxies on the sky, in Right Ascension and Declination, is shown in the outer circular section. (Reproduced with permission from R. Kirschner.)

Andre Doroschkevich and collaborators (including those responsible for the survey) have carried out an analysis that suggests the identification of three different sets of structures in the survey. The first is 'superlarge-scale structures' accounting for 60 per cent of the galaxies, sheet-like in nature, with typical separations of 8,000 km/s. The next is 'rich filaments', with 20 per cent of the galaxies, but with characteristic separation of 3,000 km/s. The last component comprises 'poor, sparsely populated filaments' with separations around 1,300 km/s. All three sets of structures show a random (Poissonian) distribution in space. This description is somewhat subjective in nature, but still very much in line with the character of the nearby structures, as conveyed in the previous chapter. Moreover, it is a quantitative confirmation of 'repetition'.

5.5 THE SCALE OF INHOMOGENEITY

The Las Campanas Redshift Survey probably answers one of the most fundamental questions in cosmology: On what scale is the Universe homogeneous? The structures described in the previous chapter had indicated that inhomogeneities larger than 10,000 km/s existed. Moreover, those structures were almost as large as the volume of space surveyed. Had a larger volume been surveyed, would still larger features be apparent?

By contrast, the foundation of all cosmological models is known euphemistically as the 'Cosmological Principle'. It states that the Universe is homogeneous and isotropic; it is uniform in nature, and looks the same in all directions no matter where an observer is situated. Once this assumption is made, then Einstein's field equations can be reduced to a fundamental set that describes the Universe as a whole. Solutions such as the Friedmann models can be derived from them. It is the basis of modern cosmology.

While the Universe is clearly far from homogeneous on very small scales, the Cosmological Principle was devised at a time when the largest known inhomogeneities were believed to be clusters of galaxies, little more than a few hundred km/s in size. The recognition of large-scale structures has shifted the goalposts by approaching at least two orders of magnitude. If a scale of homogeneity cannot be found, then all of cosmological theory might be rendered useless.

Nevertheless, the advocates of the Cosmological Principle have always taken comfort from deep photography. Whatever direction we choose outside the Milky Way, telescopes have revealed a Universe populated by similar-looking galaxies at similar sky density, as far out as we can penetrate. Consequently, it is widely assumed that homogeneity is attained once a sufficiently large scale is reached. The question here is: how far out do we have to carry our redshift surveys to find it?

The Las Campanas survey is the first survey to show clearly an ongoing repetition of structure, and not an increasing hierarchy of ever-larger structures. As such, it does look as if the claim that it sees 'beyond bigness' is true. Granted, it looks out only in certain directions, but it does seem as if homogeneity is reached somewhere at a scale of around 20,000 km/s.

5.6 BEKS PEAKS

In parallel to the Las Campanas survey, with its indication of a general repetition of structure, has come an apparent indication of a 'periodic' repetition of structure that has evoked much controversy. It is derived from very deep 'pencil-beam' surveys. Such surveys are restricted to one or more 'very small' fields in the sky. Within each field, redshift observations of selected galaxies can be carried out. Since the directions to the galaxies are very much the same, the only spatial information extracted from these 'pencil beams' is one-dimensional: a histogram of the distribution of the galaxies in terms of redshift. Peaks in the distribution may therefore indicate where the pencil beam has intercepted a large-scale structure such as a great wall. Gaps in the distribution may be evidence of voids.

The approach is analogous to drilling a limited number of boreholes into the ground to determine subterranean structures. It fulfils the role of a preliminary exploration; for example, it was the technique by which the Boötes Void was first detected (as described in

Chapter 1). In that case, three sample fields, though seemingly widely spaced, all showed an underdense region at a similar depth.

As our mapping and knowledge of large-scale structures has progressed outwards, so these exploratory pencil-beam surveys have provided the deepest probes of large-scale structures. Such a strategy has for many years been followed by collaborations centred around Richard Ellis, originally based at Durham but now at Cambridge. Many of the observations have been made with the Anglo–Australian Telescope. The earlier probes intercepted overdensities of galaxies, subsequently revealed to be the large-scale structures described in the previous chapter. However, the investigations led to a stunning and controversial paper when Tom Broadhurst and Richard Ellis combined surveys made towards the North and South Galactic Poles with those of David Koo and Alexander Szalay in the United States.

Published in the journal *Nature* in 1990, the distribution in redshift of galaxies revealed a 'periodic' structure, with peaks occurring every 12,800 km/s interval in redshift. The original plot is reproduced in Figure 5.6. It goes as deep as 150,000 km/s – some three times deeper than the Las Campanas survey.

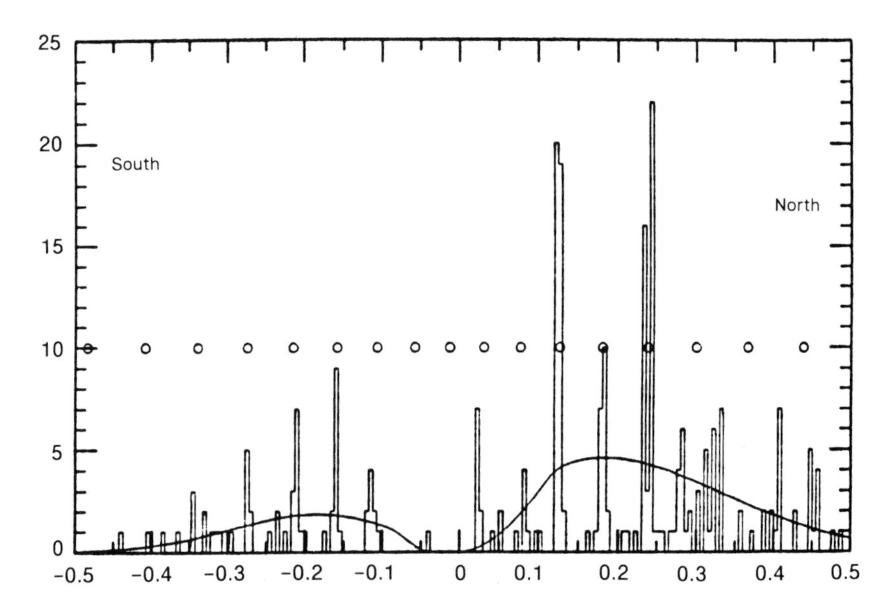

Fig. 5.6. The histogram of redshifts, in the directions of the South and North Galactic Poles, shows remarkable periodicity (as indicated by the line of open circles) in its peaks. (Reproduced with permission from T. Broadhurst and *Nature* (**343**, 726). Copyright (1994) Macmillan Magazines Ltd.)

For a moment it looked as if our Galaxy was surrounded by concentric rings of galaxies – almost a return to a geocentric Universe that overthrew all scientific sense! However, the pencil beams probed only towards the Galactic poles, and not all over the sky. Indeed, subsequent surveys (further pencil beams) indicated that the effect faded as one moved away from the Galactic poles. Even so, the discovery was sensational since it indicated repetition on a consistent scale, and, if true, a dramatic constraint on cosmology.

As can be seen from Figure 5.6, our own Galaxy was situated midway between the closest density peaks in the northern and southern Galactic hemispheres. We can reconcile this with the nearby structures described in the previous chapter. In the direction of the North Galactic Pole, the first peak (at half the periodicity, i.e. 6,400 km/s) is the Coma Great Wall. The first peak in the direction of the South Galactic Pole (at approximately 6,400 km/s) is the Sculptor Wall.

A reference was also made in Chapter 4 to the recent claim of H. di Nella and G. Paturel to there being a concentric concentration around our Galaxy at this distance, that includes the Coma Great Wall. However, the interception of the beam with the Sculptor Wall at some 6,000 km/s was fortuitous, since that wall runs at an angle to our line of sight. Had the pencil beam intercepted that wall at a different point, the peak would not have matched the periodic redshifts.

Neta Bahcall has shown that a number of the more distant density peaks matched known 'superclusters' of Abell clusters.

In the years since their original announcement, the BEKS team (Broadhurst, Ellis, Koo and Szalay) have sought to gather more redshifts by extending their original survey to a wider angle. They have added 21 new pencil beams distributed within 10-degree fields at the Galactic poles. These confirm the existence of the 12,800 km/s modulation, but put its amplitude at a more modest 12 per cent.

Not all the follow-up work has provided clear confirmation. Christopher Willmer, working with BEKS authors Koo and Szalay, together with Michael Kurtz, carried out a 'minislice' survey at the North Galactic Pole. They found that the high-density regions that produced the first five peaks had irregular shapes. Also, the third peak, that is so conspicuous in the original paper, would have been missed altogether had the original pencil beam been a degree different in direction.

Similarly, the ESO–Sculptor Faint Galaxy Redshift Survey, that obtains low resolution spectra to at least R = 20.5, covered a thin strip through the South Galactic Pole but found no evidence of the periodic structure.

Some 600 galaxies in pencil-beam surveys towards the South Galactic Pole, out to 100,000 km/s, were observed by Ettori, Guzzo and Tarenghi. They found a tendency towards 'some regularity' in the spacing of peaks in the histogram of redshifts in two of the pencil beams.

The Estonian astronomers Maret and Jaan Einasto, together with Gavin Dalton and Heinz Andernach, have sought to reconcile the distribution of Abell/ACO and APM clusters with the 12,800 km/s periodicity. They reported in 1994 that the supercluster/void scale did show a characteristic scale that agreed with the BEKS results.

Early in 1997, an extended Einasto collaboration made a quite sensational claim. They fitted the distribution of 319 clusters to a cubical lattice with 12,000 km/s periodicity. Their fit is shown in Figure 5.7. One of their plots shows their claim quite well, but the others seem less convincing.

The debate as to an exact periodicity will no doubt continue for some time, but quite how it could be interpreted may override normal cosmological sense, short of the possibility of there being some sort of periodic wave phenomenon. However, the work stands as evidence of the tendency towards a characteristic size of cellular structure.

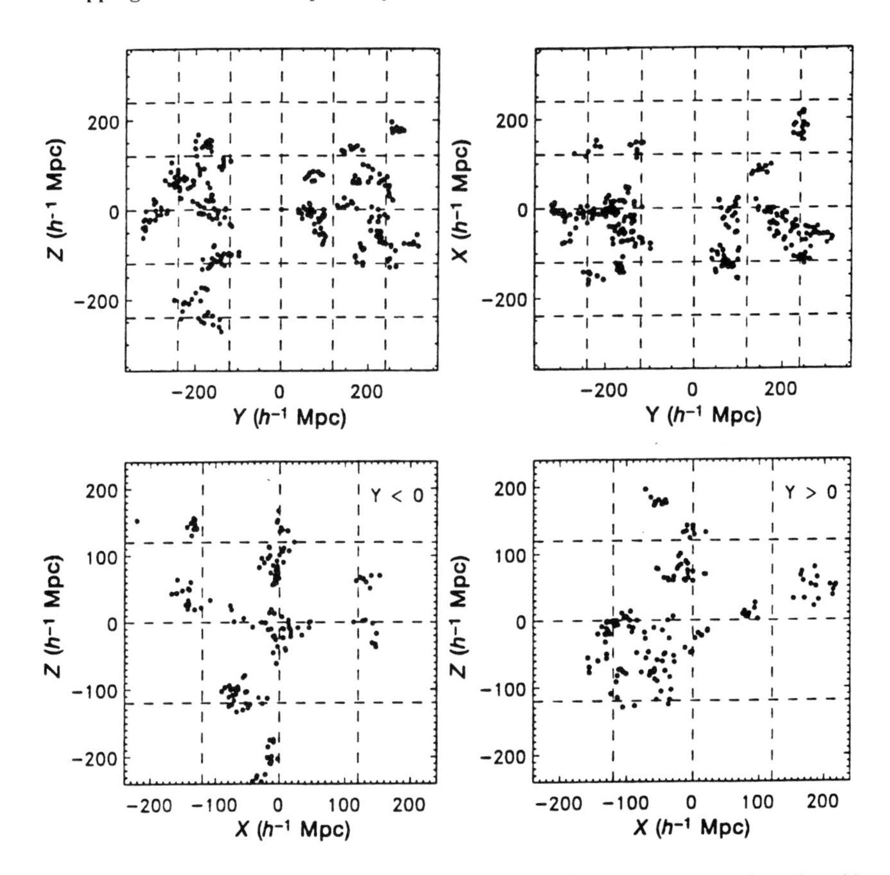

Fig. 5.7. The distribution of clusters of galaxies shows a tendency to congregate along the grid lines of a cubical lattice, with step size 120 h^{-1} Mpc in this plot by J. Einasto and colleagues. The coordinates are supergalactic; in the lower panels, clusters are plotted separately for the northern and southern Galactic hemispheres. (Reproduced with permission from J. Einasto and *Nature* (**385**, 139). Copyright (1997) Macmillan Magazines Ltd.)

Nevertheless, the periodicity of structure in the directions of the Galactic poles is highly intriguing, if not still worrying. It does suggest a repetition of what is probably cellular-like structure on a particular scale; even so, that it should occur quite so consistently and in quite such special directions, in relation to our Galaxy, remains an enigma.

5.7 OTHER IMPORTANT SURVEYS

The work so far described has been complemented by various other major surveys. A representative sample has been presented in Chapter 3. The findings, in essence, of these and other surveys are reported below.

The sparsely sampled Stromlo–APM Redshift Survey has given coverage in the southern Galactic hemisphere out to 25,000 km/s. A sample plot is shown in Figure 5.8. Some of the features there can be related to those in the nearby plots (e.g. the Atlas in Chapter 4; see Declination slices), but clearly the data is much thinner. Nevertheless, the work does

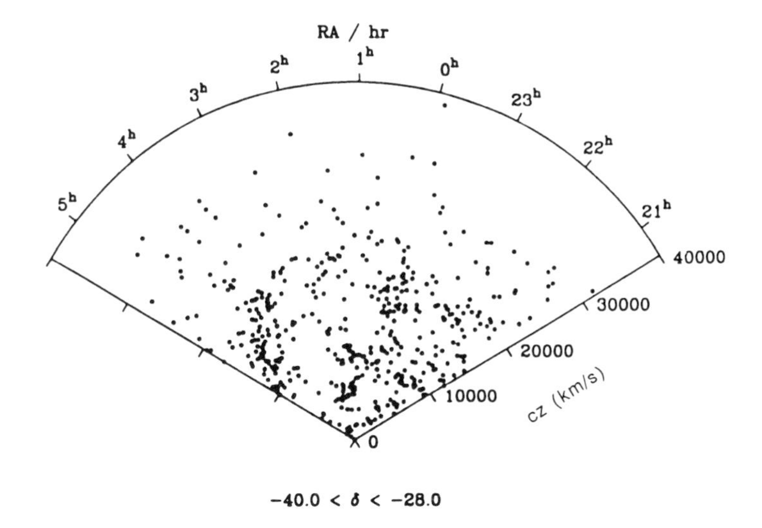

Fig. 5.8. A sample Declination slice from the Stromlo–APM Redshift Survey. (Reproduced with permission from J. Loveday and the *Astrophysical Journal* (**390**, 338, 1992).)

confirm the absence of dominant overdensities and large voids, and can also be analysed quantitatively.

Similar coverage of this region of sky is provided by the Durham/UK Schmidt Telescope Redshift Survey which has sampled some 2,500 galaxies selected at a rate of 1 in 3 from the Edinburgh–Durham *Southern Sky Galaxy Catalogue*. The FLAIR system on the UK Schmidt Telescope allows wide-angle coverage to a depth of 20,000 km/s. More global coverage, to similar depth, is provided by the *Point-Source Catalogue* survey of the QDOT collaboration.

Deep surveys to 30,000 km/s in the Virgo direction have also been attempted. Karachentsev and Kopylov have used the 6-metre telescope of the Russian Special Astrophysical Observatory to obtain redshifts of fainter galaxies. More recently, Flint and Impey have explored the region using catalogued galaxies (ZCAT) and discern a probable extended structure at 24,000 km/s.

The European Southern Observatory (ESO) Deep Slice Survey is very comparable with the Las Campanas Redshift Survey, in providing detailed coverage to some 60,000 km/s. However, it has covered only a single slice that fits between two of the six Las Campanas slices (the ESO slice covers RA 22.5 hr to 1.5 hr and Dec. –39° 45′ to 40° 45′), as shown in Figure 5.9. As with the Las Campanas slices, its importance lies in demonstrating the repetition of structures.

5.8 VERY DEEP SURVEYS

Surveys looking deeper still into redshift space cover too large a volume to allow any reasonable mapping of large-scale structures, though obviously some tentative overdensities are recorded. The main interest in these redshift surveys lies with galaxy evolution, since they look back to earlier epochs in the history of the cosmos. As these are outside the main scope of this book, we shall only summarise some main findings here.

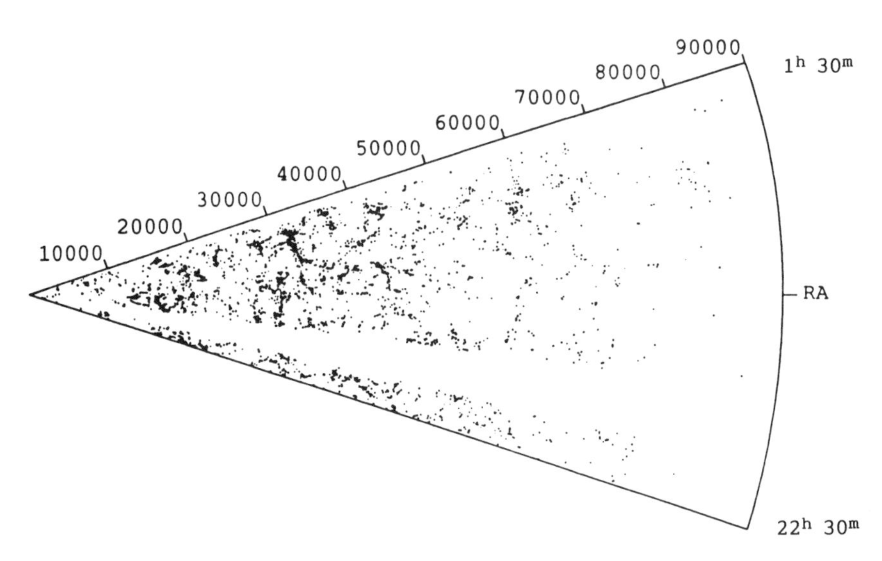

Fig. 5.9. The deep declination slice (–39.75° to –40.75°) of the European Southern Observatory Deep Slice Survey. Over 3,300 galaxies are plotted. (Reproduced with permission from G. Vettolani.)

The LDSS multi-slit spectrograph on the Anglo–Australian Telescope has been used to obtain galaxies as faint as $b_J = 22.5$ and out to redshifts of $z = 0.7$ (some 200,000 km/s). Matthew Colless and collaborators reported that although the sky counts of these faint galaxies are approximately twice the no-evolution model, the redshift distribution is much closer to a non-evolving population. It suggests that a large portion of the faint galaxies are really at low redshifts, as found by other observers.

The more recent Canada–France Redshift Survey, which covers the range $0 < z < 1$, reports a marked increase in the co-moving luminosity function (see Section 6.8) with redshift, indicating a rapid evolution of the star formation rate. As found elsewhere, blue galaxies show very marked evolutionary effects (a brightening by a magnitude by $z = 0.8$), whereas red galaxies seem to show no effects of evolution.

Other efforts with the Anglo–Australian Telescope – the AUTOFIB Redshift Survey – have pushed as faint as $b_J = 24.0$ to $z = 0.75$. A sample of 1,700 galaxies from $b_J = 11.5$ in various pencil beams has been analysed, and changes in the luminosity function with redshift detected. It confirms that the luminosity of star-forming galaxies has declined by 50 per cent or more, while quiescent galaxies are largely unchanged.

Early results with the DEEP survey on the Keck Telescope have also revealed active star-formation in galaxies that are probably conterparts of present-day dwarf spheroidal systems.

The Hubble Space Telescope has also probed deep into space and back to early epochs, but lacks the aperture for gathering redshifts.

5.9 QUASARS AS PROBES OF LARGE-SCALE STRUCTURE

As redshift surveys probe to the greatest depths, so even the most luminous conventional galaxies become almost too faint to provide usable spectra. However, quasars, known to be the active nuclei of certain galaxies, have intrinsic luminosities that often far exceed those of conventional galaxies. As such, they are the most luminous individual objects in the Universe, and can therefore be seen to the greatest distances.

Quasars are effectively extreme Seyfert 1 galaxies, the Seyfert label being reserved for the nearer cases where the host galaxy is more clearly discernible. Such host galaxies, typically spirals, appear to reflect the general spatial distribution of galaxies, and only their rarity means that one cannot see convincing filaments and voids by their distribution alone. Seyfert galaxies, being mainly spirals, avoid rich clusters; otherwise they show no discernible differences in spatial distribution to normal galaxies (as, for example, explored by this author in 1988). Consequently they should serve as markers – albeit sporadic – of large-scale structures.

That quasars can be seen to enormous distances makes them potential tracers of very distant large-scale structures. At best they will only reveal very gross overdensities. The problem in using quasars lies in selection effects, which are considerable. Nevertheless, some researchers, including Roger Clowes, have reported finding regions where significant numbers of quasars reveal similar redshifts, thereby suggesting that there could be an overdense region in redshift space. Clowes found a group of quasars at $z = 1.3$ (towards RA 10 hrs 40 min, Dec. 5°) that may span 20,000 km/s. Such regions, if they exist, are going to represent volumes of space generally much larger than those of the large-scale structures we seek. The reality of these findings has, however, been challenged; for instance, S. Croom and T. Shanks report only a weak clustering of quasars on small scales, but definitely no significant clustering above 5,000 km/s.

An alternative to quasars is their cousins, radio galaxies, since they are also seen to high redshift. C. Benn and J. Wall have shown how constraints on their isotropy may reflect the size of the largest cellular distribution.

Quasars can also be used in a different way. Some unorthodox cosmological models have entertained the idea of two or more images, seen in different directions, as originating from a common object. Individual images may not be reconcilable, but patterns of quasars may be recognisable. Although a couple of such cases have been found, they do not appear statistically significant.

Quasars, however, are more than individual objects; the spectra of high redshift quasars usually show absorption line systems of intermediate objects – presumably the extended envelopes of intervening galaxies. Groups such as G. Williger and collaborators, and J. Quashnock and colleagues, have analysed the absorption systems of adjacent quasars and found indications of clustering on scales of a few thousand km/s out to $z = 3$. These findings are obviously at too early a stage to offer convincing evidence that large-scale structures exist at high redshifts.

5.10 FURTHER READING

Specialised

Abell and other clusters

Collins, C.A. *et al.*, The Edinburgh–Durham Southern Galaxy Catalogue – VII. The Edinburgh–Milano cluster redshift survey, *Mon. Not. R. astr. Soc.,* **274**, 1071 (1995).

Dalton, G.B. *et al.*, Spatial Correlations in a Redshift Survey of APM Galaxy Clusters, *Astrophys. J.,* **390,** L4 (1992).
Ebeling, H. and Maddox, S.J., Abell Identifications of APM Clusters of Galaxies, *Mon. Not. R. astr. Soc.,* **275,** 1155 (1995).
Efstathiou, G. *et al.*, The Correlation Function of Rich Clusters of Galaxies: a Comparison of APM and Abell Clusters, *Mon. Not. R. astr. Soc.,* **257,** 125 (1992).
Huchra, J.P. *et al.*, A Deep Abell Cluster Redshift Survey, *Astrophys. J.,* **365,** 66 (1990).
Kalinkov, M. and Kuneva, I., Superclusters of Galaxies. I. The Catalog, *Astr. Astrophys. Suppl.*, **113**, 451 (1995).
Kolatt, T., Is the Density Distribution of Clusters Non-Gaussian?, [in] *Clustering in the Universe* (*Ed.* S. Maurogordato, C. Balkowski, C. Tao, J. Tran Thanh Van), p.241, Edition Fontieres (1995).
Postman, M. *et al.*, The Distribution of Nearby Rich Clusters of Galaxies, *Astrophys. J.,* **384,** 404 (1992).
West, M.J., On the Morphology of Superclusters, *Astrophys. J.,* **347,** 610 (1989).

Some individual features

Ashby, M.L.N. *et al.*, A Massive z = 0.088 Supercluster and Tests of Starburst Galaxy Evolution at the North Ecliptic Pole, *Astrophys. J.,* **456**, 428 (1996).
Batsuki, D.J. *et al.*, Redshift Observations of Abell/ACO Galaxy Clusters in Two Candidate Superclusters, *Astr. Astrophys.,* **294**, 677 (1995).
Burns, J.O. *et al.*, A 40 h–1 Mpc Diameter Void in Pisces–Cetus, *Astrophys. J.,* **355**, 542 (1988).
Ciardullo, R., The Morphology of the Rich Supercluster 1451+22, *Astrophys. J.*, **321,** 607 (1987).
Connolly, A.J. *et al.*, Superclustering at Redshift z = 0.54, *Astrophys. J.,* **473**, L67 (1996).
Dey, A. *et al.*, A Deep Redshift Survey of IRAS Galaxies towards the Boötes Void, *Astrophys. J.,* **99,** 463 (1990).
Raychaudhury, S. *et al.*, X-Ray and Optical Observations of the Shapley Supercluster in Hydra–Centaurus, *Mon. Not. R. astr. Soc.,* **248**, 101 (1991).
Szomoru, A. *et al.*, An HI Survey of the Boötes Void. I. The Data, *Astr. J.*, **111,** 2141 (1996).
Yamada, T. and Saito, M., A New Supercluster behind the Milky Way in Monoceros, *Publ. Astron. Soc. Japan,* **45**, 25 (1993).

Las Campanas

Shectman, S., *et al*, The Las Campanas Fiber-Optic Redshift Survey, [in] *Wide Field Spectroscopy and the Distant Universe* (*Ed.* S.J. Maddox, A. Aragon-Salamanca), p.98, World Scientific (1995).
Doroshkevich, A.G. *et al.*, Large- and Superlarge-scale Structure in the Las Campanas Redshift Survey, *Mon. Not. R. astr. Soc.,* **283**, 1281 (1996).

BEKS peaks

Bahcall, N.A., Superclusters and Pencil-Beam Surveys: The Origin of Large-Scale Periodicity, *Astrophys. J.*, **376**, 43 (1991).
Bellanger, C. and de Lapparent, V., Mapping the Galaxy Distribution at Large Distances, *Astrophys. J.*, **455**, L103 (1995).
Broadhurst, T. *et al.*, The Clustering of Galaxies at the Galactic poles (The 35th Herstmonceux Conference), *Ed.* S.J. Maddow and A. Aragon-Salamanca, p.178, World Scientific, 1995.
Broadhurst, T.J. *et al.*, Large-Scale Distribution of Galaxies at the Galactic Poles, *Nature,* **343**, 726 (1990).
Einasto, M., The structure of the Universe traced by Rich Clusters of Galaxies, *Mon. Not. R. astr. Soc.*, **269**, 301 (1994).
Einasto, M., A 120-Mpc Periodicity in the Three-dimensional Distribution of Galaxy Superclusters, *Nature*, **385** (1997).
Ettori, S. *et al.*, A Study of the Large-Scale Distribution of Galaxies in the South Galactic Pole Region – I. The Data, *Mon. Not. R. astr. Soc.*, **276**, 689 (1995).
Kirshner, R., The Universe as a Lattice, *Nature*, **385** (1997). (Commentary.)
Willmer, C.N.A. *et al.*, A Medium-Deep Redshift Survey of a Minislice at the North Galactic Pole, *Astrophys. J.*, **437**, 560 (1994).

Other important surveys

Bellanger, C. *et al.*, The ESO–Sculptor faint galaxy redshift survey: The Spectroscopic Sample, *Astr. Astrophys. Suppl.*, **110**, 159 (1995).
Flint, K.P. and Impey, C.D., Large-Scale Structure in the Direction of the Virgo Cluster, *Astr. J.*, **112**, 865 (1996).
Karachentsev, I.D. and Kopylov, A.I., Galaxy Redshift Survey in a Narrow Strip crossing the Coma Cluster, *Mon. Not. R. astr. Soc.*, **243**, 390 (1990).
Loveday, J. *et al.*, The Stromlo–APM Redshift Survey. I. The Luminosity Function and Space Density of Galaxies, *Astrophys. J.*, **390**, 338 (1992).
Maddox, S.J. *et al.*, Galaxy Correlations on Large Scales, *Mon. Not. R. astr. Soc.*, **242**, 43P (1990).
Oliverm S.J. *et al.*, Large-Scale Structure in a New Deep IRAS Galaxy Redshift Survey, *Mon. Not. R. astr. Soc.*, **280**, 673 (1996).
Ratcliffe, T. *et al.*, The Durham/UKST Galaxy Redshift Survey – I. Large-scale Structure in the Universe, *Mon. Not. R. astr. Soc.*, **281**, L47 (1996).

High redshift

Cole, S. *et al.*, The Spatial Clustering of Faint Galaxies, *Mon. Not. R. astr. Soc.*, **267**, 541 (1994).
Colless, M. *et al.*, The LDSS Deep Redshift Survey, *Mon. Not. R. astr. Soc.*, **244**, 408 (1990).

Ellis, R.S. *et al.*, Autofib Redshift Survey – I. Evolution of the galaxy luminosity function, *Mon. Not. R. astr. Soc.,* **280**, 235 (1996).

Lilly, S.J. *et al.*, The Canada–France Redshift Survey: The Luminosity Density and Star Formation History of the Universe to z~1, *Astrophys. J.,* **400**, L1 (1996).

Quasars

Benn, C.R. and Wall, J.V., Structure on the Largest Scales: Constraints from the Isotropy of Radio Source Counts, *Mon. Not. R. astr. Soc.,* **272**, 678 (1994).

Croom, S.M. and Shanks, T., QSO Clustering – III. Clustering in the Large Bright Quasar Survey and Evolution of the QSO Correlation Function, *Mon. Not. R. astr. Soc.,* **281**, 893 (1996).

Quashnock, J.M. *et al.*, High-Redshift Superclustering of Quasi-Stellar Object Absorption-Line Systems on 100 h–1 Mpc Scales, *Astrophys. J.,* **472**, L69 (1996).

Roukema, B.F., On Determining the Topology of the Observable Universe via Three-Dimensional Quasar Positions, *Mon. Not. R. astr. Soc.,* **283**, 1147 (1996).

Williger, G.M. *et al.*, Large-Scale Structure at $z \sim 2.5$, *Astrophys. J. Suppl.,* **104**, 145 (1996).

6

Quantitative measures of the large-scale structures

6.1 A QUALITATIVE INTRODUCTION

The way of science is to measure – on a reproducible basis and not a subjective one. This chapter concerns the various attempts to quantify the character of the large-scale structures. It seeks to ask 'what' they are, rather than 'how' and 'why', since the origin and formation of the structures will be addressed in the remaining chapters. The purpose of such measurements is, of course, to provide criteria against which theories of cosmological structure can be measured.

Numerous qualitative descriptions of large-scale structures – such as great walls surrounding voids – have filled the previous pages. A useful start may therefore be to summarise the findings so far, which are as follows:

> The visible content of the Universe is the galaxies, which, relative to the scale of distances that separate them, are effectively particles. Few galaxies, if any, exist in isolation; almost all have neighbours within 200 km/s in redshift space. The continuous large-scale structures so defined often form thick wall-like or ribbon-like structures, which contain internal cellular or filamentary substructure and have irregular boundaries. The large-scale structures are interconnected in a labyrinth. Instances of near-parallel walls and near-orthogonal walls occur.
>
> Large voids, approximately spherical and up to 5,000 km/s in diameter, are contained between the large-scale structures. These voids are seemingly interconnected, much like those of a bathsponge where the interconnections allow the air or water to be squeezed completely out.
>
> While some of the large voids are virtually empty, others look as if previous existing structures had been 'bleached out', leaving 'anaemic' remnants and almost isolated galaxies. Further superimposed on these structures are occasional clusters of galaxies, there being a tendency for rich clusters to lie near

the centroids of the great wall structures. Clusters do not necessarily occur at the intersections of wall-like structures. Quite remarkably, the adjacent substructures seem almost unaffected by the presence of the cluster.

Figure 6.1 attempts to convey the characteristics described above. This description may still prove the greatest constraint to theory, but modern theories are best assessed by mathematical analysis. We therefore seek statistical measurements of large-scale structures. The list of specialised reading at the end of this chapter reflects the growth of an industry devoted to this purpose. Whilst 'correlation functions' or 'power spectra' are the most acceptable methods for doing the job, we shall see that there is no shortage of alternatives. We shall review the main thrusts below, starting with that most widely used.

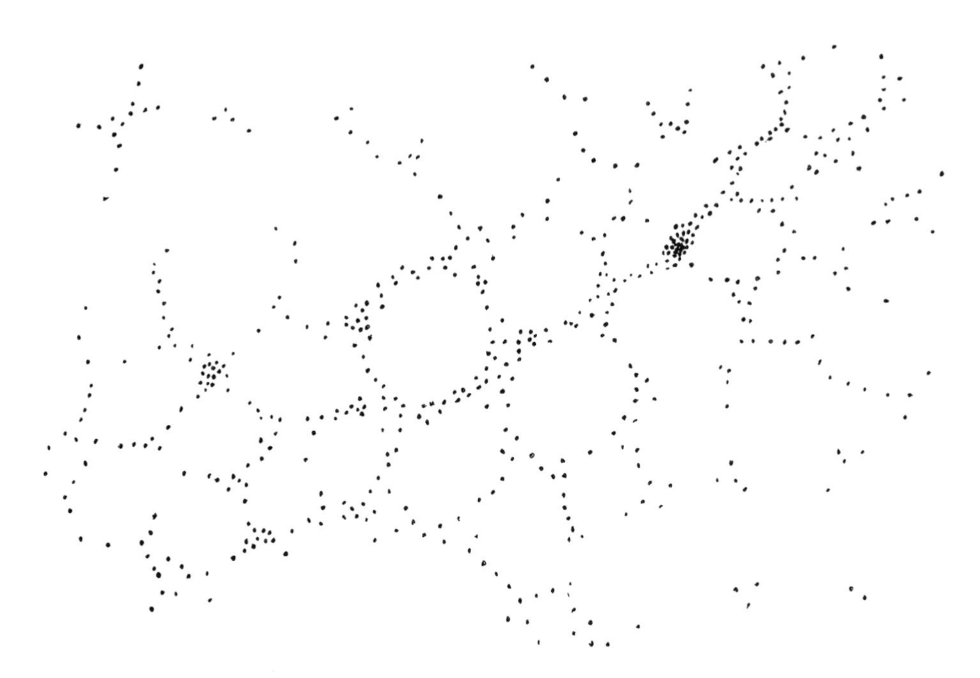

Fig. 6.1. A schematic representation, in cross-section form, through large-scale structures and voids. The figure conveys the general characteristics mentioned in the text.

6.2 THE TWO-POINT CORRELATION FUNCTION

The spatial distribution of galaxies is anything but random. The correlation function is simply a measure of the 'deviation from randomness'. First introduced by M. Totsuji and T. Kihara in 1969 and popularised by J. Peebles, the function is a measure by which the position of one galaxy is governed or influenced by others:

$$\xi(r) = \frac{n_{DD}(r)}{n_{RR}(r)} - 1$$

where $n_{DD}(r)$ is the number of pairs of galaxies with a separation of r, that can be found in the data, and $n_{RR}(r)$ is the number of pairs with the same separation had the data been distributed 'randomly', as suggested in Figure 6.2. Consequently, if $\xi(r)$ is positive, there is a degree of correlation or clustering (while a negative $\xi(r)$ would imply 'anti-clustering').

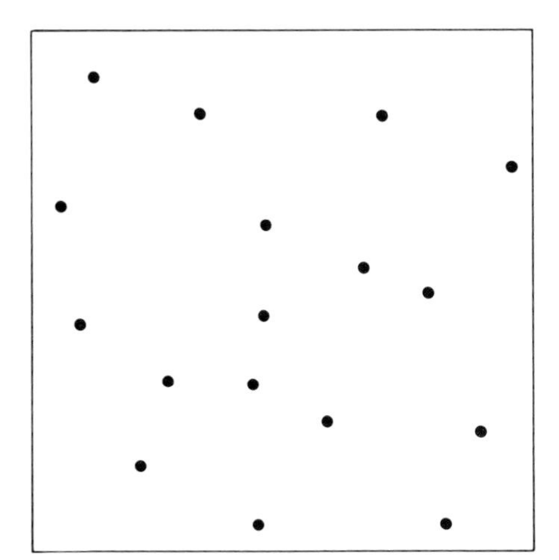

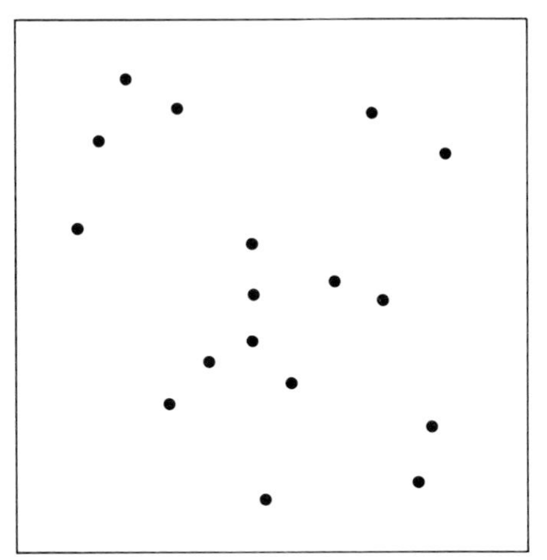

Fig. 6.2. The distribution of points in the left panel is completely random, whereas the distribution in the right panel shows a tendency to 'correlate' or cluster.

Since our definition involves separations between 'pairs' of galaxies, $\xi(r)$ above is the 'two-point' correlation function. It is possible mathematically to have higher orders – 'three point or more' correlation functions – but the simple two-point version is by far the most commonly used.

The two-point function involves pairs. It is relatively straightforward to calculate the geometric separation between any two galaxies in the sample (especially if the positions are first expressed in rectangular coordinates). The calculation is repeated for all possible pairs of galaxies within the sample volume; for example, if there were only four galaxies in the sample, there would be six possible pairings. Obviously the number of pairings increases dramatically with the number of galaxies, (and before computers, calculations of correlations were impractical). The list of all possible separations between pairs are binned and counted in intervals of the separation. The points are then rearranged in random fashion, within the same volume, and the whole process repeated. The correlation function can then be found.

The definition may seem quite simple, but in practice there are many complications. To begin with, the real data thins out with increasing distance, and may be subject to sharp cut-off boundaries. Such selection effects have everything to do with the way the data is collected, and nothing to do with the character of its distribution. Therefore, the random distribution against which it is compared must embody the same selection effects. A pure random distribution must be modified such that the density of data declines with increasing distance, in the same way as the real data. It must also have the same boundaries imposed. 'Monte-Carlo' techniques are usually used to produce a number of such random distributions, against which smoothed average values can be extracted.

The more complicated problem addresses the fact that the real data are, of course, not a complete sample, but are usually a magnitude-limited subset of the complete population. Hence the overall decline of the density of the data with distance. However, if the random distribution against which it is compared has the same overall decline, then the correlation so derived – which gives equal weight to each galaxy – is a reasonable measure. This is normally what is done. The alternative is to adopt a luminosity function and use it to extract a weighting function applied either to the individual galaxies or the distance. Then, since pairs of galaxies, or pairs of distances, are involved, each pair must take either the weighting of one of its galaxies or combine the weighting of both. The manner of weighting is much discussed in the literature; a number of authors point out that weighting both galaxies in a pair causes an overestimate of the correlation function.

A further, and even more complicated, problem is that the real data exist in redshift space and not true three-dimensional space. Peculiar velocities distort radial distances; for example, greatly increasing the separations of galaxies in clusters, and thereby affecting the correlation function. It is possible, but mathematically more complicated, to extract two correlation functions, one involving radial components, the other non-radial. Since the two functions should be the same, the difference ought to be a reflection of the distortion in redshift space. In fact the distortion can be used to constrain a parameter β relating to the 'bias' and 'mass density' (which will be taken up and discussed in Section 7.11).

Some investigations have revealed that in redshift space, the power law slope (γ defined below) is shallower than in real space, particularly if dense clusters are involved. Since, in practice, it is difficult to translate to real space, correlations are simply carried out in redshift space.

Figure 6.3 shows examples of correlation functions, using different weighting schemes to allow for a magnitude-limited sample. Happily, there are not large differences between the curves. The diagram also shows the universal characteristics of correlation functions. The correlation function is positive, but declines towards zero with increasing distance. The correlation is well defined at small separations, where there are ample pairs of galaxies, but is more poorly defined at large separations, where fewer pairs are available, and the noise is therefore much greater. In general, the correlation function can be approximated by a power law:

$$\xi(r) \propto \left(\frac{r}{r_0}\right)^{-\gamma}$$

where r_0 is the 'correlation' length, the separation corresponding to $\xi(r) = 1$, where the number of pairings is twice that of a random distribution. In Figure 6.3, slopes with $\gamma = 1.3$ and 1.9 are indicated. The most commonly extracted value, here and elsewhere, is around 1.8. The correlation length is typically in the range 500 to 1,000 km/s (5 to 10 h^{-1} Mpc).

The correlation function has been applied both to the distribution of individual galaxies and to the distribution of rich (Abell) clusters. Surprisingly, the value of γ still remains

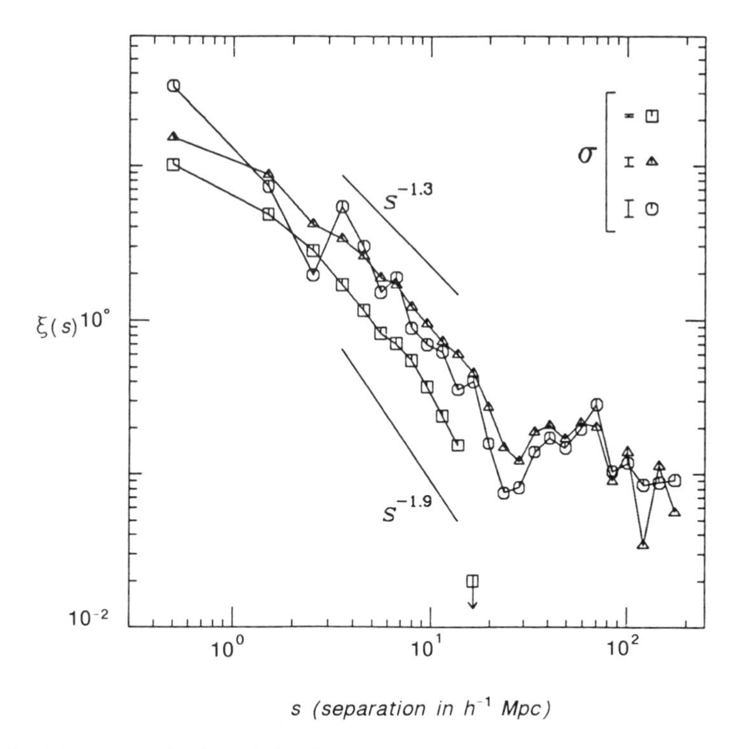

Fig. 6.3. An example of correlation function, as determined from the CfA slices. The different curves reflect different weighting schemes. (Reproduced with permission from M. Geller and the *Astrophysical Journal* (**332**, 44, 1988).)

near 1.8. The correlation length for clusters is approximately 2,500 km/s ($25h^{-1}$ Mpc). For example, the major survey by Huchra and collaborators reports 2,100 km/s, although the Oxford group report a value of 1,400 km/s for their APM clusters. The correlation function remains positive up to some 5,000 km/s.

In keeping with the scope of this book, our discussion here is restricted to correlations within three-dimensional space. It is possible to work with an 'angular' correlation function that is concerned with the angles between pairs of galaxies on the sky; there are, indeed, numerous papers in the literature. However, since such correlation functions take no account of distance – and may therefore pair a near galaxy with a distant one – they are far less informative. They may perhaps serve as a preliminary analysis.

The correlation function has been the workhorse of galaxy distributions. It is a fundamental measure of the degree of clustering. Nevertheless it is all to do with separation of galaxies irrespective of spatial directions. Consequently, it is totally insensitive to texture. The recognition of voids and walls in the large-scale distribution of galaxies has all come about without any perturbations or changes in the work on correlation functions. Useful as it is, the correlation function was conceived on the basis that there was no meaningful texture to the Universe that had to be assessed; only a degree of clustering. It is still a fundamental measure of the fashion in which galaxies are separated, but it does not reflect the character of large-scale structures.

6.3 SEGREGATION

A consistent finding reported in the literature is that the correlation function is different for early type high-luminosity galaxies than it is for late-type low-luminosity galaxies. On scales less than about 2,500 km/s, the former typically have $\gamma = 1.85$, whereas the latter have $\gamma = 1.65$. Early-type galaxies are more clustered than late types. This endorses the well known fact that rich clusters tend to contain mainly elliptical and SO galaxies, rather than spirals. However, that the effect persists to a scale larger than that of the clusters is interpreted to imply that such 'morphological segregation' is fundamental (see Section 7.11, which deals with galaxy biasing), and not an outcome of environmental effects. No luminosity segregation appears to occur on scales larger than 2,500 km/s. Nevertheless, the effect is mild, such that slightly conflicting opinions have been expressed. A. Iovino and collaborators have claimed that morphology and luminosity are independent parameters that affect the clustering properties. Magnitude-limited samples, which favour elliptical and SO galaxies are similarly found to be more clustered than diameter-limited samples that emphasise spirals.

Special investigations have focused on the characteristics of the distribution of dwarf and low surface brightness galaxies. Only mild morphological segregation has been reported, compared with bright luminous galaxies, confirming the visual impression that such galaxies do not fill in the voids.

6.4 POWER SPECTRA

The equivalent information in a correlation function can also be expressed as a power spectrum, which since the late 1980s has become more fashionable. Mathematically, the power spectrum is the Fourier pair of the correlation function (as *frequency* is the Fourier pair of *time*; for example, the Fourier transform of a segment of a piece of music would readily reveal the discrete set of frequencies and overtones of the musical instruments used). Fourier transforms have proved a powerful tool elsewhere in astronomy, for instance in revealing the discrete frequencies of vibration of variable stars. Computer routines are readily available.

The ordinate of the Fourier transform of a spatial dimension is the 'wave number', usually denoted by k. A small wave number therefore reflects the largest structures in the data, and the high wave numbers reflect the smallest structures present. An example of a power spectrum is shown in Figure 6.4. Power spectra normally peak at a certain value. In the figure, this is around a wave number of 0.05h Mpc^{-1}. At higher wave numbers, it can be seen that the spectrum slope is close to uniform. Below the peak the spectrum declines towards 0.01 Mpc^{-1}.

Different sets of data give slightly different peaks. For example, the 12,000 km/s periodicity of the BEKS peaks (described in Section 5.6) ought to give a distinct peak at 0.05h Mpc^{-1} in the power spectrum. This is in fact found by J. Einasto and collaborators in a power spectrum based on the spatial positions of 869 clusters. There they report a 'well-defined' peak at 0.052 Mpc^{-1} which corresponds to the 12,000 km/s periodicity.

The spatial dimension processed by the Fourier transform is not infinite, but is a 'window'. As with the variable star data, the window may introduce apparently spurious signals in the transform. However, these can be calculated and removed from the spectra.

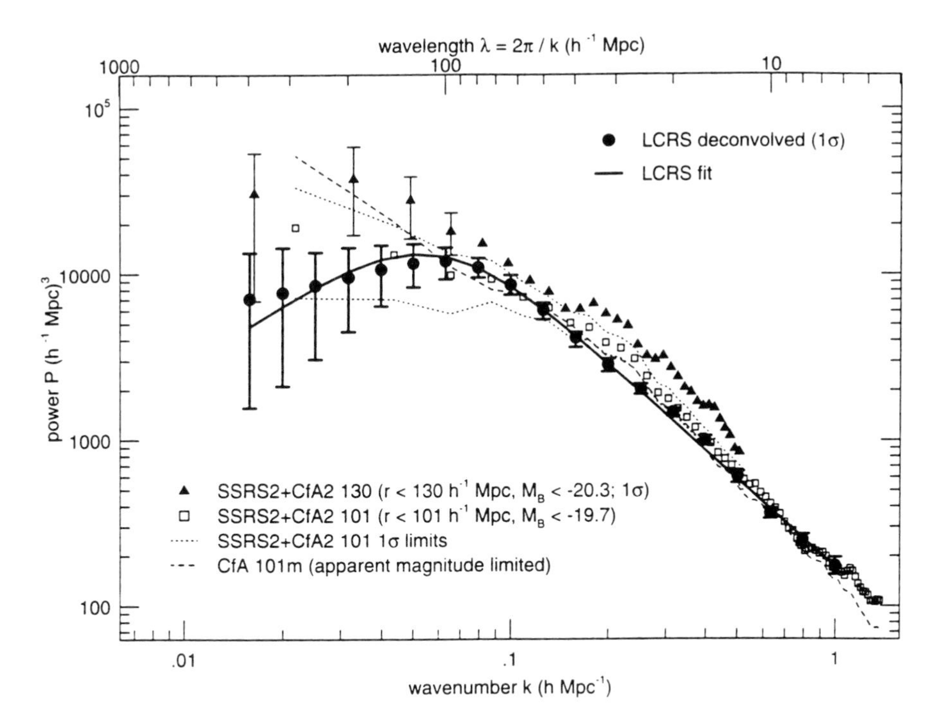

Fig. 6.4. The power spectrum of the Las Campanas Redshift Survey is shown as a solid line, in comparison with the power spectra of shallower surveys. (Reproduced with permission of R. Kirshner and the *Astrophysical Journal* (**471**, 617, 1996).)

Changbom Park and others have developed methods to calculate the power spectrum directly from the distribution of a sample of N galaxies:

$$P(k) = \left(< |\delta_k|^2 > - \frac{1}{N} \right) \Big/ \sum_k |M_k|^2$$

where

$$\delta_k = \frac{1}{N} \sum_{j=1}^{N} e^{ik.x_j} - M_k$$

The first term on the right is the Fourier transform of the galaxy distribution, and the second term is the Fourier transform of the survey region (which is then subtracted). Michael Vogeley and Alexander Szalay have also proposed alternative approaches and improvements.

Correlation functions and power spectra are the most common tools employed to compare real data with simulations, such as the variety of 'cold' or 'hot dark matter' models that will be discussed in Chapter 9.

6.5 COUNTS IN CELLS

An alternative, but roughly equivalent, method to the correlation function, is 'counts in cells'. As the name suggests, a rectilinear three-dimensional grid of cells is superimposed on the data, and the numbers of galaxies in each cell counted. The conventional statistical variance σ^2 is then calculated. The process is carried out repetitively for varying cell sizes, so that σ^2 is measured as a function of the cell size. If the cells are cubical with each side l, then the variance is related to the two-point correlation function as follows.

$$\sigma^2(l) = \frac{1}{V^2} \int_{V=l^3} \xi(r_{12}) dV_1 dV_2$$

Typical values are $\sigma^2 = 0.25$ for $l = 30\ h^{-1}$ Mpc, with a power law decline similar to the correlation function. As with the correlation function, counts in cells are also affected by the limited volume and geometry of the survey, as well as 'shot noise'

With cubical cells, this method is sufficiently similar to the correlation function as to have similar limitations; for example, not reflecting the texture of the large-scale distribution. However, because the voids that occur in that distribution are mainly spherical, the method has been found to depend on the shape of the cells. The variance is less for elongated cells, since voids are less likely to fit inside them.

The use of standard approaches of mathematical statistics opens the door to more sophisticated analyses, as might be carried out by expert statisticians. There are many papers of this nature in the literature.

6.6 THE VOID PROBABILITY FUNCTION

Another statistical tool, established in the literature, is the 'void probability function' (VPF): the probability $P_0(V)$ of finding no galaxies in a test sphere of volume V, which was derived mathematically by Simon White from the correlation functions. Again, in practice, comparison must be made against a random distribution, for which the void probability function reduces to its simplest case

$$P_0 = e^{-nV}$$

where n is the mean number density and V is the volume.

After initial applications to two-dimensional situations, S. Maurogordato, F. Bouchet and M. Lachièze–Rey extended the VPF to the three-dimensional distribution of galaxies. They and others have used this function to establish that there were more voids present in various redshift samples than would occur in a random one. B. Ryden and A. Melott have pointed out that the VPF is enhanced in redshift space, compared with normal space.

6.7 THE VOID-SIZE SPECTRUM

A straightforward comparison between theory and observational data can simply be the spectrum: a histogram of the relative numbers of voids according to size. In practice, this means that a standard way of recognising voids must be established. Since voids occur within a random data set, it is necessary to establish which of them are statistically significant, over those that would occur in the random distribution. The void sample should then be volume-limited. In reality, different-sized voids can be discerned (to the same statistical significance) out to different distances, but this can be accommodated.

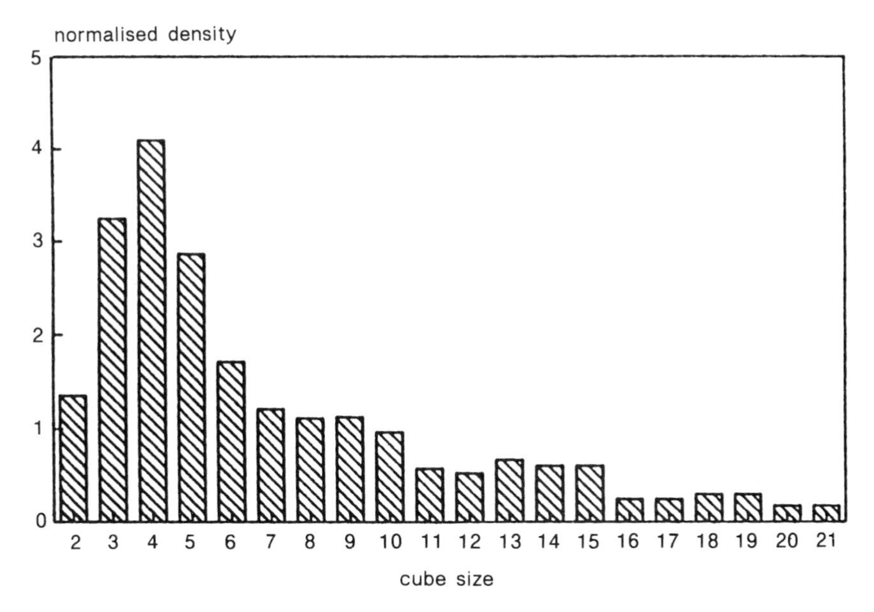

Fig. 6.5. A spectrum of the normalised density of voids against void size. The horizon scale is in units of 200 km/s, so the peak corresponds to 800 km/s.

The first measurement of such a spectrum, as illustrated here in Figure 6.5, was probably carried out by Guinevere Kauffmann and the author. For this work a large number of randomly generated databases (random, but incorporating the same radial selection effects as the real data) were generated. Then the number of voids occurring in the real sample could be compared against the random sample. This work also produced a list of statistically significant voids (included earlier in Table 4.10).

The question as to whether voids themselves cluster has also been addressed, with S. Haque-Copilah reporting a slight tendency on a characteristic scale of 5,900 km/s. However, since this is comparable to the sampling depth, some caution must be exercised, and it must also be seen in the light of the findings (see Chapter 4) that voids are in any case ubiquitous.

6.8 LUMINOSITY FUNCTIONS

The luminosity function conveys the number density of galaxies per interval of absolute magnitude (intrinsic luminosity). As introduced in Section 3.4.2, its tendency towards a universal form – the Schechter function – is the rationale behind selecting galaxies for redshift surveys according to limits in apparent magnitude. It is the interconnecting relation between the sample of galaxies observed and the general population of galaxies.

It is possible to extract the luminosity function from the data of a magnitude-limited redshift survey, where for a certain region of the sky, all galaxies down to a fixed apparent magnitude have been observed, and therefore the apparent magnitudes of those galaxies are known. Such a procedure is a check on the universality (or otherwise) of the luminosity function. While it is not in itself a measure of the spatial distribution of galaxies, and therefore appropriate to this chapter, it does, as we shall see presently, show luminosity segregation in space.

The derivation of a luminosity function is reasonably complicated, and we shall confine our discussion to essential elements. Assuming redshift to represent distance, the absolute magnitudes (at a specified colour) of each of the observed galaxies can be determined by:

$$M = m - 5\log\frac{V}{100h} - 25$$

As before, m is the apparent magnitude, V is the velocity of recession, and the Hubble constant = 100h km/s per Mpc.

The same formula implies that galaxies, of a particular absolute magnitude, are visible out to a limiting distance, as given by:

$$d = 10^{\left(\frac{m_0 - M}{5} - 5\right)}$$

where M is the absolute magnitude, m_0 is the limiting magnitude of the survey and d is the limiting distance in Mpc. Thus the number of galaxies, within an interval of absolute magnitude, can be counted. The survey volume is obtained from the limiting distance, and consequently the spatial density of the galaxies (one point on the luminosity function) is derived. The procedure is then repeated for different intervals of absolute magnitude, and a luminosity function derived.

In determining the luminosity function over a range of even five absolute magnitudes, the limiting distance varies by a factor of ten, and the volume sampled therefore varies by a thousand. Spatial inhomogeneities between the sampled volumes can therefore cause major problems; for example, a distant overdensity in the sample can cause far too many high-luminosity galaxies to be counted per unit volume. A nearby void or underdensity causes far too few low-luminosity galaxies to be counted.

Techniques have been developed to address this problem. One in particular (used by V. de Lapparent, M. Geller and J. Huchra) follows how the ratios of the counts, rather than the actual counts, of galaxies of different luminosity vary, according to

$$Y(M)\delta M = \frac{N(\le M) - N(\le M - \delta M)}{N(\le M)}$$

A differential form of the luminosity equation can then be obtained.

As introduced in Section 3.4.2, the form of the luminosity function can be fitted to a Schechter function:

$$\phi(M) = (0.4\ln 10)\phi^*\left(10^{0.4(M^*-M)}\right)^{1+\alpha} \exp\left(-10^{0.4(M^*-M)}\right)$$

The differential form allows a good determination of M* and α, but only an approximate evaluation of the space density ϕ^*. Probably the most accurate determination of the luminosity function (in the red) is that extracted from the Las Campanas Redshift Survey, which probes a volume as deep as 60,000 km/s, and has the advantage of a very large number (18,000) of galaxies. It is shown in Figure 6.6, and the following parameters are extracted:

$$M^* = -20.29 \pm 0.02 + 5 \log h$$
$$\alpha = -0.70 + 0.05$$
$$\phi = 0.019 \pm 0.01 \; h^3 \, \text{Mpc}^{-3}$$
(For $M - 5 \log h \leq -17.5$, the mean galaxy density is $0.029 \pm 0.002 \; h^3 \, \text{Mpc}^{-3}$)

As another example, working in the green J band, the Stromlo–APM Redshift Survey finds $M^* = -19.5 \pm 0.13$; $\alpha = -0.97 \pm 0.15$.

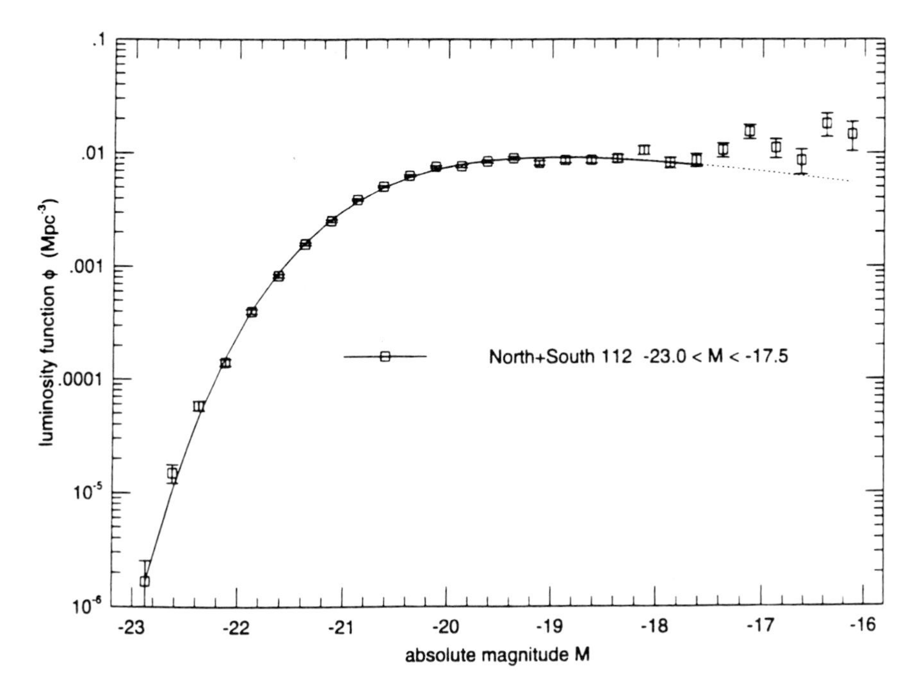

Fig. 6.6. The luminosity function of the Las Campanas Redshift Survey. (Reproduced with permission of R. Kirshner and the *Astrophysical Journal* (**464**, 60, 1996).)

A further complication in determining the luminosity function is caused by uncertainties in determining the apparent magnitudes. This is particularly applicable to the wide-angle Center for Astrophysics surveys, which have depended on the relatively unsophisticated Zwicky magnitudes. Errors of 0.35 apparent magnitudes can affect M^* by 0.3 and α by 0.1.

There are also concerns (naturally enough) with the faint end of the function, since some investigators suspect there is a major population of dwarf galaxies, with α possibly rising to −2.2. The luminosity function is also susceptible to evolutionary effects, which must be taken into account for deep surveys.

The luminosity function has also been used to examine morphological segregation. Irregular and dwarf galaxies, being of lower luminosity, have a steeper luminosity function.

By contrast, the luminosity function of only the 'bulge components' of galaxies seems more universal.

6.9 TOPOLOGY

As has been remarked, none of the quantitative measures discussed above have really addressed the 'labyrinth' of interconnected structures – the true character of large-scale structures. By contrast, 'topology' – introduced by Richard Gott and collaborators – deals directly with the nature of the labyrinth, rather than of the separations of individual galaxies.

At the outset, the observed distribution must be heavily smoothed (usually to a scale greater than the correlation length) so the large-scale structures (the walls etc.) take on three-dimensional forms. We would then find topological forms that might vary between two extremes – aptly described as 'meatball' and 'Swiss cheese'.

'Meatball' topology implies that the high-density regions – concentrations of coalesced galaxies – form isolated islands in three-dimensional space that is otherwise empty. The opposite is 'Swiss cheese', where isolated voids exist within a high-density medium.

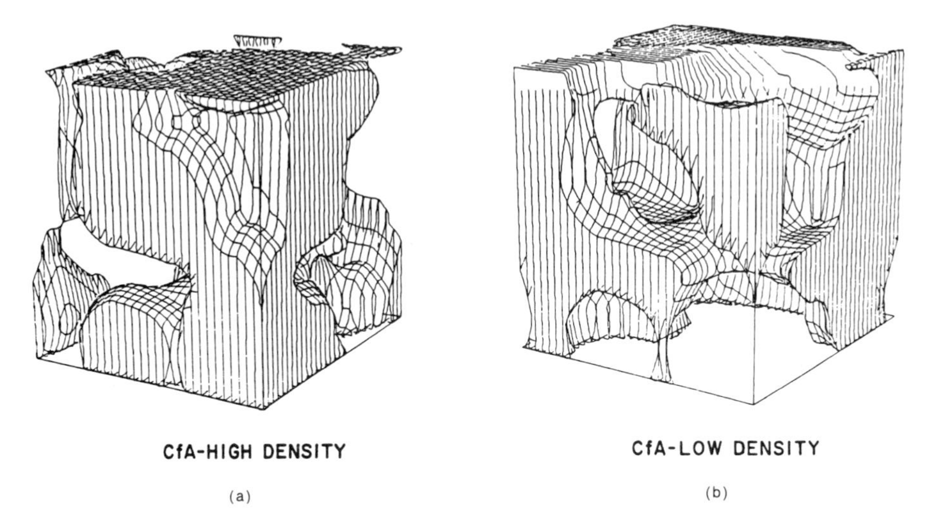

Fig. 6.7. These three-dimensional plots reveal high-density and low-density regions in a heavily smoothed version of the CfA1 survey, allowing the 'genus' to be determined. (Reproduced with permission from R. Gott III and the *Astrophysical Journal* (**306**, 341, 1986).)

The true topology of the large-scale structures – such as that shown in Figure 6.7 – falls between these extremes, and is described as 'sponge-like', in that it closely resembles a bath sponge. The elastic material that forms the sponge is completely interconnected; it has no isolated regions, or they would simply fall off. Similarly the voids are all interconnected, otherwise the bathwater could not be squeezed out. In the distribution of galaxies, it is possible to move from one high-density region to another without ever leaving such an environment. Similarly, it is possible to travel from one void through emptiness to any other void.

Gott uses a nice illustration: If the bath sponge could be filled with liquid cement that was allowed to set, and some acid then used to dissolve away the original sponge material, the resulting cement sponge would reflect the same texture as the original.

'Genus' is a quantitative measure of topology. It can be defined as the number of holes that occur minus the number of additional isolated surfaces. For example, a torus by itself has one hole (+1) and no separate isolated surfaces (0) and a genus of +1. Two separate spheres have no holes (0), but one extra surface (−1) and a genus of −1. 'Meatball' topology, having no holes, but lots of additional surfaces, has very negative genus. 'Swiss cheese' topology is full of holes and has positive genus. True sponge-like topology has zero genus.

Genus analyses have been made on most of the established controlled redshift surveys. Since almost all are magnitude-limited and not volume-limited, there is some debate as to how the smoothing ought to be done; it has to be varied with increasing redshift. There is also the problem that gross undersmoothing will create negative genus and gross oversmoothing will create positive genus. The smoothing must be scaled to the characteristic size of the large-scale structures. Such analyses have revealed genus values centred around zero, but perhaps very slightly negative. This implies 'sponge-like' topology, but with a slight tendency towards 'meatball'.

6.10 VORONOI TESSELLATION

The texture of the distribution of galaxies has often been likened to 'soapsuds'. The analogy goes further. Soapsuds normally begin as a 'wet foam' where the bubbles are spherical and there is plenty of liquid contained in the interface volumes. However, as gravity drains the liquid out, the foam changes to a 'dry foam' where the bubbles retain only approximate spherical forms as convex polyhedrons. The interfaces between bubbles are thin planar double-sided surfaces. The quantity of soapy liquid has been minimised, without destroying individual bubbles. Mathematically, such a situation is known as 'Voronoi tessellation'.

Since the voids in the distribution of galaxies are approximately spherical yet separated from their neighbours by interfaces that are roughly planar, Voronoi tessellation has been seen as a possible model for the cosmic texture. Examples are shown in Figures 6.8 and 6.9.

A tessellated structure can be constructed as follows. A set of points (usually a random distribution) can be established in three-dimensional space. For every adjacent pair of points, a planar surface is generated that is a perpendicular bisector to the line joining the two points. The surface is terminated where it meets up with those generated by neighbouring pairs of points. The vertices, where three such surfaces meet, define a set of points in space. R. van der Weygaert has found that the correlation function derived for these points shows similar power-law behaviour to that of the galaxies and clusters of galaxies. This has been seen to indicate that tessellation is a viable model. It has also led to the belief that clusters of galaxies occur at the vertices of walls, which is not generally true in the real data.

Voronoi tessellation is a popular model for cellular structures. For instance, it has been used to explore the periodicities in redshift found by Broadhurst and colleagues, as described in the previous chapter. It is found that of various one-dimensional 'pencil beams' extended from points in Voronoi foams, about 15 per cent would encounter periodic structure.

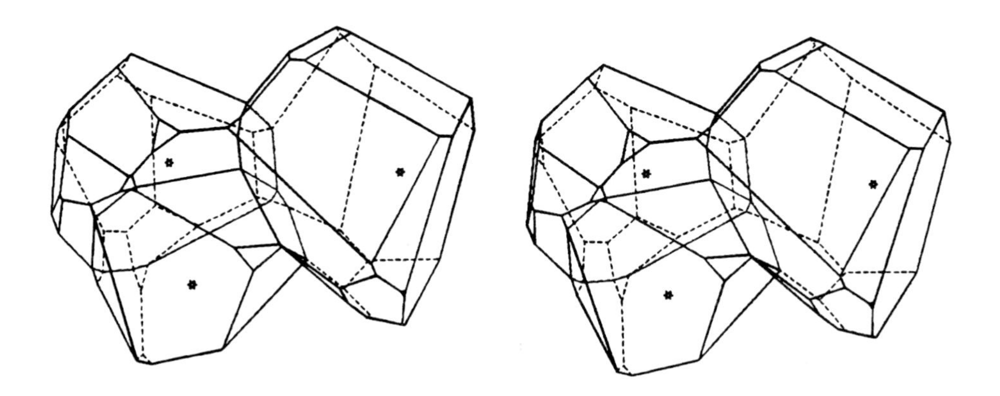

Fig. 6.8. Voronoi tessellation cells. This is a stereoscopic pair of images; if the reader's left eye is directed at the left image, and the right eye at the right image, then the cells will be seen in three dimensions. The stars represent the 'nuclei' (see text and following figure). (Reproduced with permission from R. van de Weygaert and *Astronomy and Astrophysics* (**213**, 1, 1989).)

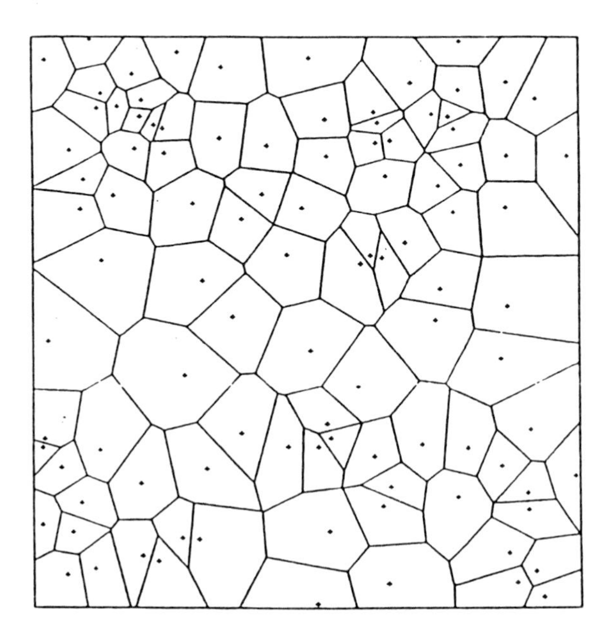

Fig. 6.9. A two-dimensional Voronoi tessellation is generated from a random set of points (dots). Cell walls are perpendicular bisectors to the lines between the points. (Reproduced with permission from P. Coles and *Nature* (**346**, 199). Copyright (1990) Macmillan Magazines Ltd.)

6.11 FRACTALS

The similarity in the character of structures on a small scale to a large scale has led some investigators to consider a fractal Universe, where the same pattern, or same structure, re-occurs 'independent' of scale. Many readers will already be familiar with fractal geometric patterns; for example, a tree-like structure that branches to ever smaller scales. An enlargement of the smallest structure would match that of the largest structure.

If fractal structure exists, then it is constrained between upper and lower limits; the lower limit provided by the average separation of the participating galaxies and the upper limit by the size of the volume sampled or the scale at which the Universe becomes homogeneous.

A fractal dimension D can be defined, according to Mandelbrot:

$$D = \frac{d(\log(N(r)))}{d(\log(r))}$$

where N(r) is the number of objects within a scale r. The value of D can vary as a function of r. As long as D holds close to a particular value, within a range of r, then the distribution has a fractal structure within that range. For example, a random distribution is fractal, since it is independent of scale and has an unchanging fractal dimension of 3.

The distribution of galaxies would be expected to show fractal dimensions in the region of 1.2 to 1.4, but various investigations have shown very different values, with very different ranges.

Whether or not fractal structure exists, it would in any case present a major problem in interpretation. We shall later discuss (in Chapter 9) the role of gravity in shaping structures, the outcome of which is very definitely different for different scales. An alternative 'physical' theory that operates identically, irrespective of scale, is quite unpalatable. Nevertheless, fractal dimension is a useful analytical tool. For instance, it has shown variation according to different galaxy luminosity – another indication of segregation effects.

6.12 DIPOLES AND QUADRUPOLES

Where samples of galaxies cover the entire sky, number counts in redshifts can also be interpreted in terms of spherical harmonics, of which dipoles and quadrupoles represent the simplest cases. For example, the distribution of IRAS galaxies has shown a dipole effect – an excess of galaxies on one side of the sky, the direction of which is extracted by mathematical analysis. Such a dipole roughly agrees with that extracted from the cosmic microwave background, the basis for discussion in Chapter 7.

6.13 PERCOLATION AND LEAST SPANNING TREES

Many of the quantitative measures so far discussed tell a different story from that which the eye seems to see in the large-scale distribution of galaxies. The eye tends to pick out individual structures such as the great walls described in Chapter 4. It would therefore be desirable to have a mathematical procedure that does much the same as the eye. Clearly, the eye connects 'dot-to-dot' until a coherent structure can be recognised. Percolation algorithms closely resemble this. Starting from an initial galaxy in the data set, a sphere is expanded to locate the nearest neighbour. A connecting line between the two galaxies is noted and the routine repeated from the neighbouring galaxy. Whilst this would soon interconnect all the galaxies in the data, imposing a limit to the size of percolation spheres limits the extent of the connections identified, such as the 200 km/s limit used in Chapter 4 to delineate large-scale structures. As the data thins with increasing distance, it is normally necessary to adjust the limiting radius of the percolation spheres with distance (though this was not done in Chapter 4). Clearly all sorts of variations are possible; for example, constraining the directions of the interconnections.

The 'least spanning tree' or 'minimal spanning tree' is an alternative means of analysis of a distribution of points, such as the distribution of galaxies. As before, the points are interconnected by lines, but by a sequence of branches radiating from a central concentration. The pattern has to be tree-like, such that no closed circuits of lines are included. A computer then explores all possible tree structures to find that which gives the minimum value to the sum of the lengths of the interconnecting lines. In this way, the minimal spanning tree finds and defines the skeletal structure. The system works very much like the human brain in picking out structures, but there is nothing subjective about it.

The minimal spanning tree serves to define a set structure, which can then be explored by mathematical analysis (moment of inertia, etc.).

6.14 OTHER APPROACHES

Further mathematical techniques are being advanced as suitable for the analysis of large-scale structures. One promising method involves 'wavelet transformations' which may give a rigorous way of detecting features, and will provide positions, dimensions and probabilities.

The use of various techniques has, over the years, seen changes as new approaches have passed in and out of fashion. Such trends show that we have by no means agreed on a common consensus as to what is best. Correlation functions, power spectra and the like are well entrenched in the literature, but cannot even begin to represent the textures described qualitatively in the opening section to this chapter.

6.15 FURTHER READING

Specialised

General

Sahni, V. and Coles, P., [Section 8 of] Approximation methods for non-linear gravitational clustering, *Phys. Rep.,* 262, 1 (1995).

Correlation functions

Alimi, J.-M., *et al.*, A cross-correlation analysis of luminosity segregation in the clustering of galaxies, *Astron. Astrophys.,* **206**, L11 (1988).

Bahcall, N.A. and Soneira, R.M., The Spatial Correlation Function of Rich Clusters of Galaxies, *Astrophys. J.*, **270**, 20 (1983).
Blanchard, A. and Alimi, J.-M., Practical determination of the spatial correlation function, *Astron. Astrophys.*, **203**, L1 (1988).
Boerner, G., *et al.*, The two-point correlation functions of galaxies with different luminosities, *Astron. Astrophys.*, **209**, 1 (1989).
Bonometto, S.A., *et al.*, Correlation functions from the Perseus–Pisces Redshift Survey, *Astrophys. J.*, **419**, 451 (1993).
Börner, G. and Mo, H.J., On the two-point correlations of galaxies in the CfA survey, *Astron. Astrophys*, **227**, 324 (1990).
Börner, G., *et al.*, Correlation functions of galaxies with different weightings according to luminosity and mass, *Astron. Astrophys.*, **221**, 191 (1989).
Boschán, P. *et al.*, On the accurate determination of the clustering hierarchy of galaxies, *Astrophys. J. Suppl.*, **93**, 65 (1994).
Cappi, A. and Maurogordato, S., The spatial distribution of nearby galaxy clusters in the northern and southern galactic hemispheres, *Astron. Astrophys.*, **259**, 423 (1992).
Davis, M., *et al.*, On the Universality of the Two-Point Galaxy Correlation Function, *Astrophys. J.*, **333**, L9 (1988).
de Lapparent, V. *et al.*, The Mean Density and Two-Point Correlation Function for the CfA Redshift Survey Slices, *Astrophys. J.*, **332**, 44 (1988).
Fisher, K.B. *et al.*, Clustering in the 1.2-Jy IRAS Galaxy Redshift Survey – II. Redshift distortions and *n*(*r*p, *p*), *Mon. Not. R. astr. Soc.*, **267**, 927 (1994).
Fisher, Karl B., *et al.*, Clustering in the 1.2-Jy IRAS Galaxy Redshift Survey – I. The redshift and real space correlations functions, *Mon. Not. R. astr. Soc.*, **266**, 50 (1994).
Ghigna, S., *et al.*, Deviations from Hierarchical Clustering in Real and Redshift Space, *Astrophys. J.*, **463,** 395 (1996).
Hamilton, A.J.S., Evidence for Biasing in the CfA Survey, *Astrophys. J.*, **331**, L59 (1988).
Hermit, S., *et al.*, The two-point correlation function and morphological segregation in the Optical Redshift Survey, *Mon. Not. R. astr. Soc.*, **283**, 709 (1996).
Klypin, A.A., *et al.*, Analysis of 'Coma strip' galaxy redshift catalog, *Mon. Not. R. astr. Soc.*, **246**, 193 (1990).
Loveday, J., *et al.*, The Stromlo–APM Redshift Survey II. Variation of Galaxy Clustering with Morphology and Luminosity, *Astrophys. J.*, **442**, 457 (1995).
Ramella, M., *et al.*, The Two-Point Correlation function for groups of Galaxies in the Center for Astrophysics Redshift Survey, *Astrophys. J.*, **353**, 51 (1990).

Segregation

Benoist, C., *et al.*, Biasing in the Galaxy Distribution, *Astrophys. J.*, **472**, 452 (1996).
Börner, G and Mo, H., Geometrical analysis of galaxy clustering: dependence on luminosity, *Astron. Astrophys.*, **223**, 25 (1989).
Domíngues-Tenreiro, R. *et al.*, Large-Scale Morphological Segregation in optically selected galaxy redshift catalogs, *Astrophys. J.*, **424**, L73 (1994).
Einasto, M., Structure and formation of superclusters – XII. Morphological and luminosity segregation of normal and dwarf galaxies, *Mon. Not. R. astr. Soc.*, **250**, 802 (1991).

Hasegawa T. and Umemura, M., Luminosity dependence of galaxy clustering in extinction-corrected CfA data, *Mon. Not. R. astr. Soc.,* **263**, 191 (1993).
Iovino, A. *et al.*, Galaxy clustering, morphology and luminosity, *Mon. Not. R. astr. Soc.,* **265**, 21 (1993).
Mo, H.J. *et al.*, Can morphological segregations of galaxies exist on 10 h^{-1} Mpc scales?, *Mon. Not. R. astr. Soc.,* **255**, 382 (1992).
Santiago, B. and Strauss, M.A., Large-Scale Morphological Segregation in the Center for Astrophysics Redshift Survey, *Astrophys. J.,* **387**, 9 (1992).
Santiago, B.X. and da Costa, L. N., Segregation Properties of Galaxies, *Astrophys. J.,* **362**, 386 (1990).
Solanes, J.M., *et al.*, The Luminosity of Galactic Components and Morphological Segregation, *Astron. J.*, **98**(3), 798 (1989).
Thuan, T. X. *et al.*, Northern Dwarf and Low Surface Brightness Galaxies. IV. The Large-Scale Space Distribution, *Astrophys. J.,* **370**, 25 (1991).

Power spectrum

Baugh, C.M., Large-scale fluctuations in the distribution of galaxies, *Mon. Not. R. astr. Soc.,* **282**, 1413 (1996).
da Costa, L.N. *et al.*, The Power Spectrum of Galaxies in the Nearby Universe, *Astrophys. J.,* **437**, L1 (1994).
Fong, R. *et al.*, Correlation function constraints on large-scale structure in the distribution of galaxies, *Mon. Not. R. astr. Soc.,* **257**, 650 (1992).
Gramann, M. and Einasto, J., The Power Spectrum in Nearby Supercluster, *Mon. Not. R. astr. Soc.,* **254**, 453 (1992).
Labini, F.S. and Amendola, L., The Power Spectrum in a strongly inhomogenous Universe, *Astrophys. J.,* **468**, L1 (1996).
Landy, S.D. *et al.*, The Two-Dimensional power spectrum of the Las Campanas Redshift Survey: Detection of Excess Power on 100 h–1 Mpc Scales, *Astrophys. J.,* **456**, L1 (1996).
Lin, H. *et al.*, The Power Spectrum of Galaxy Clustering in the Las Campanas Redshift Survey, *Astrophys. J.,* **471**, 617 (1996).
Newman, W.I. *et al.*, Redshift Data and Statistical Inference, *Astrophys. J.,* **431**, 147 (1994).
Park, C. *et al.*, Large-Scale Structure in the Southern Sky Redshift Survey, *Astrophys. J.,* **392**, L51 (1992).
Park, C. *et al.*, Power Spectrum, Correlation function, and tests for luminosity bias in the CfA Redshift Survey, *Astrophys. J.,* **431**, 569 (1994).
Peacock, J.A. and West, M.J., The power spectrum of Abell cluster correlations, *Mon. Not. R. astr. Soc.,* **259**, 494 (1992).
Schuecker, P., *et al.*, The Muenster Redshift Project. I. Simulations of power spectra and analytical corrections, *Astrophys. J.,* **459**, 467 (1996).
Schuecker, P. *et al.*, The Muenster Redshift Project. II. The Redshift Space Galaxy Power Spectrum, *Astrophys. J.,* **472**, 485 (1996).
Tegmark, M., A method for extracting maximum resolution power spectra from galaxy surveys, *Astrophys. J.,* **455**, 429 (1995).

Vogeley, M.S. and Szalay, A.S., Eigenmode Analysis of Galaxy Redshift Surveys. I. Theory and Methods, *Astrophys. J.*, **465**, 34 (1996).
Vogeley, M.S. *et al.*, Large-Scale Clustering of Galaxies in the CfA Redshift Survey, *Astrophys. J.*, **391**, L5 (1992).

Counts in cells

Alimi, J-M. *et al.*, Nonlinear Clustering in the CfA Redshift Survey, *Astrophys. J.*, **349**, L5 (1990).
Bouchet, F.R. *et al.*, Moments of the Counts Distribution in the 1.2 Jansky IRAS Galaxy Redshift Survey, *Astrophys. J.*, **417**, 36 (1993).
de Lapparent, V. *et al.*, Measures of Large-Scale Structure in the CfA Redshift Survey Slices, *Astrophys. J.*, **369**, 273 (1991).
Efstathiou, G. *et al.*, Large-scale clustering of IRAS galaxies, *Mon. Not. R. astr. Soc.*, **247**, 10P (1990).
Efstathiou, G., Counts-in-cells comparisons of redshift surveys, *Mon. Not. R. astr. Soc.*, **276**, 1425 (1995).
Elizalde, E. and Gaztanaga, E., Void probability as a function of the void's shape and scale-invariant models, *Mon. Not. R. astr. Soc.*, **254**, 247 (1992).
Gaztanaga, E., *N*-Point Correlation Functions in the CfA and SSRS Redshift Distribution of Galaxies, *Astrophys. J.*, **398**, L17 (1992).
Loveday, J. *et al.*, Large-scale structure in the universe: results from the Stromlo–APM redshift survey, *Astrophys. J.*, **400**, L43 (1992).
Maurogordato, S *et al.*, The large-scale galaxy distribution in the Southern Sky Redshift Survey, *Astrophys. J.*, **390**, 17 (1992).
Szapudi, I. and Colombi, S., Cosmic Error and Statistics of Large-Scale Structure, *Astrophys. J.*, **470**, 131 (1996)
Szapudi, I., *et al.*, Higher Order Statistics from the Edinburgh/Durham Southern Galaxy Catalogue Survey. I. Counts in Cell, *Astrophys. J.*, **473**, 15 (1996).

Void probability

Betancort-Rijo, J., Probabilities of Voids, *Mon. Not. R. astr. Soc.*, 246, 608 (1990).
Bouchet, F.R. and Lachièze-Rey, M., Voids in the Center for Astrophysics Catalog, *Astrophys. J.*, **302**, L370 (1986).
Einasto, J. *et al.*, Structure and formation of superclusters – XIII. The void probability function, *Mon. Not. R. astr. Soc.*, **248**, 593 (1991).
Ghigna, S. *et al.*, Size of Voids as a Test for Dark Matter Models, *Astrophys. J.*, **437**, L71 (1994).
Ghigna, S. *et al.*, Void Analysis as a Test for Dark Matter Composition?, *Astrophys. J.*, **469**, 40 (1996).
Lachiéze-Rey, M. *et al.*, Void Probability Function in the Southern Sky Redshift Survey, *Astrophys. J.*, **399**, 10 (1992).
Maurogordato, S. *et al.*, Void probabilities: behaviour with depth of survey in 2D catalogs, *Astron. Astrophys.*, **206**, L23 (1988).
Ryden, B.S. and Melott, A.L., Voids in Real Space and in Redshift Space, *Astrophys. J.*, **470**, 160 (1996).

Sheth, R.K., Random dilutions, generating functions, and the void probability distribution function, *Mon. Not. R. astr. Soc.*, **278**, 101 (1996).
Vogeley, M. *et al.*, Void Statistics of the CfA Redshift Survey, *Astrophys. J.*, **382**, 44 (1991).
Vogeley, M.S. *et al.*, Voids and Constraints on Nonlinear Clustering of Galaxies, *Astron. J.*, **108,** No. 3, 745 (1994).
Watson, J.M. and Rowan-Robinson, M., The void probability function for flux-limited samples, *Mon. Not. R. astr. Soc.*, **265**, 1027 (1993).

Void sizes

Buryak, O.E. *et al.*, Deep Galactic Surveys and the Structure of the Universe, *Astrophys. J.*, **383**, 41 (1991).
Haque-Copilah, S. and Basu, D., Do Voids Cluster?, *Publ. astr. Soc. Pacific*, **106**, 67 (1994).
Kauffmann, G. and Melott, A.L., The void spectrum in two-dimensional numerical simulations of gravitational clustering, *Astrophys. J.*, **393**, 415 (1992).
Kauffmann, G., and Fairall, A.P., Voids in the distribution of galaxies: an assessment of their significance and derivation of a void spectrum, *Mon. Not. R. astr. Soc.*, **248,** 313 (1990).
Ryden, B.S., Measuring q_0 from the distortion of voids in redshift space, *Astrophys. J.*, **452**, 25 (1995).

Luminosity function

de Lapparent, V. *et al.*, The Luminosity Function for the CfA Redshift Survey Slices, *Astrophys. J.*, **343**, 1 (1989).
de Propris, R. *et al.*, Evidence for steep luminosity functions in clusters of galaxies, *Astrophys. J.*, **450**, 534 (1995).
Efstathiou, G. *et al.*, Analysis of a complete galaxy redshift survey – II. The field-galaxy luminosity function, *Mon. Not. R. astr. Soc.*, **232**, 431 (1988).
Franceschini, A. *et al.* Analysis of a complete sample of galaxies at mz < 14.5: the optical, radio and far-infrared luminosity functions, *Mon. Not. R. astr. Soc.*, **233**, 157 (1988).
Kirshner *et al.*, A Deep Survey of Galaxies, *Astron. J.*, **88**, 1296 (1983).
Lilly, S.J. *et al.*, The Canada–France redshift survey: the luminosity density and star formation history of the Universe to z ~ 1, *Astrophys. J.*, **460**, L1 (1996).
Lin, H. *et al.*, The Luminosity Function of Galaxies in the Las Campanas Redshift Survey, *Astrophys. J.*, **464**, 60 (1996).
Marzke, R.O. *et al.*, The Luminosity Function of the CfA Redshift Survey, *Astrophys. J.*, **428**, 43 (1994).
Mobasher, B. *et al.*, A complete galaxy redshift survey – V. Infrared luminosity functions for field galaxies, *Mon. Not. R. astr. Soc.*, **263**, 560 (1993).
Treyer, M.A. and Silk, J., The Faint End of the Galaxy Luminosity Function, *Astrophys. J.*, **436**, L143 (1994).

Topology

Coles, P. *et al.*, Quantifying the topology of large-scale structure, *Mon. Not. R. astr. Soc.*, **281**, 1375 (1996).

Gott III, J.R. *et al.*, The Sponge-like Topology of Large-Scale Structure in the Universe, *Astrophys. J.*, **306**, 341 (1986).

Gott III, J.R. *et al.*, The Topology of Large-Scale Structure III. Analysis of Observations, *Astrophys. J.*, **340**, 625 (1989).

Park, C. *et al.*, The Topology of Large-Scale Structure VI. Slices of the Universe, **387**, 1,(1992).

Vogeley, M.S. *et al.*, Topological Analysis of the CfA Redshift Survey, *Astrophys. J.*, **420**, 525 (1994).

Tessellation and cellular patterns

Coles, P., Understanding recent observations of the large-scale structure of the universe, *Nature,* **346**, 446 (1990).

Goldwirth, D.S. *et al.*, The two-point correlation function and the size of voids, *Mon. Not. R. astr. Soc.*, **275**, 1185 (1995).

Kurki-Suonio, H. *et al.*, Deviation from Periodicity in the Large-Scale Distribution of Galaxies, *Astrophys. J.*, **356**, L5 (1990).

Subba Rao, M.U. and Szalay, S., Statistics of Pencil Beams in Voronoi Foams, *Astrophys. J.*, **391**, 483 (1992).

van der Weygaert, R. and Icke, V., Fragmenting the universe II. Voronoi vertices as Abell clusters, *Astron. Astrophys.*, **213**, 1 (1989).

van der Weygaert, R., Quasi-periodicity in deep redshift surveys, *Mon. Not. R. astr. Soc.*, **249**, 159 (1991).

Fractals

Campos, A. *et al.*, Scaling Analysis of the Galaxy Distribution in the SSRS Catalog, *Astrophys. J.*, **436**, 565 (1994).

Domínguez-Tenreiro, R. and Martínez, V.J., Multidimensional Analysis of the Large-Scale Segregation of Luminosity, *Astrophys. J.*, **339**, L9 (1989).

Martínez, V.J. and Coles, P., Correlations and Scaling in the QDot Redshift Survey, *Astrophys. J.*, **437**, 550 (1994).

Wen, Z. *et al.*, The fractal dimension in the large-scale distribution of galaxies with different luminosities, *Astron. Astrophys.*, **219**, 1 (1989).

Percolation and minimum spanning trees

Babul, A and Starkman, G.D., A Quantitative Measure of Structure in the Three-Dimensional Galaxy Distribution: Sheets and Filaments, *Astrophys. J.*, **401**, 28 (1992).

Bhavasar, S.P. and Splinter, R.J., The superiority of the minimal spanning tree in percolation analyses of cosmological data sets, *Mon. Not. R. astr. Soc.*, **282**, 1461 (1996).

Bhavsar, S.P. and Ling, E.N., Are the Filaments Real?, *Astrophys. J.*, **331**, L63 (1988).

Börner, G. and Mo, H., A percolation analysis of cluster superclustering, *Astron. Astrophys.*, **224**, 1 (1989).
Krzewina, L.G. and Saslaw, W.C., Minimal spanning tree statistics for the analysis of large-scale structure, *Mon. Not. R. astr. Soc.*, **278**, 869 (1996).
Martínez, V.J. and Jones, B.J.T., Why the Universe is not a fractal, *Mon. Not. R. astr. Soc.*, **242**, 517 (1990).

Wiener filtering

Zaroubi, S. *et al.*, Wiener Reconstruction of the Large-Scale Structure, *Astrophys. J.*, **449**, 446 (1995).

Alternative approaches

Escalera, E. and MacGillivray, H.T., Topology in galaxy distributions: method for a multi-scale analysis. A use of the wavelet transform, *Astron. Astrophys.*, **298**, 1 (1995).
Fang, F. and Zou, Z-L., Structure in the Distribution of Southern Galaxies described by the distribution function, *Astrophys. J.*, **421**, 9 (1994).
Kerscher, M. *et al.*, Minkowski functionals of Abell/ACO clusters, *Mon. Not. R. astr. Soc.*, **284**, 73 (1997).
Matsubara, T., Diagrammatic Methods in Statistics and Biasing in the Large-Scale Structure of the Universe, *Astrophys. J. Suppl.*, **101**, 1 (1995).
Moles, M. *et al.*, Taxonomical analysis of superclusters – I. The Hercules and Perseus superclusters, *Mon. Not. R. astr. Soc.*, **213**, 365 (1985).
Nusser, A. *et al.*, Evidence for Gaussian Initial Fluctuations from the 1.2 Jansky IRAS Survey, *Astrophys. J.*, **449**, 439 (1995).
Slezak, E. *et al.*, Objective Detection of Voids and High-Density Structures in the First CfA Redshift Survey Slice, *Astrophys. J.*, **409**, 517 (1993).
Taylor, A.N. and Rowan-Robinson, M., Reconstruction analysis – I. Redshift-space deprojection in the quasi-non-linear regime, *Mon. Not. R. astr. Soc.*, **265**, 809 (1993).

7

Large-scale motions

7.1 LARGE-SCALE STRUCTURES IMPLY LARGE-SCALE MOTIONS

Large-scale structures in the Universe, such as those so far described in this book, are gravitationally 'unrelaxed'; that is, they are still collapsing and drawing in surrounding material. The present-day texture of the Universe shows a very different picture from the almost-smooth early Universe revealed by the cosmic microwave background (as explained in the following chapter). The formation of large-scale structures could have only come about through large-scale movements of matter, and we should be able to measure such motions.

All extragalactic measurements of motion can only be done radially, along our line of sight. Velocities of recession from (or velocities of approach towards) our Galaxy, can be measured accurately according to the Doppler shift (as outlined in Chapter 2). Unlike the proper motions of nearby stars, we can never perceive the motion of a galaxy perpendicular to our line of sight. The distances of galaxies are so great that even if a galaxy were moving sideways at almost the speed of light, its proper motion would still be imperceptible to our measurements. However, the radial motions of galaxies – the galaxies' own 'peculiar' motions – are superimposed upon the radial motions caused by the cosmological expansion of the Universe. Earlier we pointed out how the peculiar motions blurred, or otherwise interfered with, our measurements of distance by the Hubble law. Now we can point out that cosmological expansion interferes with our measurements of peculiar velocities!

Ideally, we would like to measure the peculiar motions and the cosmological motions quite separately, but this is not easy. In the case of clusters, where all galaxies are assumed to be at the same distance, a common cluster velocity can be found and subtracted to reveal the peculiar velocities of individual members within the clusters. However, these are not the peculiar velocities we seek, since clusters of galaxies are relaxed structures (and fall outside the concerns of this book). The common cluster velocity might be its cosmological velocity, but it could also include the peculiar motion of the cluster as a whole. Similarly, outside clusters, one has to separate the cosmological and peculiar velocities of individual galaxies – a *seemingly* impossible task, but not quite so. What is needed are methods of determining distances of galaxies independent of the Hubble velocity relation. Such methods are available through the Tully–Fisher and modified

Faber–Jackson relations, which will be described later in this chapter. Unfortunately, the large inherent scatter in these relations make them only accurate for statistical samples (such as a cluster of galaxies) and not for individual galaxies.

Only very close to home, within the Local Group of galaxies and in its immediate neighbourhood, do peculiar motions dominate over the cosmological expansion. For example, the distance to the Andromeda Galaxy is gradually decreasing – not increasing – at about 80 km/s.

We can also use primary distance indicators such as Cepheid variable stars. The Local Group is somewhat 'anaemic' and its internal velocities are rather modest – always less than 100 km/s. However, all measurements are made relative to our Galaxy, or relative to a 'local standard of rest' within the Local Group. The question that remains is whether the entire Local Group is moving. The answer is that it is.

7.2 THE FIRST INDICATIONS OF LARGE-SCALE MOTIONS

Until the 1970s, large-scale motions in the Universe were as unexpected as large-scale structures. On the basis that the expansion of the Universe 'cooled' the 'thermal' velocities of galaxies, galaxies outside clusters were expected to have negligible peculiar velocities (< 25 km/s), and be subject only to cosmological expansion.

However, in 1976 Vera Rubin and collaborators made the claim that by using ScI galaxies, whose overall properties suggested they were more or less equidistant from our Galaxy, they had detected a motion of our Galaxy of some 450 km/s (in a direction towards l = 163°, b = −11°). Their finding was considered sensational, but many doubted its validity. Soon afterwards, a conflicting but nevertheless similar result was extracted from the cosmic microwave background, and was interpreted as a flow of some 600 km/s in a totally different direction from that determined by Rubin. In hindsight, it can now be seen that the Rubin result was somehow erroneous, while the extraction of a velocity from the cosmic microwave background is now well established and considered the foundation of large-scale motions.

Einstein drew on the word 'relativity' in the belief that all motions could only be made relative to one body, or relative to another, rather than to an absolute frame of rest. However, a by-product of the cosmic microwave background (discussed in detail in the next chapter) is that it can define a local standard of rest. Accurate measurements of this background have revealed a dipole effect: one side of the sky is brighter than the other by one part in a thousand. This is interpreted as the absolute motion of our Galaxy.

The most accurate measurements carried out at the time of writing are undoubtedly those from the COBE satellite. They reveal a dipole of 3.343 mK towards l = 264.4°, b = 48.4°. After correcting for the Earth moving around the Sun, the Sun moving around the Galaxy, and the Galaxy moving relative to the Local Group of galaxies, it is found that we are moving at 627 ± 22 km/s in the direction l = 276 ± 3°, b = 30 ± 3°. Such a movement must cause the apparent directions to objects in the more distant Universe to be slightly displaced from their true directions, but since we are always moving in the same direction (unlike the Earth's motion around the Sun), this is not corrected.

Since we have established velocities of nearby galaxies, relative to our Galaxy, that are almost an order of magnitude smaller than 600 km/s, the discovery from the cosmic microwave background must mean that not only is our Galaxy moving in that direction, but so are hundreds, if not thousands of neighbouring galaxies. We have stumbled on a large-scale streaming of galaxies. But has this streaming been caused by large-scale structures? The only force considered to operate on a large-scale in the Universe is gravity. It is therefore believed that if such streaming motions occur, they have come about because of localised mass overdensities that exert gravitational influences on their surroundings. Our velocity, relative to the cosmic microwave background, must come about due to the presence of one of more relatively nearby overdensities.

7.3 VIRGO INFALL

The nearest of the overdensities is the Virgo Cluster (discussed and displayed in Chapter 4), at a distance of about 15 Mpc and with a redshifted velocity of 1,100 km/s. Although, as has been frequently stated earlier, this cluster is the most conspicuous concentration of galaxies seen in the sky, it is nevertheless not a particularly rich cluster.

As with other clusters of galaxies, the 'virial' mass of the cluster can be obtained from the velocity dispersion of the cluster members. We have seen earlier that this is in the region of 5×10^{14} solar masses. However, irregularities in the sky distribution of its galaxies (see Section 4.4.1) may imply that it is not properly gravitationally relaxed. Nevertheless, the value is such that it implies a high mass/luminosity ratio within the cluster, i.e. the apparent presence of dark matter.

In 1982, a pioneering study of the Virgo Supercluster was carried out by Marc Aaronson and collaborators, using the Tully–Fisher relation (see below). It gave the first indication that there was a general infall throughout the supercluster towards its centre. The magnitude of this flow suggested the presence of dark matter, contributing mass but not luminosity. Perhaps the term 'infall' is misleading, since it does not subdue the cosmological expansion, but merely retards it slightly. Nevertheless, should we choose to work in 'comoving coordinates' which expand along with the cosmos, then the infall would be significant.

A classic investigation by Gustav Tammann and Allan Sandage drew on using groups of galaxies in the vicinity of the Virgo Cluster to measure the infall towards the cluster. Their findings suggest that the dark matter is concentrated towards the core of the cluster. At the distance of our Local Group, the infall velocity was determined to be 200 ± 50 km/s. Again, the infall is insufficient to counteract the cosmological velocity of recession of the Virgo Cluster. The distance between our Galaxy and the Virgo Cluster is nevertheless increasing, in spite of our 'falling' towards it.

This value is therefore far too low to account for the motion towards the cosmic microwave background dipole, and in any case not in quite the right direction. Figure 7.1 shows that if we correct the motion relative to the cosmic microwave background for the infall towards Virgo, the result is a velocity of around 500 km/s towards $l = 274°$, $b = +11°$, in the general direction of the constellation of Hydra. We still need to explain how that came about.

The Sandage and Tammann work is not the only analysis, as a number of alternative approaches have been made. Needless to say, there is a variety of different values for the infall. John Huchra and collaborators found an infall of 250 ± 64 km/s; other determina-

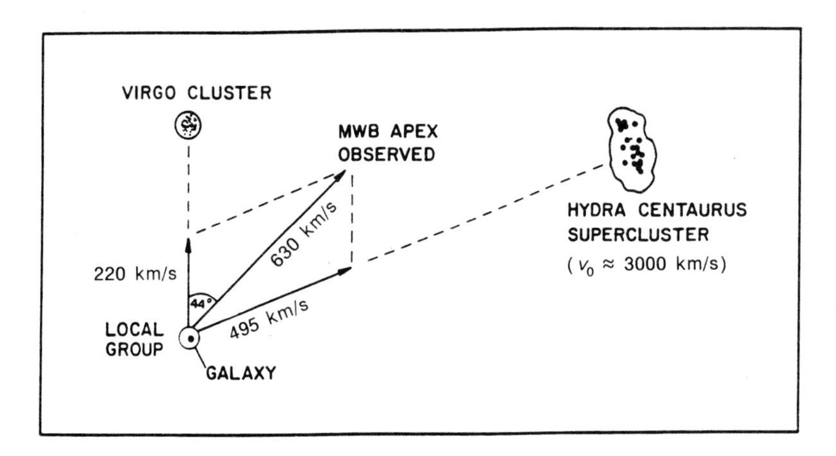

Fig. 7.1. A velocity vector diagram shows that if our infall towards the Virgo Cluster is taken as 220 km/s, then in order to account for the motion implied from the microwave background (MWB), our Galaxy would have to be moving towards Hydra–Centaurus at 495 km/s. (Reproduced with permission from G. Tammann and the *Astrophysical Journal* (**294**, 81, 1985).)

tions are lower – even as low as 38 km/s. The residual velocity, from the cosmic microwave background dipole, and its direction, vary accordingly.

7.4 MOTIONS IN CLUSTERS AND DARK MATTER

Although the emphasis of this book is generally deflected away from clusters of galaxies, it is a relatively simple procedure to determine their mass according to the virial theorem. Virial masses are derived on the assumption that clusters are collapsed to a relaxed state whereby the overall gravitational force that tries to draw the galaxies to the cluster centre is counterbalanced by their kinetic motions. The latter can be assessed as the dispersion in line-of-sight velocities for cluster members. Then, following the recipe of M. Schwarschild:

$$M = \frac{rv^2}{G}$$

where v is the velocity dispersion and r is the 'effective radius' of the cluster as determined from 'strip' counts of the galaxies in the cluster:

$$r = 2\frac{\left(\int_0^R Sdq\right)^2}{\int_0^R S^2 dq}$$

It is assumed the cluster is spherically symmetric.

Fritz Zwicky was the first to apply this technique to the Coma Cluster in 1937. The result was surprising, giving a much higher mass than would have been expected from the sum of the individual members. It was the first indication of the existence of substantial amounts of dark matter in the Universe. The masses of individual galaxies – especially spirals – are determined from the rotation velocities of their outer regions (in the same way as the mass of our Sun can be determined from the orbital velocities of the planets). Following chiefly from the work of Vera Rubin, we now know that the velocities of the outer regions of spiral galaxies fail to decline with increasing distance from their centres. This is seen as direct evidence for considerable amounts of dark matter in the envelopes of galaxies. It suggests that galaxies, as we see them, are somewhat like icebergs, in which most of the mass is not visible.

The virial masses are not without complications. The most obvious is the accidental inclusion of foreground or background galaxies as cluster members, so stretching the 'fingers of God' and producing a higher velocity dispersion than the true value. The other is the question as to whether clusters are 'virialised' (gravitationally relaxed), which surfaced, for example, in our earlier discussion regarding the Virgo Cluster.

7.5 INDEPENDENT INDICATORS OF DISTANCE – MODIFIED FABER–JACKSON AND TULLY–FISHER

The problem of separating cosmological and peculiar motions of galaxies would be solved by a method of determining distance *independent* of the Hubble relationship. The avenue to this is a measure of the intrinsic luminosity of the galaxy concerned, and a comparison between the apparent and the intrinsic luminosity in turn gives distance.

Both the modified Faber–Jackson and the Tully–Fisher relations relate luminosity to the internal dynamics of the galaxy. Luminosity depends on mass, and mass is reflected by the internal velocity dispersion. It is the manner in which that internal velocity dispersion is measured that differs between the two relations, as applied to either elliptical or spiral galaxies. Elliptical galaxies do not normally have significant amounts of free gas within them, so the best way to determine their internal velocity distribution is to measure it by their stars. Elliptical galaxies have relatively high surface brightnesses, so are reasonably easy to observe optically.

During the 1980s, a collaboration of seven US–UK astronomers (David Burstein, Alan Dressler, Sandy Faber, Roger Davies, Donald Lynden–Bell, Roberto Terlevich and Gary Wegner – since known as the 'Seven Samurai') undertook a major re-examination of elliptical galaxies with detailed photometric and spectroscopic observations. They sought to find a relationship for the intrinsic luminosity, and produced a revision of the earlier Faber–Jackson relation. They found that in a three-dimensional plot of velocity dispersion within the galaxy, total brightness and average surface brightness, the elliptical galaxies showed a nearly planar distribution. This relationship can also be expressed as follows:

$$DI^{5/6} \propto \sigma^{4/3}$$

where D is the diameter of the galaxy, I ($= L/D^2$) the surface brightness, and σ the velocity dispersion within the galaxy. The relation can be used to find the relative distances of elliptical galaxies.

The scatter is still too large to make it useful for individual galaxies; the error in distance would be around 25 per cent, which is typically of similar magnitude to that caused by the peculiar velocities. However, when applied to groups or clusters of galaxies, the error is reduced towards 10 per cent. If the distances exceed that indicated by the Hubble relationship, the group must have a peculiar motion towards us, and vice versa.

The masses and luminosities of spiral galaxies are, by contrast, best reflected by the rotation of the disk components – both stars and gas. While the optical surface brightness of the disk is low and difficult to observe, precise velocity measurements can be made via the 21-cm radio emission of neutral hydrogen (as detailed in Section 2.8). The velocity 'width' of the motions in the disk can be measured, and corrected for the apparent inclination of the disk, as though the galaxy is then viewed edge on. The velocity width is then a measure of the speed of rotation of the disk, and thereby a reflection of the mass present and the resulting luminosity.

Brent Tully and Richard Fisher were the first to show that a relation between the intrinsic optical luminosity and the 21-cm velocity width did exist, with the form

$$L \propto \sigma^{\beta}$$

Unlike elliptical galaxies, spiral galaxies contain a significant amount of dust that obscures and diminishes the light output from the disk. Since spirals are more transparent at infrared wavelengths, the present trend is to use an infrared Tully–Fisher relation, particularly in the I-band where $\beta = 3$. Again there is considerable scatter, though somewhat less than with the modified Faber–Jackson relation. As with the latter, the Tully–Fisher relation is therefore best applied to groups or clusters and not to individual galaxies. At best, there is still a scatter of some 0.3 magnitudes, which corresponds to an error in distance of 15 per cent for individual galaxies.

The Faber–Jackson and Tully–Fisher relations have opened the doors to the study of large-scale motions in relation to various large-scale structures, as we shall now discuss.

7.6 THE GREAT ATTRACTOR

The Seven Samurai applied their revised Faber–Jackson relation to their sample of elliptical galaxies, and made a major announcement at a conference held on Hawaii in 1986. They reported that a comparison between the distances established by the Faber–Jackson relation and the velocities of recession suggested a large-scale streaming towards a point at $l = 307°$, $b = +9°$ and $cz = 4{,}350$ km/s (see Figure 7.2). This result was in general agreement with the streaming known from the cosmic microwave background; but what made it so dramatic was the prediction of a point of convergence and its distance. If this is so, there must be a strong mass overdensity centred on this point, apparently within the Hydra–Centaurus Supercluster. One of the seven investigators, Dressler, dubbed it the Great Attractor, and the name has endured.

The nature and position of the Great Attractor, responsible for the large-scale streaming motion that includes our own Galaxy, has been a matter of considerable interest and debate ever since. The work of the collaboration that discovered it has been similarly followed; it was in the aftermath that the investigators were christened the Seven Samurai, in the spirit of the gentle sense of humour that operates in our community of researchers.

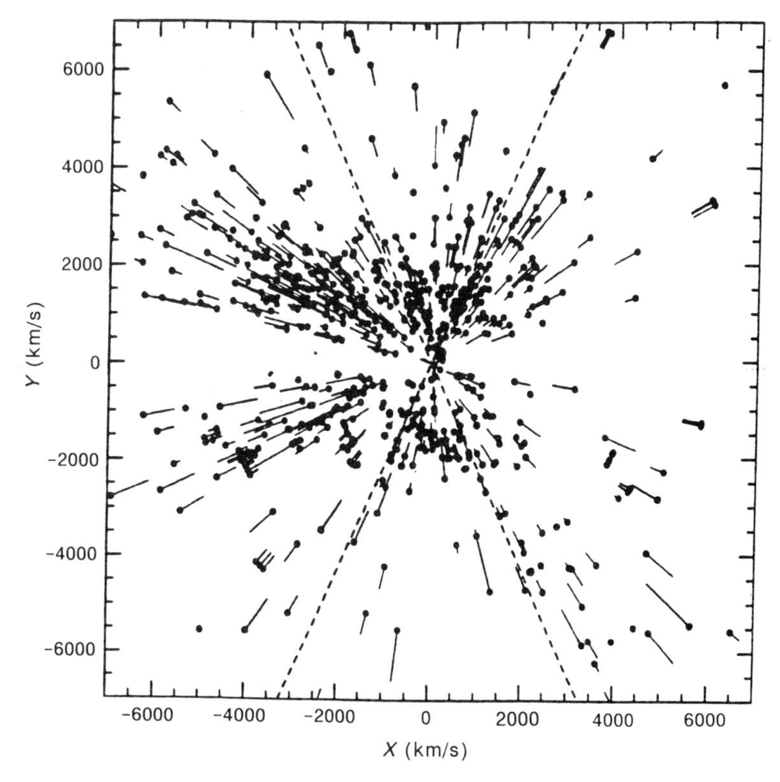

Fig. 7.2. A map showing the peculiar motions of elliptical galaxies, as aligned in the general direction of local flow. The short lines extending from the data points represent the magnitude and directions of the peculiar velocities. A tidal distortion, systematic convergence towards X = −400 km/s, is apparent. (Reproduced with permission of A. Dressler and *Nature* (**350**, 391). Copyright (1991) Macmillan Magazines Ltd.)

If the distance and effect of the Great Attractor is known, then its mass is predictable: about 10^{16} solar masses, an order of magnitude greater than the greatest cluster mass known. But what sort of mass concentration could the Great Attractor then be?

The Seven Samurai were fond of displaying the diagram shown here as Figure 7.3. They pointed out the large general concentration of galaxies in the Centaurus direction. We have seen from Chapter 4 that this overdensity, in numbers of galaxies at least, is at a redshift of around 4,500 km/s – the correct distance for the Great Attractor. Yet, impressive as the concentration may be, the numbers of galaxies were not adequate to account for the mass required. In any case, a similar concentration of galaxies exists in Perseus–Pisces, at similar redshift, yet in the opposite direction in the sky. Why should we be pulled towards the one and not the other? Murphy's law could, of course, have taken a hand. The concentration lies close to the obscuring band of the Milky Way. Perhaps there is an even greater concentration of galaxies in that portion of the Great Attractor region hidden behind the foreground contents of our own Galaxy.

Embedded in the concentration suspected of being the Great Attractor is the Centaurus Cluster of galaxies. Originally suspected of being the Great Attractor itself, it was soon

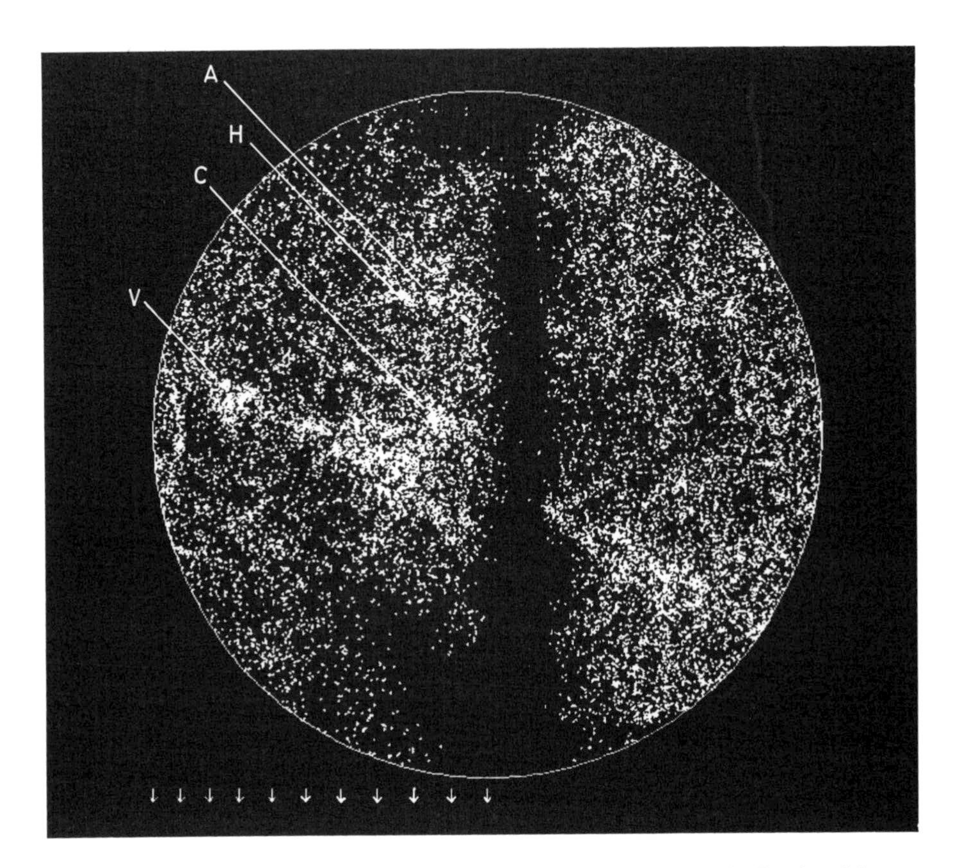

Fig. 7.3. The distribution of galaxies over a hemisphere of the sky centred on the direction of the dipole of the cosmic microwave background, as prepared by O. Lahav from optical catalogues. The letters identify the Virgo, Centaurus, Hydra and Antlia Clusters. The greatest concentration of galaxies is below and left of the Centaurus Cluster. The Centaurus Wall is apparent as the linear feature stretching across the plot. (Reproduced with permission from O. Lahav.)

found to have insufficient mass (around 10^{14} solar masses) and moreover was also being drawn towards the gravitational overdensity. The cluster has also shown an anomalous double peak in its redshift distribution. Its main peak is at 3,000 km/s, and there is a secondary peak at 4,500 km/s – the predicted redshift of the Great Attractor. Although in the past certain investigators have made the claim that there were two dynamical components intermixed at a common distance, the more likely explanation is that the 3,000 km/s peak is the true redshift (as the Seven Samurai find the cluster generally streaming away) and the 4,500 km/s peak is contamination from a much more extensive background concentration – the supposed Great Attractor region itself.

The Seven Samurai have produced a sequence of papers as they were able to add more data, and they continued to fit their data to a simple spherical infall model. One of their papers (Dressler and Faber, 1990) reported some inconclusive evidence for 'backside infall' – galaxies on the far side of the Great Attractor streaming towards us. They admitted that the evidence was tentative, particularly due to the uncertainty of the Malmquist bias (galaxies streaming inwards are more likely to be included in their sample), and their claim was soon contested by D. Mathewson and collaborators. This Australian group was using the Tully–Fisher relation on a very large number of spirals. They reported no evidence of 'backside infall', and rather went along with 'bulk' streaming, without any discernible point of convergence. Their later paper (1994) again claimed that their data could not be fitted to any infall models – whether Great Attractor, Centaurus Cluster or otherwise – and they went so far as to question whether peculiar velocities even arose from infall into dense regions.

There has been a wealth of papers from other authors on the possible causes of the streaming. Many point out that the general distribution of galaxies in optical catalogues (to magnitude or diameter limits) reveals a dipole effect in their sky distribution – likewise galaxies catalogued for their infrared luminosity (IRAS) – and these dipoles point in the same general direction as the cosmic microwave background dipole and the Great Attractor.

The truth as to whether there is infall into the Great Attractor or bulk flow to beyond may be a compromise between the two. Michael Hudson, for example, concluded that part of the motion was Great Attractor infall and part was a bulk flow (towards $l = 316°$, $b = 10°$).

What is the origin of the bulk motion? – an even Greater Attractor? A consortium of Italian researchers has pointed out the existence of the Shapley region; the 'metagalaxy' in Centaurus discovered by Shapley during the 1930s (see Section 1.3). It is one of the densest concentrations in the sky when one looks to a redshift of some 15,000 km/s – a deeper view than taken by the Seven Samurai. As described in Section 5.2.1, it is centred on 13,000 km/s. Could this be responsible for the more general streaming, the convergence of what appears to be bulk flow? If so, then its greater distance must be compensated by its having still greater mass to achieve the similar gravitational pull to the nearer Great Attractor region. Since gravity works by the well-known inverse square law, then the mass would have to be an order of magnitude greater, some 10^{17} solar masses. For many researchers, that is stretching things too far. However, Raychaudhury *et al* examined the rich clusters in the Shapley concentration and suggested that they were merging at a high rate.

In 1991, Manolis Plionis and Riccardo Valdarnini analysed Abell and ACO clusters, finding a cluster dipole within 10 degrees of the cosmic microwave background dipole. Further analysis suggested that most of the Local Group motion was caused by matter fluctuations within 5,000 km/s, but that the Shapley concentration could be responsible for 20 per cent of the motion.

Furthermore, POTENT (described below) clearly indicates streaming towards the nearer Great Attractor region. The most recent investigations, at the time of writing, reinforce the idea of the Great Attractor being an extended region of overdensity. Perhaps we are also hampered by the existence of dark matter, as the mass-to-luminosity ratio always seems to increase as we go to larger and larger scales.

A collaboration, involving the author, may have made a significant breakthrough. It involves the work (described earlier in Chapter 4) towards mapping structures partially obscured by the Milky Way. In 1995, we recognised the existence of a large rich cluster, comparable with the Coma Cluster yet somewhat closer. The cluster had already been catalogued and was known as ACO 3627; our group was, however, the first to obtain large numbers of redshifts in the cluster, which revealed a large velocity dispersion, and consequently a high mass for the cluster. The position of ACO 3627 closely matches the centre of the Great Attractor as predicted by POTENT, but the mass of the cluster is still an order of magnitude too low for the cluster to be the Great Attractor itself. The Great Attractor remains a more extended overdensity, but it is possible that the massive cluster holds a central position: the bottom of the extended gravitational well, much as does the Coma Cluster within the northern Great Wall. ACO 3627 is fairly centrally situated within the structure called the Centaurus Wall, described earlier in Chapter 4.

We still look for a clear reason as to why we are pulled towards the Great Attractor, rather than other relatively nearby overdensities such as Perseus–Pisces and even Coma. However, there is a continuous if irregular distribution of galaxies between our Galaxy and the Great Attractor region, making us a part of the Centaurus Wall. By contrast, there are mainly voids between Perseus–Pisces and ourselves, and in front of Coma. Just as concentrations of galaxies attract, so do voids repel.

7.7 TULLY–FISHER AND THE GENERAL FIELD

Numerous researchers have applied the Tully–Fisher relation to clusters scattered across the sky. Inevitably, the bulk flow in the frame of the cosmic microwave background is detected. Often, such as with Jeremy Mould and collaborators in 1993, the data do not allow distinction between convergence on the Great Attractor or simple bulk flow. Similarly, the necessity or otherwise of invoking a 'local anomaly' to deal with deviations of the Local Group region towards the Great Attractor, compared with the bulk flow towards the cosmic microwave background dipole, is debated.

A controversy has only recently been settled regarding the Perseus–Pisces region. Early indications were that it too shared the bulk motion, or even that it too might be falling towards the Great Attractor. As described in Chapter 4, the Perseus–Pisces 'Wall' or 'Supercluster' is a major structure, comparable in magnitude with the Centaurus Wall that supposedly contains the Great Attractor. The idea, then, of such massive structures sharing in massive bulk motions would have been a severe challenge to cosmological modelling, to say the least. Earlier attempts to discern infalls to both the Great Attractor and to Perseus–Pisces were frustrated by the limitations of the data. The added complication is the Malmquist effect – the bias towards selecting more infalling galaxies (with slightly lower redshifts) than outflowing.

This unsatisfactory situation has now been remedied by an analysis by Luiz da Costa and colleagues that complemented the sample of spiral galaxies assembled by Giovanelli and Haynes with those of Mathewson and collaborators. The near infrared Tully–Fisher relation has been used, together with a reconstruction algorithm similar to that described in Section 7.8 below. Figure 7.4 shows the resulting velocity map in the Supergalactic plane. For the first time, clear bifurcation of the flows towards the Great Attractor region and Perseus–Pisces is apparent (in comparison to Figure 7.5). In spite of past doubts, it is obvious that flows towards major overdensities do occur at the expense of flows away from regions dominated by voids.

The map is impressive in its identification of major overdensities and voids, from galaxy flow alone. The Great Attractor is, however, put much closer – at only 2,000 km/s

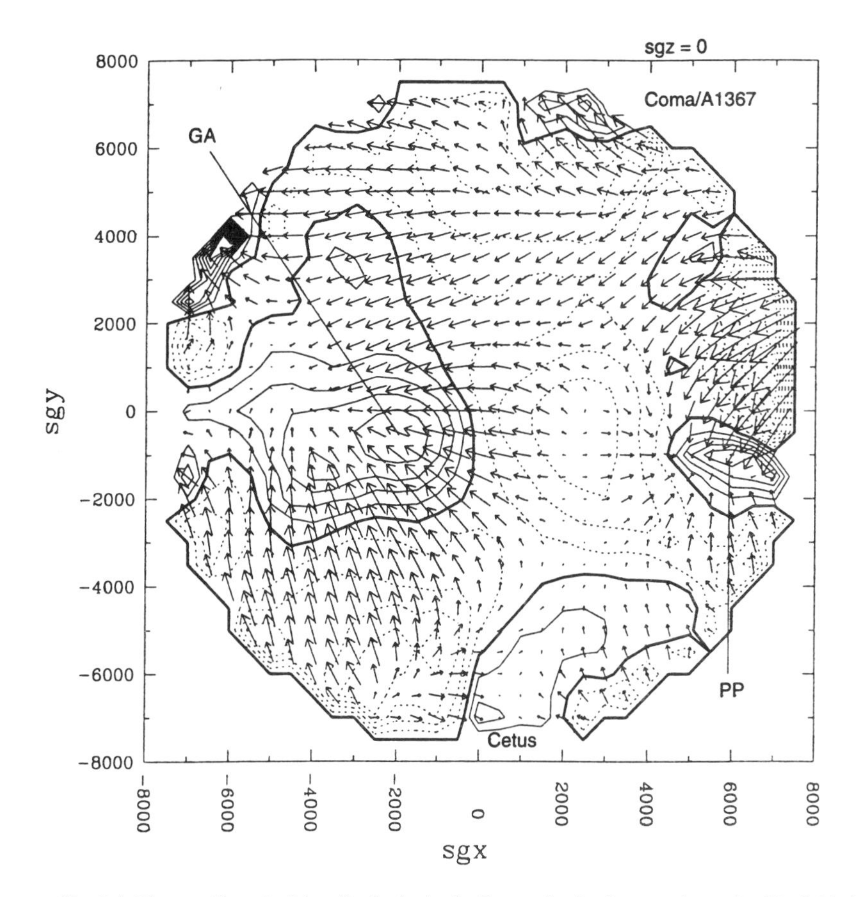

Fig. 7.4. The peculiar velocities of galaxies in the Supergalactic plane, as determined by L.N. da Costa and collaborators. The implied mass densities are conveyed by contour lines. GA = Great Attractor, and PP = Perseus–Pisces. (See discussion in text.) (Reproduced with permission from L.N. da Costa and the *Astrophysical Journal* (**468**, L5, 1996).)

(close to the intersection of the Centaurus and Hydra Walls) – than has been previously considered. Furthermore, the flow towards it is weaker; the upper left hand portion of the map shows bulk flow, possibly to a more distant mass overdensity. There is a distinct void beyond the Great Attractor.

7.8 DENSITY FIELDS – POTENT MAPS

A velocity map, such as that in Figure 7.4, may be used to construct a mass density map. One of the most successful derivations of local large-scale mass distribution has been undertaken by a collaboration of theoreticians centred on Ed Berschinger and Avishai Dekel. The technique depends on an elegant piece of software known as POTENT.

In POTENT, it is assumed that the motions of galaxies have come about as a result of the gravitational fields of the underlying mass distribution. Whether that mass is luminous or dark makes no difference, since the observed galaxies act only as test particles. In the same way as the orbital velocities of stars supposedly on the fringe of our Galaxy indicated the presence of a massive halo, so the dynamics of the galaxies must reveal the presence of massive structures, quite independent of whether or not the structures are visible.

POTENT works on the assumption that the velocity field is irrotational; that is, that the flows of galaxies are effectively laminar, and free of vortex motion. Mathematically, this can be expressed as

$$\nabla \times \mathrm{v} = 0$$

With this assumption, the velocity field can be derived from a scalar potential

$$\mathrm{v}(\mathrm{x}) = -\nabla \phi(\mathrm{x})$$

Only the 'radial' velocity field is of course available, so the potential is obtained by integrating the radial velocity field.

Whilst the principles are straightforward, the practice is complicated by realities of noise, errors and incompleteness of the data. The velocity field is sampled in an irregular fashion instead of a uniform dense manner, and the data therefore have to be suitably smoothed. Numerous mathematical complications arise (and are described by Dekel in his 1994 review). Spurious data points also tend to cause upsets. Considerable efforts have been made to assess errors in the recovered scalar field by contrived and Monte–Carlo simulations. One useful technique in recovering smooth velocity fields, known as 'Wiener' filtering, is particularly suited to dealing with data that are sparse, noisy and incomplete.

The success of POTENT is in its product. Over the years, its maps have gradually improved as more data have become available. A recent example, shown in Figure 7.5, reveals the matter distribution in the Supergalactic plane. Like the da Costa *et al* plot in the previous figure, it is impressive in how it confirms the presence of various nearby large-scale structures, entirely from the peculiar motion of galaxies. The Great Attractor (Centaurus Wall) is the most dominant feature, and the maps support its extended nature. Perseus–Pisces is clearly seen, and its milder extension into the Cetus Wall. Unlike the da Costa map, it does not as yet show the bifurcation of flow towards Perseus–Pisces. The other surprise is the weak showing of the Coma region. The maps also show the underdense regions as much as they do the overdense regions. The conspicuous valley in the figure is formed by the Sculptor and Eridanus Voids.

The great advantage of POTENT is that it should be – and has been shown to be – independent of the sample of galaxies used, because the galaxies used serve only as test particles, and do not in themselves necessarily form the mass overdensities. For example, both IRAS and Tully–Fisher samples are mainly spiral galaxies, which are not usually found in dense clusters. Nevertheless, the scalar potential maps derived from them do reveal the dense concentrations.

7.9 TULLY–FISHER ON INDIVIDUAL STRUCTURES

The Tully–Fisher relation has also been applied to individual structures. Mention has already been made of the early work of Aaronson and collaborators who detected infall within our Local Supercluster. Baffa and co-investigators have applied Tully–Fisher to the

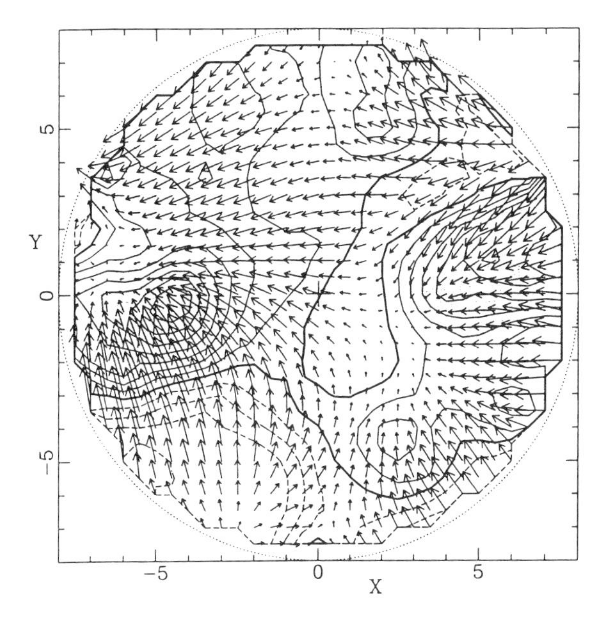

POTENT MASS DENSITY

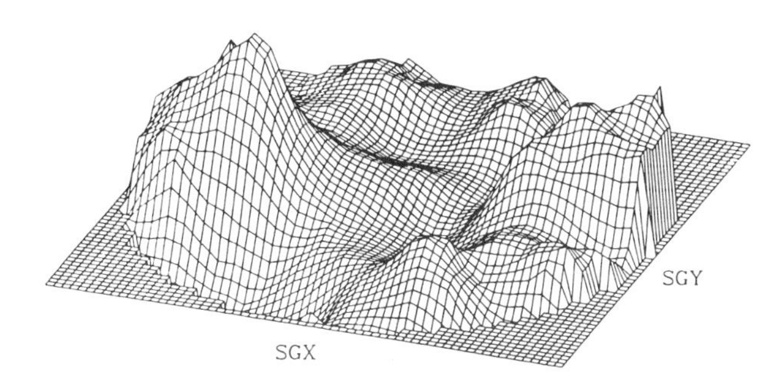

Fig. 7.5. The peculiar velocities of galaxies in the Supergalactic plane as extracted by POTENT software. The derived mass density distribution is conveyed by contours and shown as a three-dimensional plot in the lower panel. The scale in the upper panel is in 1,000s km/s. (Reproduced with permission from A. Dekel.)

Perseus–Pisces Supercluster and detected a shrinking in the main structure, indicative of a total mass of 10^{16} solar masses.

A similar analysis of the Coma Great Wall has recently been carried out by Ian Dell'Antonio, Margaret Geller and Gregory Bothun. They find a best fit for a mean infall velocity less than 150 km/s, and a 90 per cent confidence upper limit of 500 km/s. This is significant, since the Coma Great Wall is seen more or less flat on. Were the infall velocities very large, then the wall would appear much thinner seen in redshift space than it would in reality be in true three-dimensional space. Earlier, in Chapter 4, we reported that great walls typically have thicknesses of some 1,000 km/s – seen both sideways on and edge-on – implying that their thicknesses are similar in both redshift space and real three-dimensional space, and a low infall velocity confirms this. As with the Virgo Supercluster, the infall flow is nearly an order of magnitude lower than the spread in cosmological velocity.

7.10 EVEN LARGER-SCALE MOTIONS

POTENT and its counterparts let us map peculiar flows over a surrounding region out to some 6,000 km/s. This also represents the effective current range of the Tully–Fisher and modified Faber–Jackson relations. The question is whether we will find a repetition of similar flows over larger regions, or that the entire region so far covered has a bulk motion, as suggested by the residuals in the cosmic microwave background. To examine such a possibility, we need to look for a reference frame even further afield. The Abell clusters offer a possibility.

A major survey of Abell clusters by John Huchra and colleagues reported that peculiar velocities were less than 1,000 km/s. A similar result was obtained by Elena Zucca and collaborators, from finding no preference for systems to be elongated in the radial direction. The Oxford APM group put the figure even lower, at less than 500 km/s. By contrast, Neta Bahcall *et al* suggest peculiar motions of around 2,000 km/s.

In 1994, Tod Lauer and Marc Postman used 119 Abell (and ACO) clusters out to 15,000 km/s. As a distance indicator, they employed a relation between the intrinsic luminosity of the brightest galaxy in the cluster and the slope of its brightness profile – a relationship found earlier by J. Hoessel that supposedly gives distances with errors of only 16 per cent. If the Abell clusters defined a reference frame, then the Local Group of galaxies was moving at 561 km/s towards l = 220° and b = –28°. This finding was sensational, for although the magnitude of the velocity was familiar, the direction was very different from the accepted motion relative to the cosmic microwave background (l = 274° b = +11°). If the motion to the cosmic microwave background was correct, then the interpretation was that the reference frame of the Abell clusters was in motion, at several hundred km/s towards l = 343° and b = 52°. If such massive bulk flow were to exist, it would downgrade all motions discussed above to local effects. If gravitational in origin, it implies the existence of mass overdensities far exceeding those of the large-scale structures we have encountered in Chapters 4 and 5.

Lauer and Postman, well aware that the immediate response would be to shoot down their finding, did their best to eliminate the result, but reported that it seemed extremely robust, being quite insensitive to variations they tried on biases, sample compositions and geometry. As expected, a flurry of papers have since examined the Lauer–Postman effect, though it should also be mentioned that a 1992 paper by Rood suggested that Abell clusters in groups might have large peculiar motions. In 1995, two papers that used Type Ia supernovae as distance indicators produced results in disagreement with the Lauer–Postman effect.

In 1996, Alister Graham used a $R^{1/4}$ relation, as an alternative relation, on the Lauer-Postman sample, and confirmed their result. However, Riccardo Giovanelli and collaborators extended their Tully–Fisher galaxies within cones directed towards the apex and antapex of the Lauer–Postman motion. Their findings strongly disagreed with the very large-scale motion. By contrast, some mild support was found by Dell'Antonio and collaborators for their work (already mentioned) of using the Tully–Fisher relation on the Coma Wall.

Clearly then, it is still too early to be sure about the existence of very large-scale motions. We await future results and interpretations.

7.11 MEASURING THE COSMOLOGICAL DENSITY FROM LARGE-SCALE MOTIONS

Unlike clusters, large-scale structures are not virialised (not gravitationally relaxed), so one cannot employ the virial theorem to extract their mass. While velocity maps such as POTENT trace a mass density field, they show fluctuations rather than absolute values, and these also become increasingly more inaccurate with increasing redshift. This is a disappointment, because the great hope was that large-scale structures could be used to measure the cosmological density.

The cosmological density is perhaps the most crucial cosmological parameter. If the density of the Universe surpasses a critical value (expressed as $\Omega = 1$), the expansion of the Universe would be arrested and reversed. It is the difference between an infinite Universe that expands forever, and a finite Universe that reaches maximum volume and then collapses. Furthermore the inflationary Universe theory (discussed in Chapters 9 and 10) is interpreted by many as predicting $\Omega = 1$ exactly. Therefore measuring Ω observationally would be an important test of this theory.

In spite of our dealing with ill-shaped irregular large-scale structures, it appears that there are nevertheless ingenious ways that may allow us to use them to assess the cosmological density.

Peculiar velocities represent the growth of fluctuations. A dimensionless linear growth rate can be defined as

$$\beta = \frac{\Omega^{0.6}}{b}$$

Here b is the linear bias factor – the variation in the density of galaxies as a fixed fraction b of the variation in the density of total mass. (β is not the same as that used earlier in this chapter.)

While the variation in the density of galaxies is subject to this bias, the peculiar velocities of galaxies are not. The general magnitude of the peculiar velocities can be established by the general 'distortion' they impose upon redshift space (see Section 3.2), and it has been shown that β can be measured as the amplitude of this distortion. Consequently it is possible to measure β from the magnitudes of peculiar motions, and many investigators have sought to do so (see references). The values reported in the literature vary with the authors and the surveys used, but the general range is 0.4 to 0.85, though there is at least one as low as 0.21. The favoured value seems to be just below 0.6 (Stromlo–APM and APM clusters, Durham/UKST have $\beta = 0.55 \pm 0.12$), though all investigators admit to large uncertainties.

Of course, the problem in interpretation lies with the value of b, the bias factor. If we assume the bias factor is unity (galaxies represent the mass field), then the values of β suggest a range in Ω of 0.2 to 0.75, but with the preferred value around 0.3. Whilst this is generally seen as observational evidence favouring a low density Universe (an infinite Universe that will expand forever), the large errors and the unknown bias factor have led to it not being accepted by cosmologists as a definitive measurement.

Avishai Dekel and Martin Rees have also devised a clever way of assessing Ω, on the basis that large outflows are not expected in low-Ω universes. A preliminary application of their approach to the Sculptor Void has also produced a value $\Omega \leq 0.3$.

Hopefully, the future will see refinement of the techniques to accompany improved data. Perhaps it will confirm that we do, after all, live in a low-density Universe.

7.12 FURTHER READING

Specialised

Historical

Rubin, V.C. *et al.*, Motion of the Galaxy and the Local Group determined from the velocity anistropy of distant ScI galaxies. II. The analysis for the motion, *Astron. J.*, **81**, 719 (1976).

Virgo infall

Tammann, G.A. and Sandage, A., The infall velocity toward Virgo, the Hubble constant, and a search for motion toward the microwave background, *Astrophys. J.*, **294,** 81 (1985).

Tully, R.B. *et al.*, Nearby galaxy flows modelled by the light distribution: distances, model, and the local velocity anomaly, *Astrophys. J. Suppl.*, **80**, 479 (1992).

Teerikorpi, P. *et al.*, Investigations of the Local Supercluster velocity field I. Observations close to Virgo, using Tully–Fisher distances and the Tolman–Bondi expanding sphere, *Astr. Astrophys.* **260**, 17 (1992).

Great Attractor

Burstein, D. *et al.*, Evidence from the motions of galaxies for a large-scale, large-amplitude flow toward the great attractor, *Astrophys. J.*, **354**, 18 (1990).

Dressler, A. and Faber, S.M., Confirmation of a large-scale, large amplitude flow in the direction of the great attractor, *Astrophys. J.*, **354**, 13 (1990).

Dressler, A., The Great Attractor: do galaxies trace the large-scale mass distribution?, *Nature*, **350**, 391 (1991).

Dressler, A., The large-scale streaming of galaxies, *Scientific American*, **257,** 38, (Sept. 1987).

Hudson, M.J., Optical galaxies within 8000 km s^{-1} – III. Inhomogeneous Malmquist bias corrections and the Great Attractor, *Mon. Not. R. astr. Soc.*, **266**, 468 (1994).

Kaiser, N., Bulk flows, shear and the Great Attractor, *Astrophys. J.*, **366**, 388 (1991).

Mathewson, D.S. *et al.*, No back-side infall into the great attractor, *Astrophys. J.,* **389**, L5 (1992).
Mathewson, D.S. and Ford, V.L., Large-scale streaming motions in the local universe, *Astrophys. J.,* **434**, L39 (1994).
Plionis, M. and Valdarini, R., Evidence for large-scale structure on scales ~300 h^{-1} Mpc, *Mon. Not. R. astr. Soc.,* **249**, 46 (1991).
Scaramella, R. *et al.*, The Distribution of Clusters of Galaxies within 300 Mpc h^{-1} and the Crossover to an Isotropic and Homogeneous Universe, *Astrophys. J.,* **376**, L1 (1991).

Tully–Fisher

Baffa, G. *et al.*, Peculiar motions in superclusters: Perseus–Pisces, *Astron. Astrophys.,* **280**, 20 (1993).
Bahcall, N.A. and Oh, S.P., The Peculiar Velocity Function of Galaxy Clusters, *Astrophys. J.,* **462**, L49 (1996).
Courteau, S., Faber, S.M. *et al.*, Streaming motions in the local universe: evidence for large-scale, low-amplitude density fluctuations, *Astrophys. J.,* **412**, L51 (1993).
da Costa, L.N. *et al.*, The Mass Distribution in the Nearby Universe, *Astrophys. J.,* **468**, L5 (1996).
Dell'Antonio. I.P. *et al.*, Peculiar Velocities for Galaxies in the Great Wall. II. Analysis, *Astron. J.*, **112**, 1780 (1996)
Han, M. and Mould, J.R., Peculiar Velocities of Clusters in the Perseus–Pisces Supercluster, *Astrophys. J.,* **396**, 453 (1992).
Han, M., The Large-Scale Velocity Field beyond the Local Supercluster, *Astrophys. J.,* **395**, 75 (1992).
Ichikawa, T. and Fukugita, M., Hubble Flows in the Pisces–Perseus Region from the Giovanelli–Haynes Galaxy Sample, *Astrophys. J.,* **394**, 61 (1992).
Mould, J.R. *et al.*, The Velocity Field of Clusters of Galaxies within 100 Megaparsecs. II. Northern Clusters, *Astrophys. J.,* **409**, 14 (1993).
Shimasaku, K. and Okamura, S., A Study of the Velocity Field in the Local Supercluster based on a new peculiar-velocity sample, *Astrophys. J.,* **398**, 441 (1992).

Mass distribution – POTENT

Dekel, A., Dynamics of Cosmic Flows, *Ann. Rev. Astron. Astrophys.*, 371 (1994).
Newsam, A., Simmons, J.F.L. and Hendry, M.A., Bias minimisation in Potent, *Astron. Astrophys.,* **294**, 627 (1995).

Lauer–Postman

Giovanelli, R. *et al.*, A test of the Lauer–Postman bulk flow, *Astrophys. J.,* **464**, L99 (1996).
Graham, A.W., Another look at the Abell cluster inertial frame bulk flow, *Astrophys. J.,* **459**, 27 (1996).
Lauer, T.R and Postman, M., The motion of the local group with respect to the 15,000 kilometre per second Abell cluster inertial frame, *Astrophys. J.,* **425**, 418 (1994).

Riess, A.D. *et al.*, Determining the motion of the local group using type Ia supernovae light curve shapes, *Astrophys. J.*, **445**, L91 (1995).

Rood, H.J., Peculiar motions of Abell clusters in compact groups, *Mon. Not. R. astr. Soc.*, **254**, 67 (1992).

Watkins, R. and Feldman, H.A., Interpreting new data on large-scale bulk flows, *Astrophys. J.*, **453**, L73 (1995).

Zucca, E. *et al.*, All-sky catalogs of superclusters of Abell–ACO clusters, *Astrophys. J.*, **407**, 470 (1993).

Beta

Branchini, E. and Plionis, M., Reconstructing positions (and peculiar velocities) of galaxy clusters within 25,000 kilometers per second: the cluster real space dipole, *Astrophys. J.*, **460**, 569 (1996).

Davis, M. *et al.*, Comparison of velocity and gravity fields: the mark III. Tully–Fisher catalog versus the IRAS 1.2 Jy survey, *Astrophys. J.*, **473**, 22 (1996).

Dekel, A. and Rees, M.J., from velocities in voids, *Astrophys. J.*, **422**, L1 (1994).

Dekel, A., Measuring independent of galaxy biasing, [in] *Clustering in the Universe* (*Ed.* S. Maurogordato, C. Balkowski, C. Tao, J. Tran Thanh Van), p.89, Edition Fontieres, 1995.

Freudling, W. *et al.*, The peculiar velocity field in the Hercules region, *Astrophys. J.*, **377**, 349 (1991).

Hamilton, A.J.S. and Culhane, M., Spherical redshift distortions, *Mon. Not. R. astr. Soc.*, **278**, 73 (1996).

Hudson, M.J., Optical galaxies within 8000 km s^{-1} – II. The peculiar velocity of the Local Group, *Mon. Not. R. astr. Soc.*, **265**, 72 (1993).

Kaiser, N. *et al.*, The large-scale distribution of IRAS galaxies and the predicted peculiar velocity field, *Mon. Not. R. astr. Soc.*, **252**, 1 (1991).

Plionis, M. *et al.*, The QDOT and cluster dipoles: evidence for a low-Omega Universe?, *Mon. Not. R. astr. Soc.*, **262**, 465 (1993).

Shaya, E.J. *et al.*, Action principle solutions for galaxy motions within 3000 kilometers per second, *Astrophys. J.*, **454**, 15 (1995).

Strauss, M.A. *et al.*, A redshift survey of IRAS galaxies. V. The acceleration on the local group, *Astrophys. J.*, **397**, 395 (1992).

Miscellaneous

Chengalur, J.N. *et al.*, Galaxy Pairs, Redshift Catalogs, and the cosmic peculiar velocity, *Astrophys. J.*, **461**, 546 (1996).

Lonsdale, C. *et al.*, Galaxy evolution and large-scale structure in the far-infrared. II. The IRAS faint source survey, *Astrophys. J.*, **358**, 60 (1990).

Pellegrini, P.S. and Nicolai da Costa, L., Peculiar velocity of the local group from nearby redshift surveys, *Astrophys. J.*, **357**, 408 (1990).

Scharf, C. *et al.*, Spherical harmonic analysis of IRAS galaxies: implications for the Great Attractor and Cold Dark Matter, *Mon. Not. R. astr. Soc.*, **256**, 229 (1992).

8

The cosmic microwave background

8.1 INTRODUCTORY EXPLANATION

The discovery of the cosmic microwave background has been one of the most important breakthroughs of the twentieth century, and some even see it as the biggest advance in cosmology since Copernicus. However, unlike the Copernican shift from a geocentric to a heliocentric model, the concepts involved are far beyond the comprehension of most members of the general public. Even students of science may become confused, so we shall open this chapter with an explanation of the nature of the cosmic microwave background.

We begin by the use of a simple analogy. Looking out my window, I see the weather is partly cloudy. Some parts of the sky are clear and blue and completely transparent, since I can even see the Moon at last quarter. But there are clouds; they are opaque, and they occasionally hide the Moon. The clouds are sufficiently opaque that not only can I not see through them, I can't even seem to see into them! Like fish in water, they seem like solid three-dimensional forms floating in a transparent medium. Of course, that is not quite true, as they are not so solid that a plane could not fly through them. If I were in a plane, then once inside the cloud I would find I could see a short but very limited distance, which physics would label the 'optical depth'. Were I flying just outside a cloud, it would look as if it were enclosed by a surface, which physics would label the 'surface of last scattering'. Furthermore, it would not be abrupt, but would only seem like a surface, because the optical depth – the distance I could see into the cloud – is so short.

Now consider the Sun. It looks like a ball, and appears as if one could land on its surface, save for the problem of its intense heat. Yet there is no solid surface to land on, since the Sun is gaseous throughout. However, it is opaque; you cannot see 'into' its interior. It only looks as if it has a surface, and the apparent surface is known as the 'photosphere'. Furthermore, the photosphere is not the boundary of the Sun, as there is still gas above it, just as there is gas below it. Why, then, should the gas below the photosphere appear opaque, whilst the gas above the photosphere is quite transparent? Atomic physics gives us the answer. The gas below the photosphere is hotter, denser and partially ionised. Ionised gas has free electrons that readily scatter the photons of light attempting to escape from the Sun. (The corona surrounds the Sun, and although highly ionised, it has far too low a density to have any effect.) The photosphere is the 'surface of last scattering', and as with the cloud, it is not abrupt. The optical depth – the distance one can see into the Sun – is around 100 km, which is minute in comparison with its 700,000-km radius, so it still looks as if the Sun has a distinct surface. It also tells us that gas which is sufficiently dense and hotter than a few thousand degrees is ionised and opaque, and that gas cooler than a few thousand degrees is not ionised and therefore transparent.

So too with the Universe. The standard cosmological model is the hot Big Bang, implying that the Universe was once in a much denser and therefore, by the laws of physics, a much hotter place than we find it now. The very early Universe must have resembled the present interior of our Sun – sufficiently hot to be gaseous throughout, and sufficiently hot and dense to be ionised and opaque throughout. As the Universe expanded, so the temperature and density decreased. There came a time, some half a million years after the Big Bang itself, when the temperature dropped below 3,000 degrees and, like fog lifting, the Universe changed from being opaque to being transparent. It has remained transparent ever since.

Figure 8.1 is a schematic representation of our perspective of the entire 'observable' Universe. For almost the whole of its 13 or so billion-year lifetime, the Universe has been transparent. In that time, a light ray could have travelled 13 or so billion light years. No light ray could have travelled further, because the Universe is no older. Consequently, the furthest distance one can now see is some 13 billion light years. Since we can look out to such a distance in all possible directions, the entire observable Universe can therefore be represented as a sphere of that radius, as in the Figure. The surface of the sphere forms a horizon – the limit to our visibility. In cosmology, this limit is close to that known as the 'particle horizon'.

What is visible to us within the sphere is not the Universe at the present time. We are restricted by the finite speed of light, and as we look out towards the boundary of the sphere, so we are compelled to look back in time. At any significant distance, we can see the Universe only as it was, and not as it is. Furthermore, what we see may be distorted. We may be seeing back in time, but the size of the sphere is based on the current age of the Universe. We are imposing the current scale on the distant past, when the Universe was really much smaller. As a result, our representation greatly exaggerates the volume of the more distant Universe, the more so as we look further out and further back in time. Nevertheless, the diagram correctly represents what we will see, even if it is distorted.

As long as the Universe is transparent, we can look further out and further back in time, but the very early Universe was opaque, not transparent. That early Universe forms an opaque shell on the surface of the sphere, and the inside of this shell is the limit of our visibility – the point beyond which, or before which, the Universe was opaque. As with the photosphere of our Sun, it is the 'surface of last scattering'.

This 'surface of last scattering' exists in time, not in a place. The change from opaqueness to transparency occurred throughout the whole Universe at a particular time, approximately at an age of half a million years. It was a time of recombination of ions and electrons, and the end of ionisation. Like the surface that surrounds a rain cloud, or the photosphere of the Sun, the surface of last scattering is diffuse, not abrupt. It was a gradual clearing, not an instantaneous event.

At the time of the clearing, the contents of the Universe were at a temperature of thousands of degrees, and like the photosphere of the Sun, the contents emitted blazing light. If we can see back to such a time, why then is such blazing light not apparent? Of course,

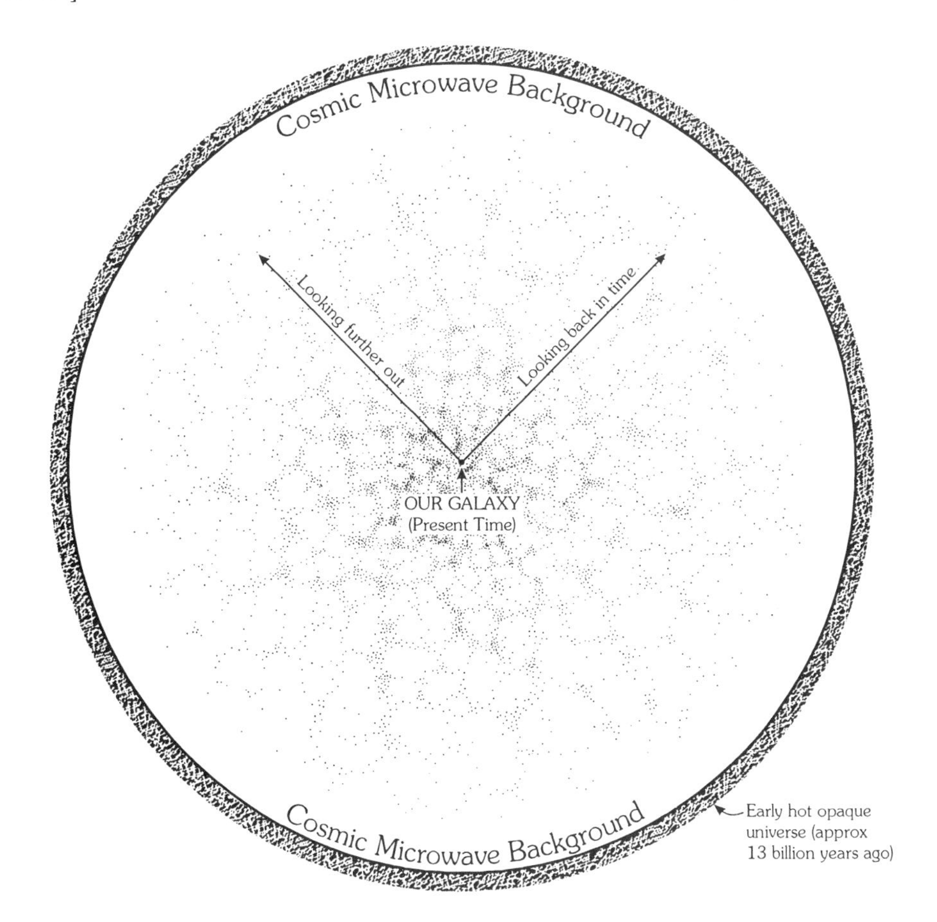

Fig. 8.1. A schematic representation of our apparent view of the Universe. As we look further out, so we look back in time, until an opaque curtain forms the 'ultimate horizon'. The cosmic microwave background originates from its inside surface, making us appear to lie at the centre of a hollow shell. (See further explanation in text.)

were it spread over the entire sky, our existence would be jeopardised. The answer comes from the extreme distortion in looking out so far; both the surface of last scattering and the light rays that should come from it have been grossly stretched. The surface has been stretched to encircle the entire visible Universe, and the light has been stretched to microwaves. The result is the cosmic microwave background.

The cosmic microwave background – the radiation received from the surface of last scattering – is an image of the embryonic Universe. It is also confirmation of the existence of a hot early phase, as physics would have predicted. At the time of its serendipitous discovery by Arno Penzias and Robert Wilson in New Jersey in 1965, physicists elsewhere in the same state (at Princeton) were even preparing an experiment in an attempt to detect it. Penzias and Wilson, whose persistence in tracking down extraneous noise from a horn antenna, led to the breakthrough, were later rewarded with the Nobel Prize.

Our view of the visible Universe (in both optical and microwaves) is therefore very weird. Not only do we look back at a parade of the history of the Universe, with images ever distorted, but everything is contained within a 'hollow spherical shell'. If the Earth had proved to be hollow inside, and we found ourselves floating at its centre, we might be similarly surprised! The shell that surrounds us, enclosing us in our own observable Universe, may appear as though it is a physical structure, whereas in reality it is a temporal structure.

Were we to shift our position in the Universe, with the freedom of a god, and try to approach the shell, it would recede from us as fast as we approached it. It is a horizon and, short of time travel, could no more be reached than could the horizon on the surface of the Earth. At this cosmological time, wherever our position in the Universe, we would have a similar view.

We can relate the discussion above to the redshift parameter, z, introduced earlier (in Section 2.6). Since the radius of the (outer) surface of the sphere in Figure 8.1 is set by the age of the Universe, it is expanding at the speed of light, or $z = \infty$. The inner surface (the surface of last scattering) is therefore expanding at very close to the speed of light; the era of recombination is put at just over $z = 1{,}000$. It is this extreme redshift that causes the original abundant visible light to be shifted all the way to the microwave region.

The extreme redshift has shifted the peak wavelength of the emitted thermal radiation. Today we know that the cosmic microwave background radiation has a Planck black-body spectrum. The most precise measurement (at the time of writing), from J. Mather and colleagues using the FIRAS instrument on the COBE (COsmic Background Explorer) satellite, shows an almost perfect fit to a black-body spectrum, that enables the temperature of the cosmic microwave background to be derived with remarkable precision:

$$T = 2.726 \pm 0.010 \text{ K}$$

If the redshift of the spectrum were to be removed, the shape would still retain its black-body nature, only the derived temperature would be just more than a thousand times higher. The Universe would be close to isothermal at the time of recombination.

The early cosmic origin of the microwave background is considered convincing, but a handful of researchers have nevertheless challenged the interpretation. One concern is the unlikely possibility of the Universe being re-ionised (and again becoming opaque) at much later epochs. In this case, much of the argument remains the same, but the whole time-scale would be disrupted. A small group of eminent cosmologists (Sir Fred Hoyle, Geoffrey Burbidge and Jayant Narlikar) have even suggested that iron whiskers, in the vicinity of our Galaxy, could mimic the cosmic microwave background. We shall, however, accept the conventional interpretation in the discussions that follow.

8.2 ANISOTROPIES IN THE COSMIC MICROWAVE BACKGROUND

The cosmic microwave background provides an image of the early Universe, which is of course very relevant to the development of large-scale structures. However, even at the time of its discovery, it was apparent that the background was remarkably isotropic (the same in every direction). In the years that followed, the search for temperature variations, $\Delta T/T$, in the background was pushed to ever-lower limits in the expectation of seeing structure in the early Universe. In the mid-1970s an overall dipole anisotropy was detected; one side of the sky was 0.1 per cent brighter than the opposite side. This, however, was soon interpreted as not being due to any structure in the distant early Universe, but rather to the motion of our Galaxy at some 600 km/s relative to a local standard of rest. This result has already formed the foundation of our discussion on the peculiar motions of galaxies (Chapter 7).

The attempts to detect anisotropies on a smaller angular scale than the dipole have proved frustrating, even pushing theoretical predictions down to at least the 0.01 per cent level. The understanding was that any anisotropies would be weak precursors of what is apparent today. Eventually, anisotropies, on scales of 10 degrees or larger, were detected by the DMR instrument on the COBE satellite at a level around 0.001 per cent, as reported by George Smoot and co-workers in 1992. These COBE results received enormous publicity and recognition, but it is only fair to point out that tentative anisotropies had also been reported by the Jodrell Bank experiment on Tenerife a few years earlier, and that several independent claims of detection of anisotropies followed within a year or so.

Figure 8.2 conveys the extent of the apparent anisotropies in the first year of the COBE DMR data. For good reason, many researchers expressed initial scepticism. Substantial contributions from the Milky Way had first to be subtracted from the all-sky image. The fluctuations that remained were comparable in magnitude to the predicted noise, so much so that one cannot say whether any individual feature was due to noise or to a true anisotropy. What George Smoot and colleagues showed, by very careful analysis, was that the character of the fluctuations was such that they could not have been produced by noise alone.

Scepticism has waned as supporting results have come in, and it is now widely accepted that we are probably seeing primordial fluctuations in the early Universe.

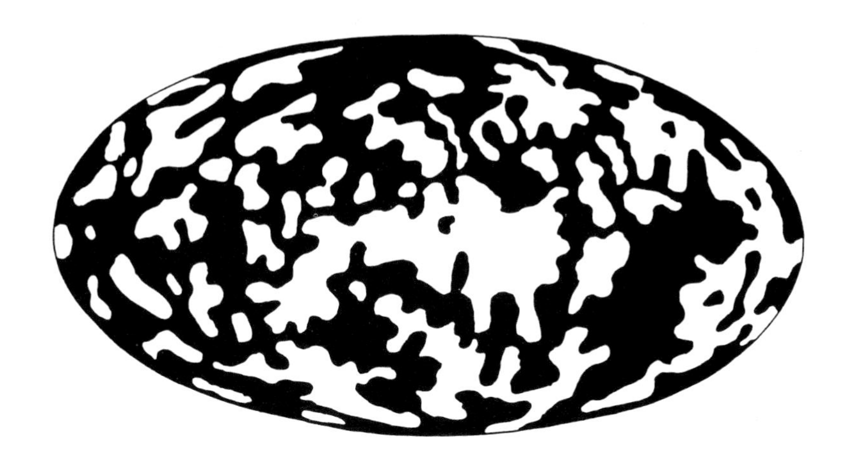

Fig. 8.2. An all-sky representation of the cosmic microwave background, where the white areas are marginally brighter than the black areas, thus revealing anisotropies (see text).

8.3 LARGE-SCALE STRUCTURE AND THE COSMIC MICROWAVE BACKGROUND

The excitement surrounding the discovery of the anisotropies in the cosmic microwave background is because we may be seeing the beginnings of the large-scale structures. The anisotropies may literally be 'fossils' of the structures visible today, and it may therefore be possible to reconstruct a complete scenario showing how the structures developed.

How do the scale of the anisotropies relate to that of the present-day large-scale structures, such as those depicted in the maps that accompany Chapter 4? We can make a simple reconciliation. The representation of the observable Universe, shown as Figure 8.1, looks back in time, but disregards the expansion of the volume of the Universe. It shows things as they were, but on a scale that they would have today, rather than on the scale they would have had at the time of their existence. For example, the surface of last scattering has been stretched over the inner surface of an enormous sphere. Figure 8.1 is therefore a portrayal of the visible Universe in co-moving coordinates (as introduced in Section 3.3) which are independent of the expansion of the Universe. The radius of the sphere, expressed in light years, has been set to match the age of the Universe in years (less the short interval of half a million years when the Universe was opaque). While the age of the Universe is of cosmological debate, the figure of 13 billion years, mentioned earlier, is taken as a reasonable accepted value. If so, then the radius of the sphere is 4,000 Mpc and the circumference is 25,000 Mpc. The angular scale would then be 25,000/360 = 70 Mpc per degree subtended, and if the Hubble parameter is around 70 km/s per Mpc, then a degree on the the cosmic microwave background corresponds to about 5,000 km/s of present-day redshift space, such as shown in the many maps that accompany earlier chapters in this book.

The resolution of the now famous COBE maps is, of course, not a degree, but is towards an order of magnitude greater. Consequently, the anisotropies of the COBE maps are suggestive of the seeding of large-scale structures some 50,000 km/s or larger. That is, of course, very much larger than any of the structures that appear in this book. Chapter 4, for instance, was concerned with structures out to 10,000 km/s and its detailed mapping of structures petered out by 7,000 km/s. However, a more detailed comparison will be possible once the local mappings of large-scale structure are extended outwards (e.g. the Las Campanas survey) and the mappings of the anisotropies on the cosmic microwave background are carried out with smaller angular resolution.

Plans are already afoot for the latter. NASA intends to launch the Microwave Anisotropy Probe in the year 2000, and this satellite should have a resolution some thirty times better than COBE. The European Space Agency has a similar mission planned for 2004, and a number of balloon-based instruments are also planned. Moreover, one of these has produced an early result. Figure 8.3 shows how observations by the airborne Millimeter-wave Anisotropy eXperiment (MAX), when treated by the 'maximum entropy' method, produce a resolution of 0.5 degrees over a very limited region of the sky. The area is most densely sampled over the central 5 × 2 degrees. At this resolution, various peaks and depressions are seen, with a general scale that matches that of the walls and voids presented through much of this book. The fluctuations are, of course, of incredibly low amplitude – nothing like which we find today – but they are probably the seedings of the fabric.

It should be remembered that, like a photograph of a baby that has since grown to an adult, the image frozen in the cosmic microwave background is a 'snapshot' of the early Universe, complete with its primordial structures. With reference to the latter, it is as

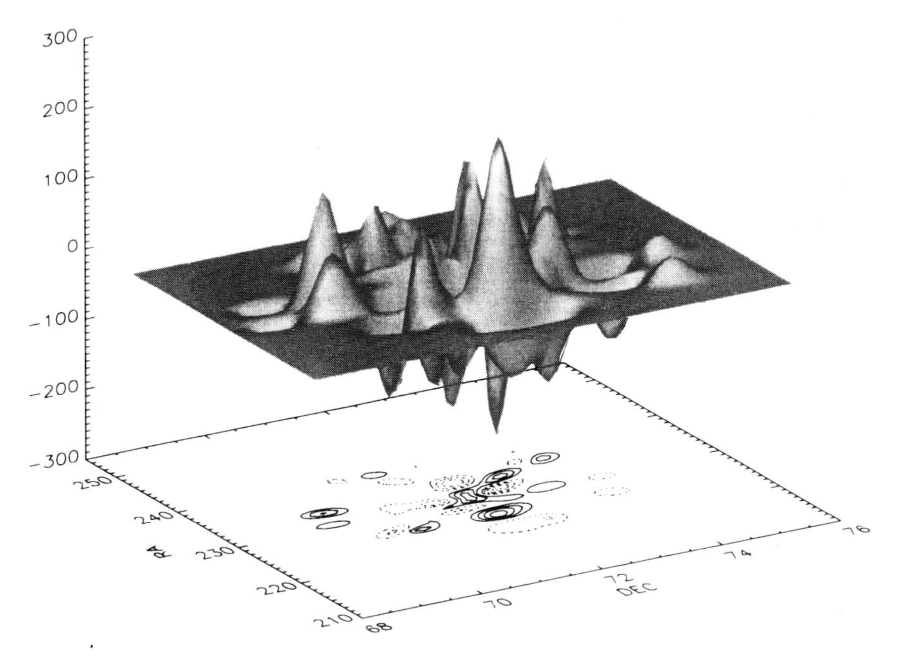

Fig. 8.3. A first map of the cosmic microwave background at 0.5-degree resolution (see text). (Reproduced with permission from M. White and the *Astrophysical Journal* (**443**, L53, 1995).)

though a sphere were to be expanded, from the point where we are now situated, and at a crucial moment, everything registered on the surface of the sphere were to be recorded for posterity. It is a once-only snapshot; that registered previously on the sphere is lost, and that registered after the snapshot is also lost. At the time the snapshot was made, the expanding sphere would have seemed minute compared to the radius of the hollow shell of Figure 8.1. Since the snapshot was made, both the structures it shows, and the snapshot itself, have grown enormously. We have literally inherited a gross 'photographic' enlargement that is so big that it takes the size of the observable Universe just to store it!

The photograph is a little out of focus. The surface of last scattering is not precise, but as suggested earlier, has a certain thickness. While the surface itself is put at a redshift in the region of $z = 1,000–1,100$, estimates of the thickness put Δz from 80 to 110. Ultimately, it will be this blurring, and not the technology, that may limit its usefulness. Clearly, the finer details are already lost.

It may also be like the photograph of a baby; but while the baby has since grown up, we have no photograph of the adult it became. The seedings of large-scale structures on the cosmic microwave background have presumably since grown to fully-fledged walls and voids, but we have no way of seeing what those particular structures are like today. The recorded snapshot of the cosmic microwave background is not a record of the early appearance of the very same structures we examine today, and we can only presume that the structures it gave rise to are of the same character as those we see today. If that is so, then a comparison is feasible.

8.4 QUANTITATIVE ASSESSMENTS OF THE ANISOTROPIES

Many of the quantitative measures described in Chapter 6 can also be applied or adapted to the cosmic microwave background. The main difference is that the background is a continuous two-dimensional image, whereas the distribution of galaxies is that of points in three-dimensional space. This book, however, is concerned with the salient points of the character of the cosmic microwave background, and not its detailed analysis, since that is a thesis in itself, for which there are excellent, if not highly technical, reviews available (such as that by M. White, D. Scott and J. Silk; see list at the end of this chapter).

Given the obvious limitations of the data at this stage, there is flexibility in the manner in which an analysis can be made. As an example, the analysis by E. de Gouvela Dal Pino and co-authors extracts evidence for very large fractal structure from the COBE data.

However, the power spectrum is more generally seen as the best comparative tool. At this stage, it is necessary to make assumptions – such as assuming a cold dark matter model (see next chapter) – to be able to extract meaningful points. By comparison with existing power spectra of large-scale structures – in particular that shown in Figure 6.4 – there is broad reconciliation and general overlap at low wave numbers, with k in the range 0.001–0.01 Mpc^{-1}. This is simple agreement, as already suggested in Section 8.3 above, with the probable existence of very large structures of 10,000 km/s or larger.

Collective results also suggest the power spectrum of the cosmic microwave background rising towards the expected turnover at k = 0.05 Mpc^{-1}. Calculated models also predict that the power spectrum of the cosmic microwave background might show so-called 'Doppler' peaks rising at values of k above 0.01 Mpc^{-1}, but these have not yet been observed.

There are, however, good prospects for the future, particularly if the anisotropies in the cosmic microwave background are mapped at increasingly higher resolution. It might then be possible to extract a true 'primordial' spectrum of fluctuations, to be compared with the present power spectrum of large-scale structures, and thereby impose much tighter constraints on theory and interpretations – the topic of the next chapter.

8.5 FURTHER READING

Efstathiou, G., Anisotropies of the Microwave Background Radiation and the Large-Scale Structure of the Universe, [in] *Clustering in the Universe* (*Ed.* S. Maurogordato, C. Balkowski, C. Tao, J. Tran Thanh Van), p.381, Edition Fontieres, 1995.

Silk, J., Recent Results of the Cosmic Microwave Background, [in] *Clustering in the Universe* (*Ed.* S. Maurogordato, C. Balkowski, C. Tao, J. Tran Thanh Van), p.373, Edition Fontieres, 1995.

White, M. and Bunn, E.F., A First Map of the Cosmic Microwave Background at 0.5 Resolution, *Astrophys. J.,* **443**, L53 (1995).

White, M. *et al.*, Anisotropies in the Cosmic Microwave Background, *Ann. Rev. Astron. Astrophys.*, **32**, 319 (1994).

de Gouvela Dal Pino, E.M. *et al.*, Evidence for a very large scale fractal structure in the Universe from COBE Measurements, *Astrophys. J.,* **442**, L45 (1995).

9

Simulations and interpretations

9.1 GENERAL CONSIDERATIONS

Large-scale structures in the Universe have been an astonishing revelation; no scientist or layperson, no theory or speculation, predicted their nature in advance of discovery. There is something quite bizarre about finding the Universe, on its largest observable scale, to contain a maze-like labyrinth that might be more at home in the cyberspace of one's personal computer than out there in the cosmos. Yet scientific intrigue and excitement thrive on the unexpected. Like the discovery of a Rosetta Stone, new and exciting interpretations are possible.

The emphasis of this book is on 'observational evidence', such that theory and interpretation have been reserved for this chapter. That it should be almost confined to a single chapter is hardly fair, since theory alone could easily form a book. It also indicates that this chapter is but a summary of an immense topic. Here we shall review the interpretations arising from large-scale structures, and see how computer simulations have contributed to much of the understanding.

Were it not for the existence of large-scale structures, the Universe would be seen as a pure gravitational hierarchy. Gravity causes matter to condense and form stars and planets, and assembles stellar systems, galaxies and even clusters of galaxies. Presumably, gravity must also cause condensation of the large-scale structures, but in itself it could not fabricate the maze-like labyrinth.

In the wake of the publicity surrounding the release of the Center for Astrophysics' 'Slice of the Universe' in the mid-1980s, the only theory that seemed to fit the bill was 'explosions' – giant explosions in the early Universe had blown material aside to leave the enormous cavities, much like supernova explosions have done within the interstellar material of our Galaxy. That theory, though not discarded, has today fallen out of favour. Instead, most investigators have looked for the seedings that grew to today's large-scale structures in an inflationary Universe theory. This theory, originated by Alan Guth in 1980, sought to explain the remarkable homogeneity of the early Universe, especially as seen in the almost perfect uniformity of the cosmic microwave background.

Guth's theory holds that the present era of sedate expansion of the Universe was preceded by an extraordinarily rapid exponential expansion. Roughly speaking, the Universe became 10^{30} times larger in only 10^{-30} of a second. The theory involves a satisfactory merging of field theory (which successfully explains particle behaviour at very high energies, e.g. at the very high temperatures of the extremely early Universe) and General Relativity (used to model the cosmos). Though it still seems quite bizarre to mere astronomers like myself, it has the great advantage of explaining why the Universe should now look so homogeneous (as seen earlier in Chapter 5), even when looking far out in opposite directions. In particular, it helps the understanding of why the cosmic microwave background (the distant Universe in all directions) should appear so uniform.

Inflation allows the possibility that quantum fluctuations that exist on the smallest scale possible could be so enormously magnified that they would become manifest in the Universe. They would become the seedings for the labyrinth of large-scale structures; the barely discernible anisotropies in the image of the embryonic Universe conveyed by the cosmic microwave background.

One assumption must be made from the outset: it is that the only force significant on a very large scale in the Universe is gravity. Once the initial 'seedings' were in place, gravity alone so enhanced them that they evolved into the present large-scale structures.

Suppose matter were to be absolutely uniformly spread through the early Universe; then, in theory, every particle of matter would experience equal gravitational force in every direction, and then nothing would form. However, this would be a delicate equilibrium, and the slightest deviation from absolute uniformity would upset everything. The slightest increase above average density at any point would start gravity pulling other material towards it, and the slightest decrease from average density would have the opposite effect. As time ran its course, so the overdensities would gain more material and more gravitational pull. The underdensities would lose material and effectively repel material due to their loss of gravitational pull. The process would escalate, so that in time most mass would gather where the overdensities are situated, and the underdensities would be gradually emptied of matter.

This general scenario has almost universal agreement. Where there is the greatest difference in opinion is in the initial 'sizes' of the inhomogeneities. Let us recall (from the opening chapter of this book) the two 'rival' theories from the 1970s. One was the 'American' or 'Western' view of 'hierarchical clustering'. It held that gravity first formed the galaxies, and that thereafter the galaxies assembled into groups that assembled into clusters which gradually settled into large-scale structures – if time permitted.

By contrast, the Soviet school favoured the theory of Ya. B. Zel'dovich. In what is now considered classic theory, Zel'dovich held that a large overdensity would collapse progressively under its mutual gravity, but would be most rapid along its smallest dimension and so assume a 'pancake' shape. 'Pancaking' is the signature of the Zel'dovich theory. In time, the next shortest axis would collapse the pancake to a filament, and eventually the filament would drain towards a central cluster. The discovery that galaxies were arrayed in sheets surrounding voids, and not just in clusters, was interpreted by the Soviet researchers as favouring 'pancaking'. Zel'dovich also leant strongly towards 'biased' galaxy formation, where only condensations that exceeded a threshold density actually formed into galaxies.

This initial difference in views across the Iron Curtain shows a fundamental separation of theory that is still only partially resolved today. We can pose the questions as follows: Which came first, the galaxies or the large-scale structures? Did the galaxies form first and afterwards assemble into the large-scale structures, or did the structures form first and then the galaxies condense within them? These two basic alternative scenarios, conveyed schematically in Figure 9.1, are known as 'bottom-up' and 'top-down' respectively.

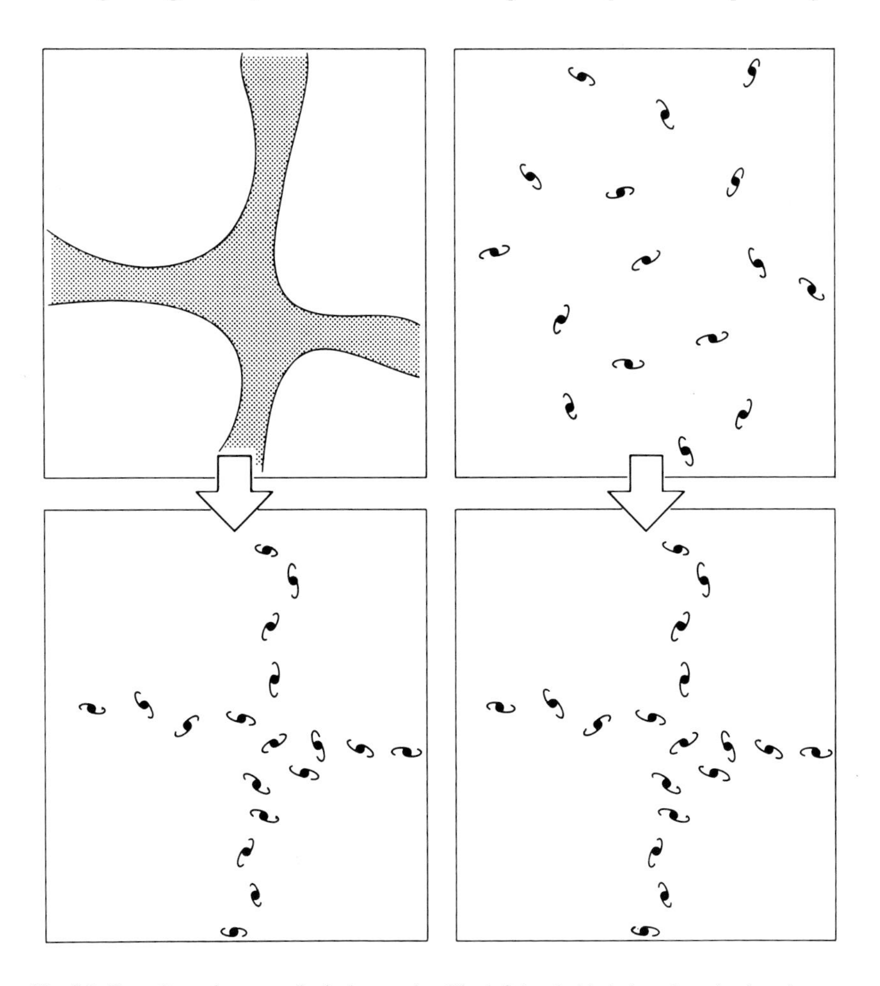

Fig. 9.1. Two alternative cosmological scenarios. The left-hand side is 'top-down', where large-scale structures form first and then condense into galaxies. The right-hand side is 'bottom-up', where galaxies form first and thereafter congregate into large-scale structures.

Since gravity is governed by an inverse-square law, the timescale on which it operates must increase dramatically with physical scale. For example, in our Solar System, the orbital periods of the planets increase according to their distance from the Sun, raised to the power of 3/2 (Kepler's third law). Orbital periods about the centre of the Galaxy are some 10^3–10^5 times longer than those of the Solar System's planets, and the same again is true in stepping up to the scale of a cluster of galaxies. The 'crossing time' for a galaxy to pass from one side of a moderately rich cluster to the other is typically a significant fraction of the age of the Universe. Were the condensation any larger, the crossing time would far exceed the age of the Universe. Consequently, large-scale structures are totally 'unrelaxed' gravitationally. The dramatic increase in the timescale with increasing physical scale means that gravity very decidedly operates 'bottom-up'.

Yet the galaxies are now seen to be arrayed in formations reminiscent of a foam-like fluid like soapsuds (Voronoi tessellation; Section 6.10), or, as some might prefer, the froth on top of a tankard of beer! Surely, the substance of the Universe must first have behaved like a fluid – to form the frothy texture – before that substance condensed into swarms of galaxies. Common sense, in the light of the remarkable fluid-like/foam-like texture, seems to endorse a 'top-down' scenario.

Common sense or not, the 'top-down' picture runs contrary to observational evidence. By looking ever deeper into space, we look back into the early history of the Universe. We have long been able to see quasars – the superbright nuclei of active galaxies – to redshifts of $z = 4$ or 5. Currently, the Hubble Space Telescope is showing us regular galaxies at similar redshifts, probably to at least $z = 7$. A current claim for the most distant galaxy ever observed (thanks to the Hubble and Keck telescopes, and gravitational lensing) suggests we may be seeing much further back, with star formation in a galaxy when the Universe was only a billion years old. If so, galaxies were already being formed by that time. If the 'top-down' scenario were to hold, then the whole process has to be complete within a billion years, and the time scale does not allow it.

Though we can now see galaxies at this early epoch of the Universe, we cannot of course discern whether or not they are arrayed into the large-scale structures. Perhaps, in time, that may be possible. Suffice to say, for now, that the related observational evidence does not seem to support the 'top-down' picture.

If this gravitational scenario is correct, then the great walls and labyrinth of the large-scale structures that we see today must first have existed as a pattern of weak fluctuations from an otherwise uniform density, as suggested in Figure 9.2. Gravity did not in itself create the pattern, but it would have enhanced the pattern enormously.

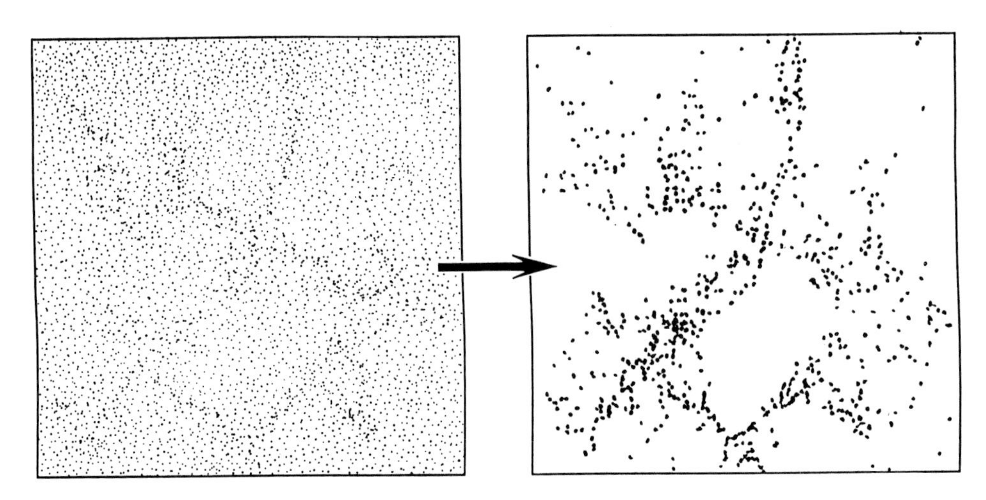

Fig. 9.2. The weak patterns in the cosmic microwave background provide the seeds for the condensation of the large-scale structures apparent today.

We shall presently see how simulations have enabled researchers to model the growth of density fluctuations. They have also revealed the need for there to be substantial dark matter involved. Since the 'inflationary Universe' theory also helped explain why the Universe is so 'flat' (that is, it does not show marked relativistic curvature), it has been widely interpreted – though not necessarily correctly so – as implying that the Universe should have a density close to the critical value (see also Sections 7.11 and 10.2). If so, then the bulk of the mass present may be dark matter.

While many researchers have moved towards a common scenario – the standard 'cold dark matter' model – there is no shortage of ideas. Even if we restrict ourselves to papers dealing with the origin and development of the large-scale structures themselves, as opposed to cosmology in general, then there are still far more theoretical than observational papers. As an observer, I am sorely tempted to think of Mark Twain's famous quote of science being "wholesale return of conjecture for a trifling investment of fact", but let us hope for better than that. Nevertheless, even the 'selected' reading listed at the end of this chapter runs towards a hundred references. Consequently, it will not be possible here to mention every approach, but rather to review the mainstream trends.

9.2 GRAVITATIONAL OVERDENSITIES

The growth of gravitational overdensities was first calculated by Sir James Jeans. He found that perturbations larger than a particular length (the Jeans length) grow exponentially with time, whereas those smaller are rather stabilised by internal pressure. His work, however, assumed a static Universe; in an expanding Universe, the growth is not exponential, but is at most a power of the time.

It is usual to express the perturbations in dimensionless fashion. If the mean value of the density of the Universe is ρ_0, then regions of overdensity can be expressed as a dimensionless $\delta(x,t) = (\rho(x,t) - \rho_0)/\rho_0$, where x is a spatial coordinate and t is time. We expect condensations to grow wherever δ is significantly greater than 0. The growth can be divided into two distinct epochs. In the first, $\delta(x,t) = D(t)\delta(x)$, i.e. all such overdensities grow according to a universal function in time. This is known as the 'linear' epoch.

The second epoch involves non-linear growth, for which there is no full theoretical basis. Present-day trends are to steer through the theory by means of approximations, the most famous of which is the 'Zel'dovich' approximation.

While Jeans' approach was essentially Eulerian – since it looks at density perturbations at fixed spatial coordinates – Zel'dovich used a Lagrangian approach, and saw the perturbations in the trajectories of fluid elements. The Zel'dovich approximation is therefore a linear Langrangian perturbation theory. It can be expressed as follows:

$$x(t) = q - b(t)\psi(q)$$

where x is the Eulerian coordinate and q is the Lagrangian coordinate. $\psi(q)$ describes the density fluctuations and b(t) the linear growth.

We shall see that the Zel'dovich approximation has also proved very useful in setting initial conditions for N-body simulations.

9.3 STANDARD N-BODY SIMULATIONS

The complexity of large-scale structures is such that they cannot be easily interpreted by theory alone. Simulations that 'test drive' the theory are necessary. Their outcomes can be compared to reality.

In a convenient twist of history, the recognition of large-scale structures only came about after the advent of relatively powerful computers. Such machines allow us to explore the possible outcome of various initial conditions, quantities of matter and so on. Some researchers have labelled the pursuit 'experimental cosmology', and others refer to 'toy universes'.

Such endeavours are known as N-body simulations. The matter in the Universe is represented by a large number of particles (though obviously not nearly enough). The particles are distributed in uniform fashion and then perturbed so as to reflect an initial power spectrum of fluctuations. The particles are contained in a limited volume; usually a cube. The data are then folded in each of the three dimensions, so that the simulated Universe consists of repetitions of the cube, and there are no edge effects. The cube also grows with the expansion of the Universe, but is a constant size in co-moving coordinates (see Section 3.3 for a definition).

As the cube grows – and as the Universe evolves – so gravitational forces act on the particles, thereby causing them to move continuously. The changing configuration of the particles in turn modifies the gravitational field, which must be repeatedly recalculated. The simulation can be halted at any desired time and compared to real distributions, or even viewed as an animated sequence. Some readers may have seen the Imax movie *Cosmic Voyage* which includes an animated N-body simulation (by Frank Summers) that depicts the evolution of the Universe from early inhomogeneities to the condensation of large-scale structures and clusters of galaxies.

The manner in which the initial inhomogeneities are created, and the different possibilities, will be discussed in the following section. For the rest of this section, we shall look at how the simulation proceeds thereafter.

The greater the number of particles, obviously the better the resolution of the simulation. But the use of a limited number of particles is only for convenience; the 'true' mass distribution would be a smoothed version of the particle distribution, and similarly the gravitational field must also be smoothed.

The basic recipe for N-body simulations can be found in the (frequently cited) 1985 paper by George Efstathiou, Marc Davis, Carlos Frenk and Simon White. This collaboration – sometimes called the Gang of Four – has laid many of the ground rules, and their simulations generally pioneered and established some fundamental results that we discuss below. There are, of course, a large number of researchers in the field (as indicated in the list of specialised reading at the end of this chapter).

As an insight into an N-body simulation, let us look into a little of what is involved. The initial procedure we shall describe is usually termed a 'particle mesh' (PM). Then, according to Efstathiou *et al*, one works as follows. Divide the volume into mesh cells where M is the number of cells in one dimension. Let n be an integer triple that defines the centre of each mesh cell. If x_i is the position vector of the *i*th particle, and W is a weighting scheme reflecting the assignment of particle masses, then the mass density at the grid points is given by:

$$\rho(n/M) = \frac{M^3}{N}\sum_{i=1}^{N} W(x_i - n/M)$$

From this, the gravitational potential at the grid points can be derived:

$$\phi(n/M) = \frac{1}{M^3}\sum_{n'} \mathcal{G}[(n-n')/M]\rho(n'/M)$$

Here $\mathcal{G}$ is an approximation to Green's function of Poisson's equation. Then the gravitational forces at the grid points can be derived:

$$F(n/M) = -\frac{D_n\phi}{N}$$

where the operator D_n is the differencing scheme used to extract forces from the potentials. Finally, the forces on each of the particles can be calculated:

$$F(x_i) = \sum_{n} W(x_i - n/M)F(n/M)$$

Efstathiou *et al* point out that the choice of W, G, D and M is a trade-off between accuracy and the size of the computational task. Smoothness of the extracted potential and forces is also a necessary requirement, and can be obtained by utilising mass assignment functions of increasing order. The potential may also be calculated using a technique employing fast Fourier transforms.

An improvement on the basic PM scheme is known as P^3M (particle–particle–particle mesh) that allows for far greater spatial resolution. In these simulations, local 'corrections' are added to the force derived from the mesh. The correction is based on the usual Newtonian $1/r^2$ attraction between particles, and is calculated for all particle pairs within the immediate neighbourhood of each particle. Particles outside the neighbourhood are more economically represented by the smoothed mesh forces.

Newton's law, in comoving coordinates, can be expressed as

$$\dot{\upsilon}_i + 2\frac{\dot{a}}{a}\upsilon_i = -a^{-3}\sum_{i \neq j}\frac{Gm_j x_{ij}}{|x_{ij}|^3} = a^{-3}F_i / m_i$$

(see Efstathiou *et al*) where

$$x = r/a(t)$$

and the cosmological expansion parameter a(t) is derived from the standard Friedmann equation of cosmology:

$$(da/dt)^2 = 8\pi G\,\rho_0\left(\Omega_i^{-1} - 1 + a^{-1}\right)/3$$

where ρ_0 is the mean mass density and Ω is the value of the cosmological density parameter.

An N-body simulation involves a complex mathematical procedure. Our coverage here, however, is restricted to very simple indications, and the reader is referred to the original paper of Efstathiou *et al* for the full discussion.

Straightforward Newtonian attraction and motion is collisionless. An obvious modification that could be incorporated is to accommodate collisions, such as happen in a real gas when its particles approach close enough. Ideally, we would like to see the simulation progress from a collisional gas to an eventual collisionless stellar 'fluid'. However, given the enormous change in scale, the question is whether any N-body simulation can approach the behaviour of a collisional gas and see it evolve to produces galaxies, stars and large-scale structures.

The concern is, of course, focused on producing large-scale structures that simulate the observed labyrinth already described in this book. Thus the use of particles is to merely 'represent' the distribution of matter in the early Universe, and not to attempt to simulate its microscopic behaviour. Gravitational evolution can then examine how small fluctuations grow, and whether the spatial forms and character of those fluctuations can approach what the observations reveal. It would be over-optimistic to expect anything resembling galaxies to emerge from the simulations. At most, galaxy 'haloes' – precursors to galaxies – might emerge.

In any case, as overdensities draw in matter, so a complication may arise. As long as $\delta\rho/\rho$ is less than 1, so perturbation theory is held in a linear regime. Higher overdensities which occur on small physical scales involve non-linear theory and its associated complications, with which most simulations cannot cope, although various ways around this have been put forward.

9.4 INITIAL FLUCTUATIONS

The inflationary theory predicts that the extremely early Universe underwent enormous exponential growth. In so doing, quantum fluctuations (that are normally thought to exist in space-time on microscopic scales smaller than 10^{-30} metres) were grossly magnified, so much so that they provided the initial fluctuations from which large-scale structures grow. Quantum fluctuations show no preferred scale; they are represented by a power spectrum $P(k) \propto k$. Moreover, statistically they are Gaussian in nature.

'Gaussian fluctuations' are considered the standard starting point for the growth of density perturbations. In terms of N-body simulations, the initial positions of the particles must reflect such fluctuations. This is a far from trivial task, and again the reader is referred to details in the paper by Efstathiou *et al.* Random phases must be applied, while the resulting power spectrum must be suitably monitored, in terms of the portion of the Universe represented.

9.5 THE SUCCESS OF N-BODY SIMULATIONS

The importance of the N-body simulations is their success in mimicking the appearance of large-scale structures. From a vast repertoire of such simulations displayed in the literature, we show here three representative examples. From visual appearances alone, it is obvious that they readily show most of the essential features of large-scale clusters, particularly filaments and elongated structures.

Aside from visual appearances, N-body simulations can be compared with the real redshift data by means of the quantitative measures presented as Chapter 6 of this book. Thus, correlation functions, power spectra, counts in cells, genus and fractals have all been extensively applied to the outcome of N-body simulations.

The pioneering N-body simulations by Efstathiou and colleagues not only indicated the general success of the technique but showed that the best fits were with cold dark matter (CDM), $\Omega = 1$ and with biased galaxy formation. It was seen as an important endorsement for the existence of CDM, and has been something of a standard model ever since. However, comparisons with surveys such as QDOT suggested that this canonical model nevertheless fell short of producing enough structure on very large scales.

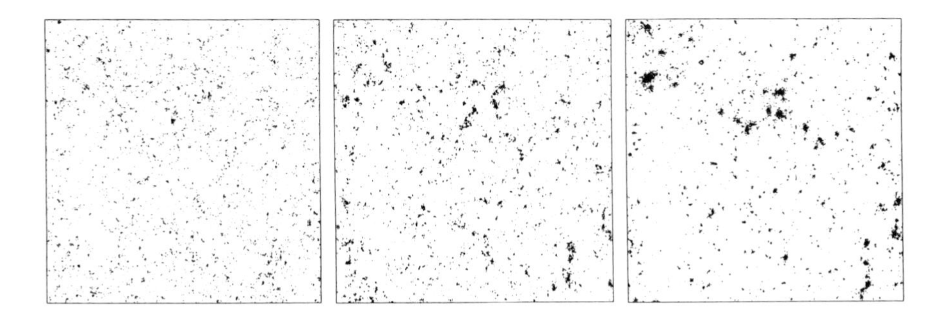

Fig. 9.3. An example of an N-body simulation seen at three different epochs. 'Large-scale structures' are seen to develop. (Reproduced with permission of G. Efstathiou and the *Monthly Notices of the Royal Astronomical Society* (**235**, 715, 1988).)

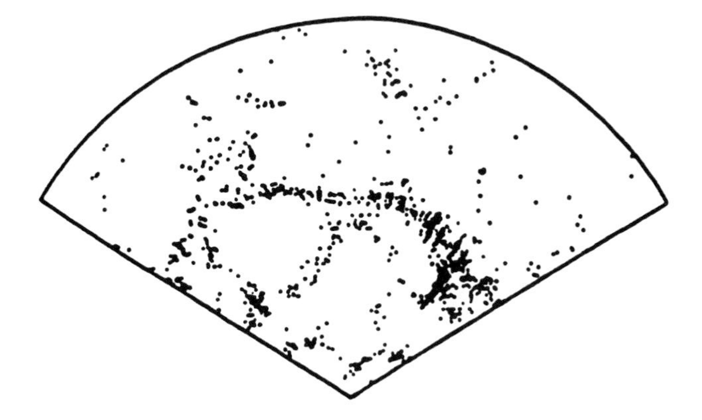

Fig. 9.4. Another N-body simulation that brings about structures resembling the Great (Coma) Wall. (Reproduced with permission from Changbom Park and the *Monthly Notices of the Royal Astronomical Society* (242, 59P, 1990).)

More recent times have seen the insistence on $\Omega = 1$ relaxed, and better fits to the data have now come from lower-density models. For example, Guinevere Kauffmann and Simon White presented an $\Omega = 0.2$ CDM model and simulation with correlation functions, counts in cells and so on, that matched observations as closely as could be expected on a scale between 100 and 2,000 km/s. Only on a smaller scale were the real galaxies found to be more strongly clustered than N-body simulation; this, however, is less a criticism of the simulation as it is probably due to the distinction between luminous and dark matter.

N-body simulations such as these have revealed that CDM is distributed much like luminous matter. The voids are just as empty of CDM as they are of galaxies. Filaments and large-scale structures are composed of CDM and not just galaxies. However, CDM is not so condensed as luminous matter is into galaxies; it rather forms haloes to individual galaxies or common envelopes.

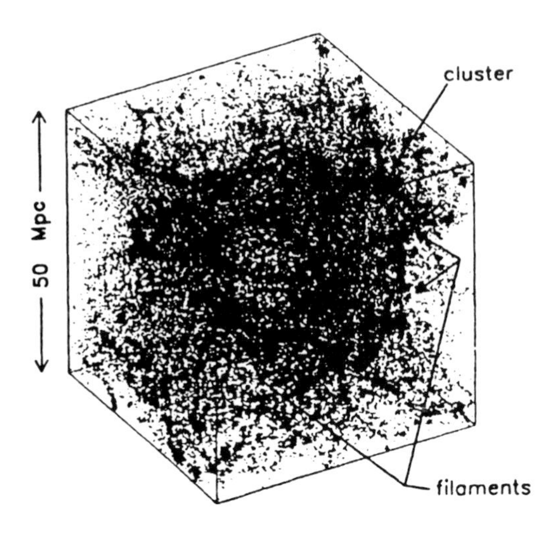

Fig. 9.5. An N-body simulation of hierarchical clustering, revealing a web of interconnecting filaments. (Reproduced with permission from L. Kofman and *Nature* (**380**, 603). Copyright (1996) Macmillan Magazines Ltd.)

9.6 COLD VERSUS HOT DARK MATTER

'Cold' dark matter implies that the particles of which it is comprised do not have significant thermal velocities. Yet, in spite of the general acceptance of CDM models, we have not moved closer to identifying what the particles actually are. 'Photinos' and 'axions' (postulated from the world of nuclear physics) have often been mentioned. Nowadays, the generic term Non-Baryonic is used, but it is nevertheless disconcerting that much of the matter in the Universe is far removed from our familiarity.

By contrast, there is a very good candidate for 'hot' dark matter – the neutrino (abundant and detectable, and well established in nuclear physics). The neutrino is, however, as close to massless as such particles can be, and even more massive versions would nevertheless have very high thermal velocities.

Hot dark matter (HDM) can be accommodated in N-body simulations and shows a quite different outcome. The high thermal velocities soon smooth out smaller density fluctuations.

Figure 9.6 shows the difference in the power spectrum. Note that for both CDM and HDM, there is a turnover due to the continuous modification of the primordial power spectrum during the radiation-dominated epoch of the early Universe (when radiation–matter coupling smoothed out small-scale fluctuations, of higher wave number).

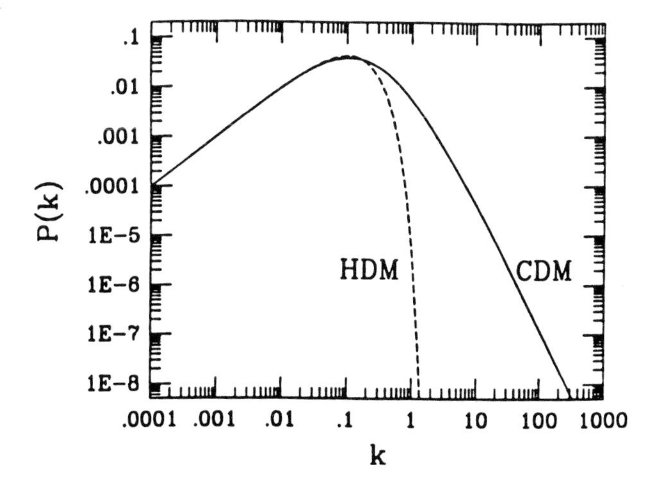

Fig. 9.6. A power spectrum of density fluctuations showing the distinction between the influences of 'hot dark matter' (HDM) and 'cold dark matter' (CDM). (Reproduced with permission from B. Ryden.)

In the HDM simulation, all but the largest scale fluctuations are smoothed away, so the condensations that form first are massive: the Zel'dovich pancakes. We have returned to the classic debate of top-down HDM versus bottom-up CDM. As in the earlier discussion, the bottom-up scenario is preferred, as similarly suggested by the CDM N-body simulations. Nevertheless, there are still some merits in top-down HDM, particularly in terms of the largest scales.

Is a compromise possible? Yes, according to the N-body simulations; many researchers today are amenable to mixed HDM and CDM, which is possibly replacing the older CDM-only standard model.

9.7 FILAMENTS, VOIDS AND CLUSTERS

N-body simulations are often seen to resemble visually the real distribution of galaxies because of the presence of filaments. In fact, filaments are so ubiquitous that some investigators have seen fit to suppose a 'cosmic web'. We have seen earlier, in the real distribution of galaxies (Chapter 4), that a number of major structures do have a filamentary nature.

An alternative approach has been to see the cosmic labyrinth as shaped by the growth of voids. Mention has already been made of the 'explosion' theory, an avenue which a number of researchers are still exploring. Figure 9.7 shows an example. Figure 9.8 displays an elegant simulation whereby the growth and merging of bubble-like voids brings about a fabric remarkably similar to that observed. Again, this is an attraction to a top-down scenario: there has to be a continuous medium trapped between the expanding bubbles. Only once the texture was generated could the material fragment and condense into galaxies. Had the walls of the expanding bubbles already been composed of galaxies, the bubbles' expansion would never have been arrested in the manner suggested in Figure 9.7. Instead, the 'bubbles' of galaxies would have merely expanded through one another, and the texture would thereafter be obliterated.

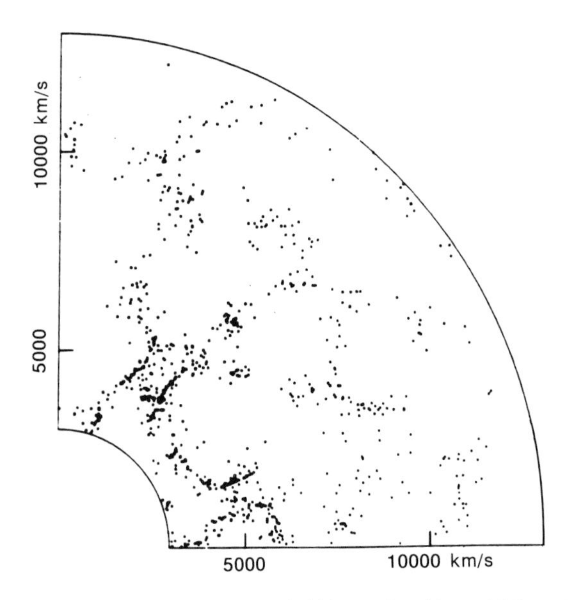

Fig. 9.7. An N-body simulation that incorporates bubbles produced by multiple explosions, giving rise to a texture very much like that seen in the observational data. (Reproduced with permission from J. Ostriker and the *Astrophysical Journal* (**338**, 579, 1989).)

Others have sought to use the spacings of filaments and voids to explain the BEKS peaks (Section 5.6). If such regularity were to exist in the Universe – particularly Einasto *et al*'s lattice (Figure 5.7) – then it must fly in the face of concepts such as the Gaussian perturbations. However, most theorists agree that as yet the evidence is far from compelling, and only mildly constraining. Perhaps a possibility for creating such regular patterns might come from 'sound' waves travelling through the material of the very early Universe.

N-body simulations have also been used to explore the spatial positions and motions of clusters. Again, correlation functions and the like can be extracted. The question of systematic motions, particularly in the light of the Lauer–Postman effect (Section 7.10) has received attention. Very large-scale bulk flows require strong power on very large scales – which is generally not accepted.

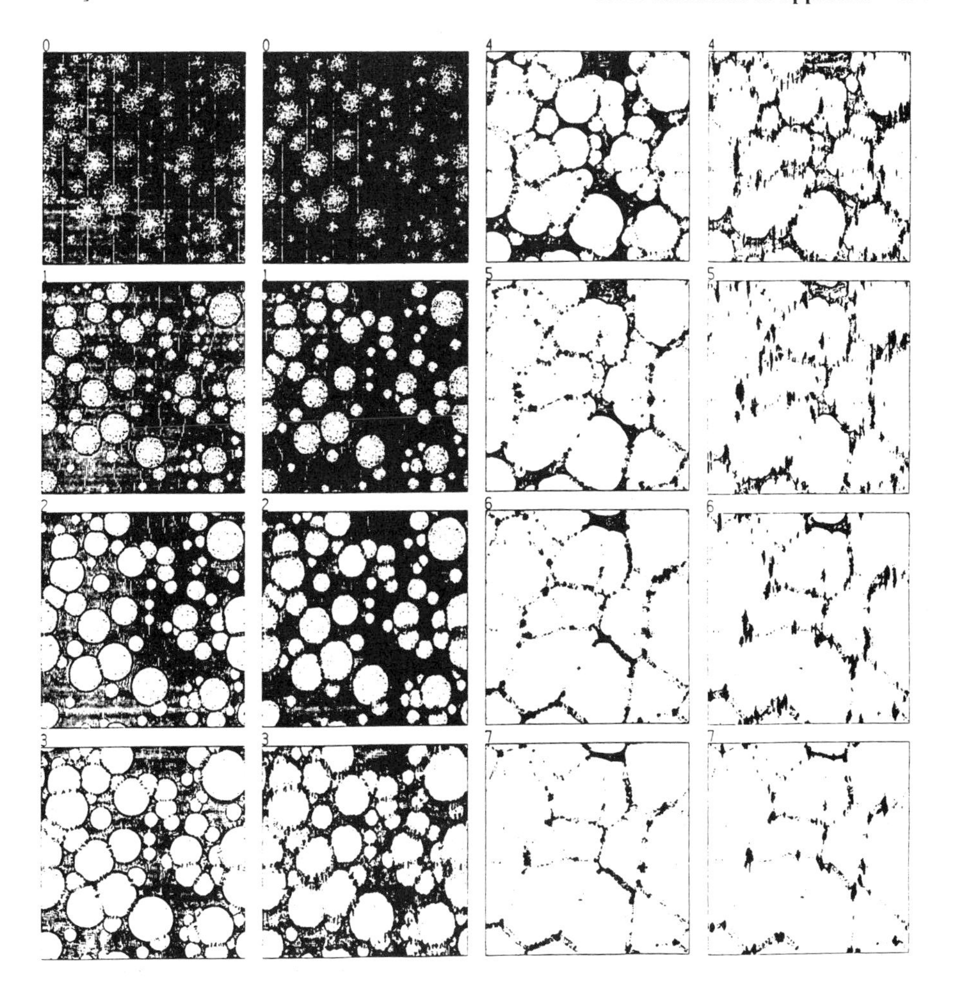

Fig. 9.8. The evolution of a two-dimensional simulation of expanding bubbles, resulting in a texture similar to that seen in the large-scale structures (by E. Regos, and M. Geller). The first and third columns show real space, and the second and fourth columns redshift space. (Reproduced with permissions from M. Geller and the *Astrophysical Journal* (**337**, 14, 1991).)

9.8 SOME VARIATIONS IN APPROACH

Finally, let us review a few interesting contributions that offer possibilities in further understanding, or which adopt quite different lines of approach.

An interesting innovation has been provided by the 'adhesion' model, an extension to the Zel'dovich model that introduces a mock viscosity term. It brings about an intriguing translation from Lagrangian to Eulerian space, that is conveyed and explained in Figure 9.9.

A number of more radical alternatives have looked at non-Gaussian perturbations as initiating the formation of large-scale structures. For a time, prominence was given to the

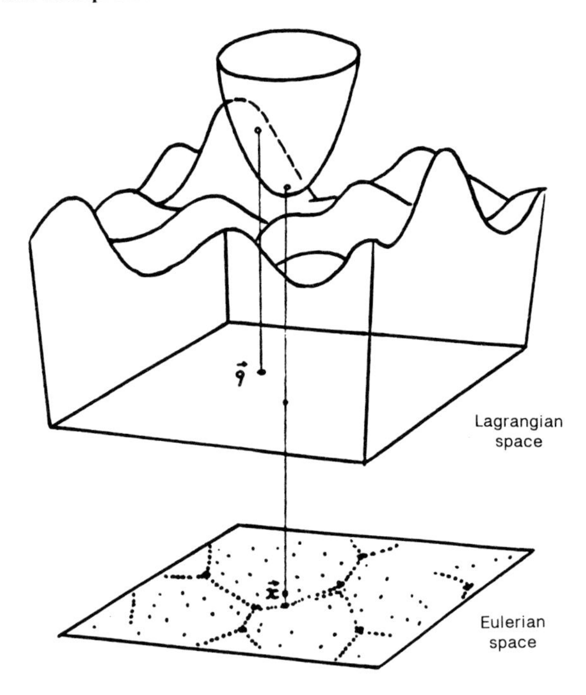

Fig. 9.9. A geometrical procedure for mapping from Lagrangian space to Eulerian space, from the 'adhesion' model. A paraboloid is lowered vertically downwards until it touches the Lagrangian surface. Where it touches at two points, a filament exists in Eulerian space. Where it touches at three points, a cluster exists. (Reproduced with permission from L. Kofman and the *Monthly Notices of the Royal Astronomical Society*.)

idea of 'cosmic strings' – topological defects from the early Universe – as creating a wake of perturbations from which galaxies might form. The occurrence of elongated structures such as great walls appeared to be compatible with cosmic strings, as suggested, for example, in Figure 9.10. The idea now seems less fashionable, particularly since the discovery of the anisotropies in the cosmic microwave background.

Though by no means perfectly satisfactory, the general success of N-body simulations seems to have provided considerable understanding. Many theorists would be happy with the state of the art. This chapter has tried to bring out the essential aspects of the mainstream background theory. In our final chapter, we shall question how good our understanding might be, and offer some speculation.

9.9 FURTHER READING

Specialised

General

Sahni, V. and Coles, P., Approximation methods for non-linear gravitational clustering, *Phys. Rep.*, **262**, 1 (1995).

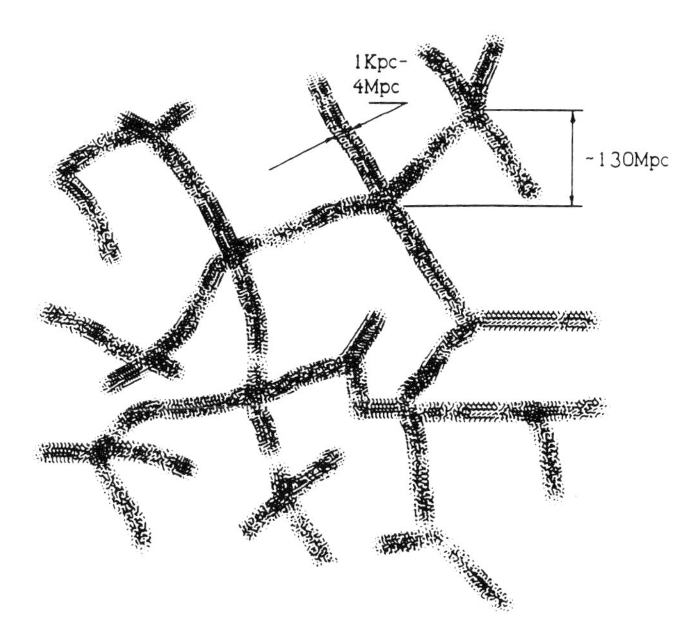

Fig. 9.10. An example of a connected string network that could fashion the geometry of large-scale structures. (Reproduced with permission from X. Luo, D. Schramm and the *Astrophysical Journal* (**394**, 12, 1992).)

Coles, P. and Sahni, V., Large-Scale Structure without N-body Simulations – The Legacy of Ya. B. Zel'dovich, *The Observatory*, **116**, 25 (1996).

Ryden, B., Cosmology. Galaxy Formation, [in] *The Astronomy and Astrophysics Encylopedia* (*Ed.* S. Maran), p.156, Van Nostrand Reinhold/Cambridge University Press, 1992.

Standard N-body CDM

Bahcall, N. A. *et al.* Redshift Space Clustering of Galaxies and Cold Dark Matter Model, *Astrophys. J.*, **408**, L77 (1993).

Baugh, C.M. and Gaztanaga, E. Testing Ansatze for quasi-non-linear clustering: the linear APM power spectrum, *Mon. Not. R. astr. Soc.*, **280**, L37 (1996).

Bouchet, F.R. *et al.* Non-linear matter clustering properties of a cold dark matter universe, *Astrophys. J.*, **383**, 19 (1991).

Brainerd, T.G. and Villumsen, J.V. On the peculiar velocity field of a CDM Universe, *Astrophys. J.*, **436**, 528 (1994).

Brainerd, T.G. *et al.* Velocity Dispersion and the Redshift-Space Power Spectrum, *Astrophys. J.*, **464**, L103 (1996).

Carlberg, R.G. Galaxy formation and clustering in an *N*-body Experiment, *Astrophys. J.*, **332**, 26 (1988).

Couchman, H.M.P. and Carlberg, R.G. Large-scale structure in a low-bias universe: *Astrophys. J.*, **389**, 453 (1992).

Doroshkevich, A.G. *et al.*, The formation and evolution of large- and superlarge-scale structure in the Universe – II. *N*-body simulations, *Mon. Not. R. astr. Soc.*, **284**, 633 (1997).
Efstathiou, G. *et al.*, Numerical Techniques for Large Cosmological *N*-body Simulations, *Astrophys. J. Suppl.*, **57**, 241 (1985). (Quoted extensively in Section 9.3.)
Efstathiou, G. *et al.* Gravitational clustering from scale-free initial conditions, *Mon. Not. R. astr. Soc.*, **235**, 715 (1988).
Gramman, M. Non-Self Similar Clustering of Galaxies, *Astrophys. J.*, **401**, 19 (1992).
Jain, B., Mo, H.J. and White, S.D.M. The evolution of correlation functions and power spectra in gravitational clustering, *Mon. Not. R. astr. Soc.*, **276**, L25 (1995).
Kates, R. *et al.* Large-scale structure formation for power spectra with broken scale invariance, *Mon. Not. R. astr. Soc.*, **277**, 1254 (1995).
Kauffmann, G. and White, S.D.M. The Observational Properties of an $\Omega = 0.2$ cold dark matter universe, *Mon. Not. R. astr. Soc.*, **258**, 511 (1992).
Lambras. D.G. *et al.* Large-Scale Structure in biased cold dark matter cosmologies, *Astrophys. J.*, **414**, 30 (1993).
Little, B. and Weinberg, D.H. Cosmic Voids and Biased Galaxy Formation, *Mon. Not. R. astr. Soc.*, **267**, 605 (1994).
Martel, H. *N*-Body Simulation of large-scale structures in $\Lambda = 0$ Friedmann models, *Astrophys. J.*, **366**, 353 (1991).
Menci, N. and Caldarnini, R. Merging Rates inside large-scale structures, *Astrophys. J.*, **436**, 559 (1994).
Nusser, A. and Dekel, A., Tracing large-scale fluctuations back in time, *Astrophys. J.*, **391**, 443 (1992).
Park, C. and Gott III, J.R. Simulation of deep one- and two- dimensional redshift surveys, *Mon. Not. R. astr. Soc.*, **249**, 288 (1991).
Park, C., Large *N*-body simulations of a universe dominated by cold dark matter, *Mon. Not. R. astr. Soc.*, **242**, 59P (1990).
Peacock, J.A. and Dodds, S.J. Non-linear evolution of cosmological power spectra, *Mon. Not. R. astr. Soc.*, **280**, L19 (1996).
Summers, F.J. and Davis, M. Galaxy Tracers and Velocity Bias, *Astrophys. J.*, **454**, 1 (1995).
Ueda, H. and Yokoyama, J. Counts-in-cells analysis of the statistical distribution in an *N*-body simulated universe, *Mon. Not. R. astr. Soc.*, **280**, 754 (1996).

Initial fluctuations and growth

Appel, L. and Jones, B.J.T. The mass function in biased galaxy formation, *Mon. Not. R. astr. Soc.*, **245**, 522 (1990).
Chodorowski, M.J. and Bouchet, F.R. Kurtosis in large-scale structure as a constraint on non-Gaussian initial conditions, *Mon. Not. R. astr. Soc.*, **279**, 557 (1996).
Henriksen, N. and Lachièze-Rey, M. The mass function of galaxies based on correlated velocity structure, *Mon. Not. R. astr. Soc.*, **245**, 255 (1990).
Hnatyk, B.I. *et al.* Great Attractor-like fluctuations: observational manifestations and theoretical constraints, *Astron. Astrophys.*, **300**, 1 (1995).

Mancinelli, P.J. and Yahil, A. Local nonlinear approximations to the growth of cosmic structures, *Astrophys. J.*, **452**, 75 (1995).

Melott, A.L. and Shandarin, S.F. Generation of large-scale cosmological structures by gravitational clustering, *Nature*, **346**, 633 (1990).

Messina, A. *et al.* Non-Gaussian initial conditions in cosmological *N*-body simulations – I. Space-uncorrelated models, *Mon. Not. R. astr. Soc.*, **245**, 244 (1990).

Muller, H-R and Treumann, R.A. Contribution of Peculiar Shear Motions to Large-Scale Structure, *Astrophys. J.*, **427**, L5 (1994).

Park, C. *et al.* Large-Scale Structure in a texture-seeded cold dark matter cosmogony, *Astrophys. J.*, **372**, L53 (1991).

Peebles, P.J.E., The Primeval Mass Fluctuation Spectrum and the Distribution of the Nearby Galaxies, *Astrophys. J.*, **473**, 42 (1996).

Ruffini, R. *et al.* The Fragmentation of Supercluster and Large Scale Structure of the Universe, *Astron. Astrophys.*, **232**, 7 (1990).

Sheth, R.K. The Distribution of Counts in Cells in the Non-Linear Regime, *Mon. Not. R. astr. Soc.*, **281**, 1124 (1996).

Stirling, A.J. and Peacock, J.A. Power correlations in Cosmology: Limits on primordial non-Gaussian density fields, *Mon. Not. R. astr. Soc.*, **283**, L99 (1996).

Cold and hot dark matter

Borgani, S. *et al.* The epoch of structure formation in blue mixed dark matter models, *Mon. Not. R. astr. Soc.*, **280**, 749 (1996).

Carr, B. Baryonic Dark Matter, *Annu. Rev. Astron. Astrophys.*, **32**, 531 (1994).

Dave, R. *et al.*, Filament and shape statistics: a quantitative comparison of cold + hot and cold dark matter cosmologies versus CfA1 data, *Mon. Not. R. astr. Soc.*, **284**, 607 (1997).

Frisch, P. *et al.*, Evolution of the supercluster-void network, *Astron. Astrophys.*, **296**, 611 (1995).

Klypin, A. and Shandarin, S.F. Percolation Technique for Galaxy Clustering, *Astrophys. J.*, **413**, 48,. (1993).

Primack, J.R. *et al.*, Cold + Hot dark matter cosmology, [in] *Clustering in the Universe* (*Ed.* S. Maurogordato, C. Balkowski, C. Tao, J. Tran Thanh Van), p.313, Edition Fontieres, 1995.

Primack, J.R. Improved ways to compare simulations to data, [in] *Clustering in the Universe* (*Ed.* S. Maurogordato, C. Balkowski, C. Tao, J. Tran Thanh Van), p.185, Edition Fontieres, 1995.

Sadoulet, B. The Nature of dark matter: the case for nonbaryonic dark matter, [in] *Clustering in the Universe* (*Ed.* S. Maurogordato, C. Balkowski, C. Tao, J. Tran Thanh Van), p.453, Edition Fontieres, 1995.

Quantitive measures of N-body simulations

Baugh, C.M. *et al.* A comparison of the evolution of density fields in perturbation theory and numerical simulations – II. Counts-in-cells analysis, *Mon. Not. R. astr. Soc.*, **274**, 1049 (1995).

Bharadwaj, S., Perturbative Growth of Cosmological Clustering. II. The Two-Point Correlation, *Astrophys. J.,* **460**, 28 (1996).

Bromley, B.C. Correlations in Cosmic Density Fields, *Astrophys. J.,* **437**, 541 (1994).

Buryak, O. and Doroshkevich. Correlation function as a measure of the structure, *Astron. Astrophys.,* **306**, 1 (1996).

Castagnoli, C. and Provenzale, A. From small-scale fractality to large-scale homogeneity: a family of cascading models for the distribution of galaxies, *Astron. Astrophys.,* **246**, 634 (1991).

Gott II, J.R., Cen, R. and Ostriker, J.P. Topology of large-scale structure by galaxy type: hydrodynamic simulations, *Astrophys. J.,* **465**, 499 (1996).

Jing, Y.P. *et al.* Three-point correlation function of galaxy clusters in cosmological models: a strong dependence on triangle shapes, *Mon. Not. R. astr. Soc.,* **277**, 630 (1995).

Matsubara, T. and Suto, Y. Nonlinear Evolution of genus in a primordial random Gaussian density field, *Astrophys. J.,* **460**, 51 (1996).

Park, C. and Gott III, J.R., Simulation of deep one- and two-dimensional redshift surveys, *Mon. Not. R. astr. Soc.,* **249**, 288 (1991).

Park, C. and Gott III, R. Dynamical Evolution of Topology of Large-Scale Structure, *Astrophys. J.,* **378**, 457 (1991).

Pierre, M. Probes for the large-scale structure, *Astron. Astrophys.,* **229**, 7 (1990).

Pons-Borderia, M.J. and Martinez, V.J. The three-point correlation function: moment method vs. direct counting, *Astron. Astrophys.,* **293**, 5 (1995).

Ueda, H. *et al.* Quantifying the pattern of galaxy clustering, *Publ. astr. Soc. Japan,* **45**, 7 (1993).

BEKS peaks

Budinich, P. *et al.*, The Spontaneous Violation of the Cosmological Principle and the possible Wave Structures of the Universe, *Astrophys. J.,* **451,** 10 (1995).

Dekel, A. *et al.*, Large-Scale Periodicity and Gaussian Fluctuations, *Mon. Not. R. astr. Soc.,* **257**, 715 (1992).

Hill, C.T. *et al.* Coherent Peculiar Velocities and Periodic Redshifts, *Astrophys. J.,* **366,** L57 (1991).

Voids, explosions and bubbles

Amendola, L. and Borgani, S. Large-scale clustering in Bubble Models, *Mon. Not. R. astr. Soc.,* **266,** 191 (1994).

Einasto, J. *et al.* The fraction of Matter in Voids, *Astrophys. J.,* **429**, 465 (1994).

Miranda, O.D. and Opher, R. The creation of Large-Scale Voids by Explosions of Primordial Supernovae, *Mon. Not. R. astr. Soc.,* **283**, 912 (1996).

More, J.G. *et al.*, Explosions in pancake models of galaxy formation, *Mon. Not. R. astr. Soc.,* **243**, 413 (1990).

Nakao, K. *et al.*, The Hubble parameter in a void universe: effect of the peculiar velocity, *Astrophys. J.,* **453,** 541 (1995).

Ostriker, J.P. and Strassler, M.J. On the generation of a bubbly universe: a quantitative assessment of the CfA slice, *Astrophys. J.,* **338**, 579 (1989).

Regos, E. and Geller, M.J. The evolution of void-filled cosmological structures, *Astrophys. J.,* **377**, 14 (1991).

Filaments

Bond, J.R. *et al.*, How filaments of galaxies are woven in to the cosmic web, *Nature,* **380**, 603 (1996).
Sathyaprakash, B.S. and Sahni, V. Emergence of filamentary structure in cosmological gravitational clustering, *Astrophys. J.*, **462**, L5 (1996).
West, M.J. Filamentary superclustering in a universe dominated by cold dark matter, *Astrophys. J.*, **369**, 287 (1991).

Adhesion

Kofman, L. *et al.*, Structure of the universe in the two-dimensional model of adhesion, *Mon. Not. R. astr. Soc.*, **242**, 200 (1990).
Nusser, A. and Dekel, A. Filamentary structure from Gaussian flucations using the adhesion approximation, *Astrophys. J.*, **362**, 14 (1990).

Strings

Hara, T. and Miyoshi, S. Comparison of three orthogonally crossed wakes with the CfA large-scale structure, *Astrophys. J.*, **405**, 419 (1993).

Clusters

Borgani, S., The cluster distribution as a test of dark matter models – I. Clustering properties, *Mon. Not. R. astr. Soc.*, **277**, 1191 (1995).
Brunozzi, P.T. *et al.*, The cluster distribution as a test of dark matter models – II. The dipole structure, *Mon. Not. R. astr. Soc.*, **277,** 1210 (1995).
Eke, V.R. *et al.* Cluster correlation functions in *N*-body simulations, *Mon. Not. R. astr. Soc.*, **281**, 703 (1996).
Gramman, M. *et al.* Large-scale motions in the universe using clusters of galaxies as tracers, *Astrophys. J.*, **441**, 449 (1995).
Jaffe, A.H. and Kaiser, N. Likelihood analysis of large-scale flows, *Astrophys. J.*, **455**, 26 (1995).
Kolokotronis, V. *et al.*, Sampling effects on cosmological dipoles, *Mon. Not. R. astr. Soc.*, **280**, 186 (1996).
Mo, H.J. *et al.*, The correlation function of clusters of galaxies and the amplitude of mass fluctuations in the Universe, *Mon. Not. R. astr. Soc.*, **282**, 1096 (1996).
Moscardini, L. *et al.* The cluster distribution as a test of dark matter models – III. The cluster velocity field, *Mon. Not. R. astr. Soc.*, **282**, 384 (1996).
Paredes, S. *et al.* The clustering of galaxy clusters: synthetic distributions and the correlation function amplitude, *Mon. Not. R. astr. Soc.*, **276**, 1116 (1995).

Alternative/miscellaneous

Anninos, W.Y. *et al.* Nonlinear hydrodynamics of cosmological sheets. II. Fragmentation and the gravitational, cooling, and thin-shell instabilities, *Astrophys. J.*, **450**, 1 (1995).
Bernardeau, F., The Effects of smoothing on the statistical properties of large-scale cosmic fields, *Astron. Astrophys.*, **291**, 697 (1994).
Fujimoto, M. A large-scale periodic clustering of galaxies as a result of hydromagnetic

ringing of gas in a recombination era of the expanding universe, *Publ. Astron. Soc. Japan*, **42**, L39 (1990).

Heavens, A., Large-scale structure in the Universe, *Mon. Not. R. astr. Soc.,* **213**, 143 (1985).

Matarrese, S. *et al.*, A frozen-flow approximation to the evolution of large-scale structures in the universe, *Mon. Not. R. astr. Soc.,* **259**, 437 (1992).

West, M.J., Groups of Galaxies and Large-Scale Structure, *Astrophys. J.,* **344**, 535 (1989).

Zaroubi, S. and Hoffman, Y., Clustering in Redshift Space: Linear Theory, *Astrophys. J.,* **462**, 25 (1996).

10

Concluding discussion

10.1 THE FABRIC OF THE COSMOS

We have come a long way since Sir John Herschel first perceived that our Galaxy was 'involved with the outlying members' of the Virgo Supercluster. And yet, it is remarkable to realise that very few investigators recognised large-scale structures until the late 1970s; and it was still many years after that before they were noticed by the wider research community.

This short closing chapter is devoted to a personal assessment – a discussion of some of the problems which, for this author, remain prominent. The content of this chapter is therefore more a matter of opinion rather than the balanced reviews of previous chapters.

Do we understand large-scale structures? The answer would have to be, "partially". There is no doubt that gravity is largely responsible for collapsing and shaping large-scale structures. That the general character of the structures is reflected by the N-body simulations is an endorsement of this. The Universe we inhabit is one governed by gravity.

The intrigue of large-scale structures, however, rests with the initial fabric; the weak patterns that initiated the collapse and formation of the large-scale structures. This fabric of the cosmos is the only 'structure' to have survived from the early Universe; the only structure not transformed by gravity into 'relaxed' entities such as the clusters, galaxies and stars. Like the cosmic microwave background, it is a representation of the early embryonic Universe. Is it enough to write off the cosmic fabric as simple statistical fluctuations? Or is there more to it than that? Majority opinion at present favours Gaussian fluctuations – quantum fluctuations blown up to gigantic proportions by the 'inflationary Universe'. But quantum fluctuations come from a scale of the Universe we have never explored, and which, by definition, we are never likely to explore.

Personally, I would be disappointed if this were really the case; there seems to me to be a greater message in the cosmic fabric than mere 'statistical fluctuations'. While I recognise that many of the N-body simulations resemble the true distribution of galaxies, I feel that too great a faith has been placed on the power spectrum as a discriminant. The power spectrum, like its parent correlation function, rests more upon the degree of coherence of galaxies than it does upon texture. As pointed out earlier in this book, correlation functions were around long before the recognition of the cellular large-scale structures, and they were largely unaffected by the finding that the Universe on a large-scale resembled soapsuds, or perhaps more accurately, a bath sponge – the astonishing revelation, or revolution, that forms the basis of this book. The topography of the Universe is not conveyed by correlation functions or power spectra.

But what can we say of the cosmic fabric that is not, in my opinion, adequately covered or accounted for by the usual quantitative measures?

Large-scale structures, as we have seen, tend to form wall-like or ribbon-like forms. Though similar features appear to exist in N-body simulations, they do not occur with the same regularity as we saw for nearby structures (in Chapter 4). There we saw that probably 80–90 per cent of nearby galaxies resided in a handful of walls or ribbons of one sort or another (those listed individually in Section 4.4). I am not convinced that the 'large-scale structures' that occur in N-body simulations are sufficiently wall-like or ribbon-like to match the real Universe.

Nature knows how to draw a right-angle on a very large-scale. We noted earlier a tendency for large-scale structures to intersect orthogonally. A number of such cases arose in Chapter 4. One was the crossing over of the Centaurus and Hydra Walls (see the Atlas of Nearby Large-Scale Structures, between pages 80 and 81, 2,000–2,999 km/s redshift shell). Moreover, in the Southern Declination slice (in the Atlas), the structures adopt an almost rectilinear grid. In this case, it is made up by the Fornax, Grus and Sculptor Walls, but with a suggestion that it could extend in similar fashion, judging by fragments of possible walls. On still larger scales, the southern Sculptor Wall is roughly orthogonal to the (Great) Coma Wall.

Nature also knows how to draw parallel lines on a very large scale. The neighbouring Fornax and Sculptor Walls in the rectilinear grid are virtually parallel (within 5 degrees of one another.) Other suggestions of parallel structures occur with Perseus–Pisces (see Atlas, RA 0 hrs Right Ascension slice).

We also encountered (in Chapter 5) the claim of Jaan Einasto and collaborators, that the spatial distribution of clusters of galaxies follows a cubical grid. Should this prove true, or even partially true, it implies that orthogonal and parallel structures exist on an enormous scale.

Do orthogonal intersections and parallel structures occur in N-body simulations? Obviously, chance alignments do bring them about, but, as far as I can see, not with the apparent frequency with which they occur in nature.

If there are orthogonal structures, then more than Gaussian fluctuations must have spun the fabric of the Universe. Perhaps there is some still hidden message that we have yet to decipher.

10.2 THE NEED FOR INFLATION

Gaussian fluctuations also rest heavily on the 'inflationary Universe' scenario. Though to a large extent, 'inflation' is well established, there has been a current trend to question it. Perhaps the biggest problem for inflation, touched on earlier, relates to observations currently favouring a cosmological density of $\Omega = 0.3$, rather than the $\Omega = 1$ as inflation supposedly predicts. However, long before the trend to $\Omega = 0.3$ emerged, my colleague George Ellis (a highly respected figure in the world of relativity and cosmology) had advocated that inflationary models did not necessarily bring about $\Omega = 1$, contrary to the widespread belief.

Without inflation, how could the Universe be so homogeneous as the cosmic microwave background suggests, and as the images of extremely distant galaxies seen in opposite directions confirm?

I cannot resist drawing an analogy. Each spring, the vast semi-desert of Namaqualand (the western portion of South Africa) bursts into flower. Tourists come from all over the world to view this spectacle. The landscape is covered by surges of colour; it is like a vast abstract painting that stretches for thousands of kilometres. In some places, one colour dominates; a single species may take over all the eye can see, from foreground to horizon. What impresses me is to examine the intricate detail in a single flower, and then to realise it is replicated trillions of times over. Even flowers a thousand kilometres away may be identical to the one in my hand. How is this possible? The answer is in the DNA codings.

Is there, then, some cosmological DNA coding that makes distantly separated galaxies look so similar, or even causes large-scale structures to be replicated? There is a 'DNA coding' in the laws of physics: given similar initial materials and conditions, they replicate the same thing. The stars in our Galaxy are like the Namaqualand flowers: those our side of the Galaxy are presumably just like those on the other side, though they may never have been in 'causal contact'. Could the same not apply to galaxies? Does one still need inflation to produce the initial conditions throughout the Universe so uniformly, or would gravity mould the same forms regardless, as it must do with stars? Is it just a case of exceeding a threshold density to get the whole thing started?

There is no doubt that without inflation we would find it very difficult to generate the cosmic lattice of fluctuations on so large a scale. If large-scale structures did not arise from the Gaussian fluctuations of quantum fluctuations, we are forced to turn to alternative cosmological scenarios that might be very different from those of the mainstream today.

Much will depend on what the newer higher-resolution observations of the cosmic microwave background reveal. It was, after all, the discovery of the anisotropies in 1992 that, by providing the 'seeds' for gravitational growth, removed the need for later 'explosions' to create the labyrinth of voids. Nevertheless, as I will discuss below (Section 10.4), there seems to me to be characteristics of voids that are not explained satisfactorily by the gravitational growth from the seeds.

10.3 THE OCCURRENCE OF CLUSTERS

Where should clusters form in the cosmic labyrinth? There is a general perception that they occur where walls intercept, but this is not the case for most of the nearby clusters, though it would seem that the galaxy density is high where walls meet.

What is apparent is that they tend to occur near the centroids of major structures. The Virgo Cluster is the central condensation in the Local Supercluster. The rich cluster Abell 3627 is the richest most central cluster of the Centaurus Wall. The Fornax Wall is centred around the Fornax Cluster. The ribbon-like Hydra Wall has the Hydra Cluster. The massive Perseus Cluster is offset from the centroid of the Perseus–Pisces Wall. The rich Coma

Cluster lies at the centroid of the Coma Wall (Great Wall). Probably the only major nearby structure to lack a conspicuous cluster is the Cetus Wall.

Over and above the rich clusters at the centroids, there are sometimes important secondary clusters. For example, the Centaurus Wall has the Centaurus Cluster, and the Coma Wall has Abell 1367. There is also a scattering of other clusters; for instance, other Abell clusters in the Centaurus Wall.

It would perhaps seem natural for rich clusters to form where the gravitational wells are the deepest: the centroids of the large-scale structures. One might also suppose that given enough time, the largest scale structures would collapse progressively along their shortest axes, from 'wall-like' to 'ribbon-like' to 'filaments' that would drain inwards towards the central cluster. If this is the pattern, then the formation and growth of central clusters must be at the expense of surrounding structures. We believe that our Local Group of galaxies is falling inward to the Virgo Cluster (Section 7.3).

The Virgo Cluster may be an ideal example to work with. Right from the start, we have seen that it possesses radial 'protuberances', rather than a uniform halo. These radial structures undoubtedly reflect tidal stretching of neighbouring clumps or condensations. The galaxies within these structures are being pulled towards the cluster, but so far have moved a part of that distance, because the fall-in time (half the 'crossing time' mentioned in the previous chapter) is far greater than the age of the Universe, and in any case their infall will never overcome the cosmological expansion.

Clusters are the largest entities that can condense, in the timescale available and in overcoming the cosmological expansion. Even the core of the Virgo Cluster has not quite reached a relaxed 'virialised' state, though it probably contains dark matter as well as the visible galaxies.

Yet why should the Virgo Cluster have protuberances and not a halo? The spaces between the protuberances are voids, effectively radially distorted by the presence of the Virgo Cluster. Is this not evidence that the voids were there in the first place, rather than being created by gravitational clearing? If the protuberances have not yet fallen into the Virgo Cluster, how did all the galaxies manage to fall into the lower-density protuberance? The time scales do not fit the geometry.

Other clusters also lack uniform haloes. If one looks again at the Atlas maps, one sees that all clusters lie in extended structures – usually massive filamentary structures that effectively form protuberances, yet with small voids between the protuberances, often pushed close up to the clusters. How could gravity have so efficiently emptied these voids, when it has barely begun to draw the protuberances into the clusters? Do we need strong biasing in the galaxy formation, where galaxies are formed only when the density passes a critical value; if so, the voids need not be so empty after all.

One of the surprising aspects of plotting where clusters lie in large-scale structures is that one does not seem able to anticipate exact positions where the clusters might be situated. If one could just see the outlines of the large-scale structures (like the stippled regions in the Atlas), could one guess where the clusters are situated? The answer is: roughly at the centroids of the large-scale structures, although the precise positions cannot be decided.

Clusters therefore have not as yet caused noticeable tidal disruption to surrounding structures. If gravity is so weak, then how could it already have shifted matter out of the voids, large or small?

10.4 VOIDS

Are voids then created by means other than gravitational clearing?

The voids in the large-scale structures, whatever their size, are mostly close to being spherical. Is this a natural occurrence in Gaussian fluctuations? One would not necessarily have thought it to be the case.

Furthermore, it is likely that even large-scale structures are permeated by smaller spherical voids, though this may simply be only my impression. Even some of the large voids (e.g. Eridanus) seem to possess fragments of fine void structure (see Figure 6.1), as these smaller structures had been 'bleached away'. Voids in the cosmos seem to me to be as ubiquitous as craters on the Moon.

I still cannot escape an intuitive attraction to some cosmological scenario that generates small voids like a sort of cosmological 'Alka Seltzer'. The compression of matter between these voids might initiate early galaxy formation. It is a top-down scenario that operates on only a small localised scale, to around $k = 1\ \mathrm{Mpc}^{-1}$.

Yet randomly scattered voids cannot generate great walls and similar structures; one would still have to superpose a lattice-like set of fluctuations. Then, given such initial geometries, gravity would be allowed to run its course, resulting in the texture of the large-scale structures that we see today.

10.5 A FINAL WORD

Large-scale structures constitute a mystery that has not as yet been completely solved. But what a fascinating revelation it has been – to discover a fabric to the cosmos. I consider myself incredibly privileged to have been around at this time.

This book has been my attempt to convey something of the excitement and fascination that I find in the subject. As emphasised at the outset, it has been written very much from the observational point of view. We have not delved deep into the mathematics behind many of the analyses and interpretations, but the accompanying reading lists and the references within them will provide avenues for the student to carry out detailed exploration of particular aspects – many of them all too briefly represented by the sections in my text.

Our exploration and understanding of the cosmos has reached a point unsurpassed in human history. For the first time, we are beginning to obtain a picture of the entire observable Universe, out to, and back to, the opaque curtain of the cosmic microwave background. In the possible (though many would consider unlikely) situation of our being the only lifeforms in this Universe, with sufficient intelligence to be able to reach this point, for the first time since the Big Bang, our Universe is coming to know itself – through us.

Index

WILEY-PRAXIS SERIES IN ASTRONOMY AND ASTROPHYSICS

Forthcoming Titles

COSMOLOGY AND PARTICLE ASTROPHYSICS
Lars Bergstrom and Ariel Goobar, Department of Physics, Stockholm University, Sweden

THE VICTORIAN AMATEUR ASTRONOMER: Independent Astronomical Research in Britain 1820–1920
Allan Chapman, Wadham College, University of Oxford, UK

EXTRASOLAR PLANETS: The Search for New Worlds
Stuart Clark, Lecturer in Astronomy, University of Hertfordshire, UK

LIFE IN THE UNIVERSE: The Solar System and Beyond
Barrie W. Jones, Physics Department, The Open University, UK

PROTOSTELLAR DISCS AND PLANETARY SYSTEM FORMATION
John C. B. Papaloizou, Astronomy Unit, Queen Mary and Westfield College, London, UK, and Caroline Terquem, Lick Observatory, University of California, Santa Cruz, USA

NEW LIGHT ON DARK STARS: Red Dwarfs, Low-Mass stars, Brown Dwarfs
I. Neill Reid, Senior Research Associate, Caltech, Pasadena, USA, and Suzanne L. Hawley, Assistant Professor, Department of Physics and Astronomy, Michigan State University, USA

COSMIC RAY PHYSICS
Todor S. Stanev, Professor of Physics, Bartol Research Institute, University of Delaware, USA